Lecture Notes in Control and Information Sciences

Edited by M. Thoma and A. Wyner

Vol. 156: R. P. Hämäläinen, H. K. Ehtamo (Eds.)
Differential Games –
Developments in Modelling and Computation
Proceedings of the Fourth International Symposium
on Differential Games and Applications
August 9-10, 1990, Helsinki University of Technology,
Finland
XIII, 292 pages. 1991

Vol. 157: R. P. Hämäläinen, H. K. Ehtamo (Eds.)
Dynamic Games in Economic Analysis
Proceedings of the Fourth International Symposium
on Differential Games and Applications
August 9-10, 1990, Helsinki University of Technology,
Finland
XIII, 311 pages. 1991

Vol. 158: K. Warwick, M. Kárný,
A. Halousková (Eds.)
Advanced Methods in Adaptive Control
for Industrial Applications
X, 331 pages. 1991

Lecture Notes in Control and Information Sciences

Edited by M. Thoma and A. Wyner

158

K. Warwick, M. Kárný, A. Halousková (Eds.)

Advanced Methods in Adaptive Control for Industrial Applications

Springer-Verlag Berlin Heidelberg GmbH

Series Editors
M. Thoma · A. Wyner

Advisory Board
L. D. Davisson · A. G. J. MacFarlane · H. Kwakernaak
J. L. Massey · Ya Z. Tsypkin · A. J. Viterbi

Editors
Prof. Kevin Warwick
Dept. of Cybernetics, University of Reading
Reading PO Box 238, RG6 2AL Berks, U.K.

Dr. Miroslav Kárný
Institute of Information Theory & Automation
Czechoslovak Academy of Sciences
Pod vodárenskou věží 4
18208 Prague 8, Czechoslovakia

Dr. Alena Halousková
Institute of Information Theory & Automation
Czechoslovak Academy of Sciences
Pod vodárenskou věží 4
18208 Prague 8, Czechoslovakia

ISBN 978-3-540-53835-6   ISBN 978-3-540-46316-0 (eBook)
DOI 10.1007/978-3-540-46316-0

61/3020-543210   Printed on acid-free paper.

PREFACE

This volume contains the selected and collected papers which resulted from a joint Czechoslovak-UK Seminar on "Advanced Methods in Adaptive Control for Industrial Applications", an event held in Praha, 14-16 May 1990. The three day seminar was intended to bring together experts in Czechoslovakia and the UK in order to provide for the presentation and fluent exchange, of ideas and problem solutions in the field of the seminar title. A major aim was to direct the papers towards the actual problems faced by industry, both at the present time and in the near future. However, a number of papers were specifically aimed at reflecting possible trends in the longer term, by looking at particular implementation issues.

The underlying hope with this text, as indeed was the case with the Praha seminar, is that the volume will make an important contribution towards the industrial take-up and usage of advanced computer control systems. The range of papers presented have is quite large, with some being of a more mathematical/theoretical nature, whilst others are very much based on practical application examples. Perhaps the norm is directed towards design and implementation of computer control systems, specifically for general aplication.

The seminar in Praha was jointly organized and sponsored by the State Committee for Scientific and Technical Development (Praha), the Department of Trade and Industry (London), the Institute of Information Theory and Automation, Czechoslovak Academy of Sciences (Praha) and the University of Reading (Reading). Our thanks go to all of those above, in particular Tony Burwood-Smith and Richard King from the Department of Trade and Industry, for chaperoning the UK delegation, to Ladislav Tichý in the Czechoslovak Embassy in London, to O. Bobko in the State Commission, Praha.

At the end of the Seminar in May, a reception was held for participants, by the British Ambassador to Praha, His Excellency Mr. P. L. O'Keeffe, and our sincere thanks go both to him and to Michael Haddock, Commercial Secretary at the British Embassy for the reception and for their help during the seminar itself.

The editors would also like to thank those concerned with Springer-Verlag, for their swift response in considering this text, particularly Prof. Dr.-Ing. M. Thoma the Series Editor, and Erdmuthe Raufelder, Editorial Assistent. Finally, the editors particularly wish to express their gratitude to those who helped to put the final version of this volume together, namely Jiřina Gajdošová at Praha and Liz Lucas at Reading.

Praha, November 14, 1990

K. Warwick
M. Kárný
A. Halousková

LIST OF CONTRIBUTORS

Dr. Josef Böhm — Institute of Information Theory & Automation
Czechoslovak Academy of Sciences
Pod vodárenskou věží 4
18208 Prague 8, Czechoslovakia
Tel:(422)8152237, Fax: (422)8584569
Telex: 122 018 ATOM C

Dr. A. Chotai — Centre for Research on Environmental Systems
Lancaster University, Lancaster LA1 4YQ, U.K.

Dr. František Dušek — Chemical University, Pardubice
Automation Department
Leninovo nám. 535, 53000 Pardubice
Czechoslovakia
Tel: (4240)513221

Dr. Pavel Ettler — Škoda Works 31600 Plzeň
Czechoslovakia
Tel: (4219)215 ext. 6161

Prof. Peter J. Gawthrop — Dept. of Mechanical Engineering
University of Glasgow, Glasgow G12 8QQ, U.K.

Prof. Mike J. Grimble — Industrial Control Unit
Dept. of Electronic & Electrical Engineering
University of Strathclyde, Glasgow
50 George Street, Glasgow G1 1QE, U.K.
Tel: (4441)5524400, Fax: (4441)5531232
Telex: 77472 UNSLIB G

Dr. Vladimír Görner — Institute of Information Theory & Automation
Czechoslovak Academy of Sciences
Pod vodárenskou věží 4
18208 Prague 8, Czechoslovakia
Tel:(422)8152310, Fax: (422)8584569
Telex: 122 018 ATOM C

Dr. Alena Halousková — Institute of Information Theory & Automation
Czechoslovak Academy of Sciences
Pod vodárenskou věží 4
18208 Prague 8, Czechoslovakia
Tel:(422)8152251, Fax: (422)8584569
Telex: 122 018 ATOM C

Dr. Vladimír Havlena — Dept. of Control, Faculty of Electrical
Engineering, Czech Technical University
Karlovo nám. 13, 12135 Prague 2,
Czechoslovakia
Tel: (422)295664

Dr. Kenneth J. Hunt — Dept. of Mechanical Engineering
University of Glasgow, Glasgow G12 8QQ, U.K.

Prof. George W. Irwin — Dept. of Electrical Engineering
Queen's University of Belfast
Belfast BT7 1NN, Northern Ireland
Tel: (44232)245133 Telex: (44232)74487

Dr. Oliver L. R. Jacobs — Dept. of Engineering Science
University of Oxford
Parks Road, Oxford OX1 3PJ, U.K.
Tel: (44865)273000 Fax: (44865)273010
Telex 83295 NUCLOX G

Dr. Jan Ježek — Institute of Information Theory & Automation
Czechoslovak Academy of Sciences
Pod vodárenskou věží 4
18208 Prague 8, Czechoslovakia
Tel:(422)815 2304, Fax: (422)8584569
Telex: 122018 ATOM C

Dr. František Jirkovský — Škoda Works 31600 Plzeň
Czechoslovakia
Tel: (4219)215 ext. 6161

Dr. Radim Jiroušek — Institute of Information Theory & Automation
Czechoslovak Academy of Sciences
Pod vodárenskou věží 4
18208 Prague 8, Czechoslovakia
Tel:(422)8152046, Fax: (422)8584569
Telex: 122 018 ATOM C

Dr. Karam Z. Karam — Dept. of Electrical & Electronic Engineering
University of Newcastle upon Tyne
Newcastle upon Tyne NE1 7RU, U.K.
Tel: Tyneside (4491)2226000 Fax: (4491)2228180
Telex: 53654 UNINEW G

Dr. Miroslav Kárný — Institute of Information Theory & Automation
Czechoslovak Academy of Sciences
Pod vodárenskou věží 4
18208 Prague 8, Czechoslovakia
Tel:(422)8152274, Fax: (422)8584569
Telex: 122 018 ATOM C

Dr. Petr Klán   Institute of Information Theory & Automation
Czechoslovak Academy of Sciences
Pod vodárenskou věží 4
18208 Prague 8, Czechoslovakia
Tel:(422)8152845, Fax: (422)8584569
Telex: 122 018 ATOM C

Prof. Pavel Kovanic   Institute of Information Theory & Automation
Czechoslovak Academy of Sciences
Pod vodárenskou věží 4
18208 Prague 8, Czechoslovakia
Tel:(422)8152230, Fax: (422)8584569
Telex: 122 018 ATOM C

Dr. Otakar Kříž   Institute of Information Theory & Automation
Czechoslovak Academy of Sciences
Pod vodárenskou věží 4
18208 Prague 8, Czechoslovakia
Tel:(422)8152232, Fax: (422)8584569
Telex: 122 018 ATOM C

Prof. Vladimír Kučera Institute of Information Theory & Automation
Czechoslovak Academy of Sciences
Pod vodárenskou věží 4
18208 Prague 8, Czechoslovakia
Tel:(422)847173, Fax: (422)8584569
Telex: 122 018 ATOM C

Dr. Rudolf Kulhavý   Institute of Information Theory & Automation
Czechoslovak Academy of Sciences
Pod vodárenskou věží 4
18208 Prague 8, Czechoslovakia
Tel:(422)8152313, Fax: (422)8584569
Telex: 122 018 ATOM C

Dr. Jaroslav Maršík   Institute of Information Theory & Automation
Czechoslovak Academy of Sciences
Pod vodárenskou věží 4
18208 Prague 8, Czechoslovakia
Tel:(422)8152254, Fax: (422)8584569
Telex: 122 018 ATOM C

Dr. Ben Minbashian   Dept. of Cybernetics, University of Reading
Reading PO Box 238, RG6 2AL Berks, U.K.
Tel: (44734)318210, Fax:(44734)318220

Dr. Gary Montague   Dept. of Chemical & Process Engineering
                           University of Newcastle upon Tyne
                           Newcastle upon Tyne NE1 7RU, U.K.

Prof. A. Julian Morris   Dept. of Chemical & Process Engineering
                           University of Newcastle upon Tyne
                           Newcastle upon Tyne NE1 7RU, U.K.

Dr. Ivan Nagy   Institute of Information Theory & Automation
                           Czechoslovak Academy of Sciences
                           Pod vodárenskou věží 4
                           18208 Prague 8, Czechoslovakia
                           Tel:(422)8152337, Fax: (422)8584569
                           Telex: 122 018 ATOM C

Dr. Petr Nedoma   Institute of Information Theory & Automation
                           Czechoslovak Academy of Sciences
                           Pod vodárenskou věží 4
                           18208 Prague 8, Czechoslovakia
                           Tel:(422)8152307, Fax: (422)8584569
                           Telex: 122 018 ATOM C

Dr. Zuzana Országhová   Dept. of Automatic Control & Measurement
                           Slovak Technical University
                           Nám. Slobody 17, 81231 Bratislava
                           Czechoslovakia
                           Tel: (427)41193

Prof. Václav Peterka   Institute of Information Theory & Automation
                           Czechoslovak Academy of Sciences
                           Pod vodárenskou věží 4
                           18208 Prague 8, Czechoslovakia
                           Tel:(422)845723, Fax: (422)8584569
                           Telex: 122 018 ATOM C

Doc. Boris Rohál-Il'kiv   Dept. of Automatic Control & Measurement
                           Slovak Technical University
                           Nám. Slobody 17, 81231 Bratislava
                           Czechoslovakia
                           Tel: (427)43 041

Dr. Richard Richter   Dept. of Automatic Control & Measurement
                           Slovak Technical University
                           Nám. Slobody 17, 81231 Bratislava
                           Czechoslovakia
                           Tel: (427)41193

Dr. Ivan Sroka   Dept. of Automatic Control & Measurement
                           Slovak Technical University
                           Nám. Slobody 17, 81231 Bratislava
                           Czechoslovakia
                           Tel: (427)41193

Dr. Ming T. Tham — Dept. of Chemical & Process Engineering
University of Newcastle upon Tyne
Newcastle upon Tyne NE1 7RU, U.K.

Dr. Wlodek Tych — Centre for Research on Environmental Systems
Lancaster University, Lancaster LA1 4YQ, U.K.

Dr. Vladimír Velička — Research Institute of Chemical Equipment
Křižíkova 70, 602 00 Brno
Czechoslovakia

Prof. Kevin Warwick — Dept. of Cybernetics, University of Reading
Reading PO Box 238, RG6 2AL Berks, U.K.
Tel: (44734)318210, Fax:(44734)318220

Prof. Peter E. Wellstead — Control System Center
EE & E,UMIST, PO BOX 88
Manchester M60 1QD, U.K.

Prof. Peter C. Young — Centre for Research on Environmental Systems
Lancaster University, Lancaster LA1 4YQ, U.K.
Tel: (44524)65201, Fax: (44524)843854

Dr. Peter Zelinka — Dept. of Automatic Control & Measurement
Slovak Technical University
Nám. Slobody 17, 81231 Bratislava
Czechoslovakia
Tel: (427)41193

Contents

X

ADAPTATION OF LQG CONTROL DESIGN TO ENGINEERING NEEDS

V. Peterka

Institute of Information Theory and Automation,
Czechoslovak Academy of Sciences
182 08 Prague 8, Czechoslovakia

Abstract

The purpose of the paper is twofold: (I) to modify the standard LQG control design procedure so that it might better meet the requirements of practicing engineers, both with respect to quality of the resulting control and with respect to its robustness, (II) to design algorithms for the control synthesis which are robust with respect to numerical errors and are suitable for real–time operation in adaptive systems based on microprocessors with reduced precision of arithmetic operations.

The control synthesis is based on the receding horizon philosophy. A suitable combination of robustness and quality of control is achieved by designing the control for the receding horizon optimally in the LQG sense but with restriction on the admissible control strategy. The process input planned for the receding horizon is restricted to be piece–wise constant for periods longer than the sampling period the control loop operates with. In addition, a modification of the quadratic criterion is recommended in order to reduce the overshoots.

The resulting control is proved to be stable and the increase of robustness is demonstrated on examples.

The algorithms are elaborated for the observable canonical state representation of the incremental ARMA or Delta model of the controlled process.

Simulated examples are given for illustration.

1   Introduction

The research reported in this paper has been motivated by the fact that in practice the LQG optimal control often cannot be applied in a straightforward way. One of the reasons is that it is difficult, or impossible, to express all requirements of the user in terms of weights entering the quadratic criterion. The other reason is that the mathematical model used in the design procedure more or less deviates from the true plant. The objective of this paper is to modify the LQG control design procedure so that the both above circumstances may be reflected.

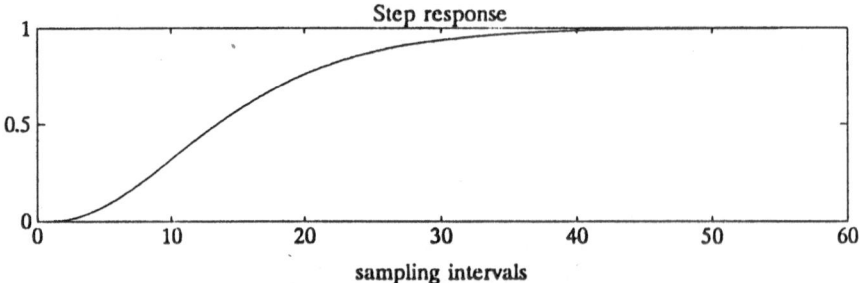

Figure 1: Step response of the system (1). Sampling interval $T_s = 0.2$ sec .

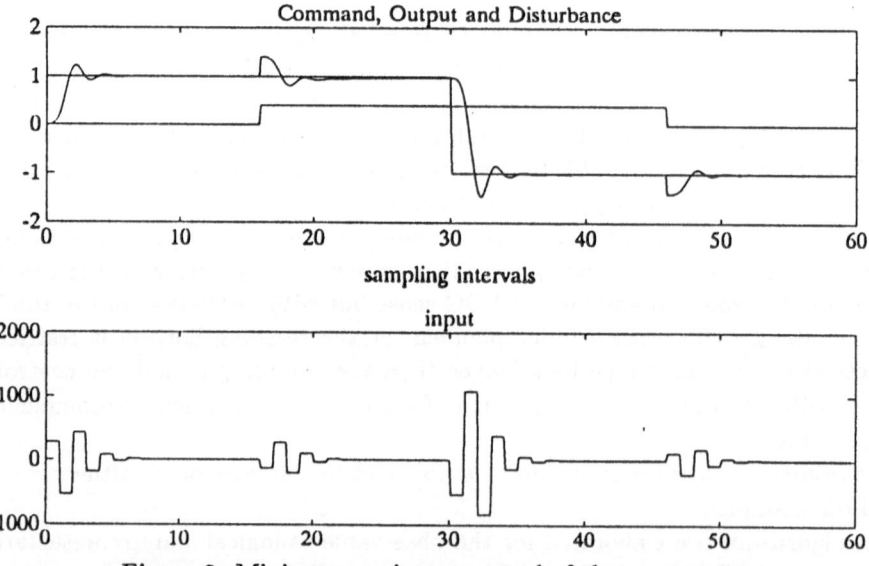

Figure 2: Minimum variance control of the system (1).

In order to make the ideas as plain as possible let us start with a simple example. Consider, for instance, the third order deterministic system

$$y = \frac{1}{(s+1)^3} \ u \tag{1}$$

the step response of which is shown in Fig.1. Suppose that this system has to be controlled with a relatively short sampling interval $T_s = 0.2$ sec. Note that the significant part of the step response is covered, approximately, by 30 sampling intervals. The simulation of the "minimum variance" control (with no penalty on the input) is registered in Fig.2. The plotted disturbance acts on the process output. It is clear that such a control is unacceptable in practice for both above reasons. Excessive movements of the actuator cannot be practically realized and require an extremely precise knowledge of the dynamics of the system.

There are two standard ways how to make the LQG control more realistic. The first possibility, shown in Fig.3, is to damp the system input by introducing a suitable weight on the input increments into the quadratic criterion. The choice of this weight,

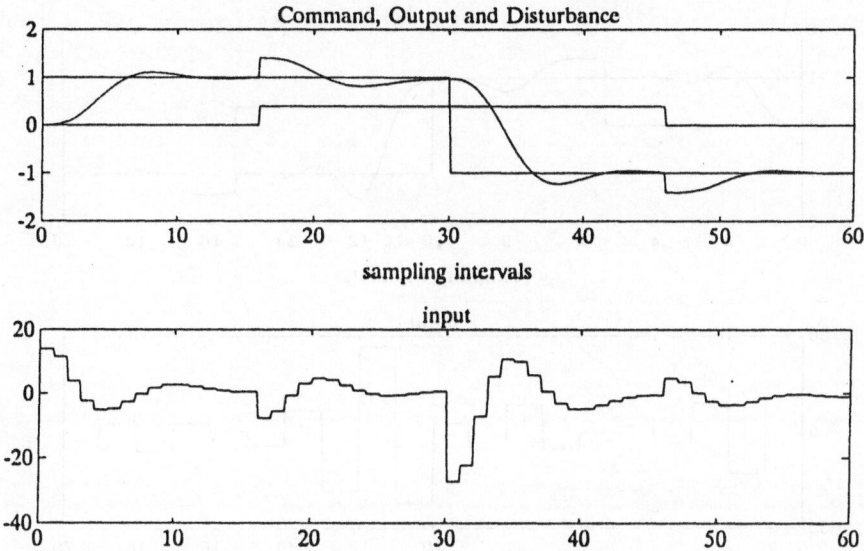

Figure 3: Damping of the control signal by penalization of the input increments

in our example $q_u = 0.00015$, is left for the user as a "performance oriented knob" but, unfortunately, no simple rule is available for this choice without much experimentation. For self–tuning control applications a procedure has been suggested [1] which introduces penalization of inputs automatically in dependence upon the observed data and the limits set on the input signal. In this paper an other possibility will be followed which has some advantages.

The other way how to reduce the movements of the actuator is to encrease the sampling interval. The rule of thumb for the choice of the sampling rate is to cover the significant part of the step response by 5 – 10 sampling intervals. The disadvantege of this method is that the system is left without supervision for relatively long periods. This may cause a slow reaction to disturbances and also undesirable overshoots may occur. This is demonstrated in Fig.4, where the sampling rate is 3 times slower than in Fig.2.

On the basis of the receding horizon philosophy, in the following Section 2 a modified LQG control problem is formulated which makes it possible to retain the original fast sampling rate and at the same time can damp the control signal without introducing the penalization of its increments. To solve the problem it is necessary to assume some mathematical model of the controlled process. For the purpose of this paper the incremental ARMA or Delta model will be assumed. A suitable canonical state representation of these models is introduced in Section 3. The solution of the modified LQG control problem is given in Section 4. Since the calculated control law has to operate in conditions which differ from those assumed in the design procedure, the stability of the closed control loop has to be investigated. This is the topic of Section 5. In Section 6 a modification of the design procedure is suggested which reduces the overshoots of the controlled output. The encrease of the robustness of the control is demonstrated

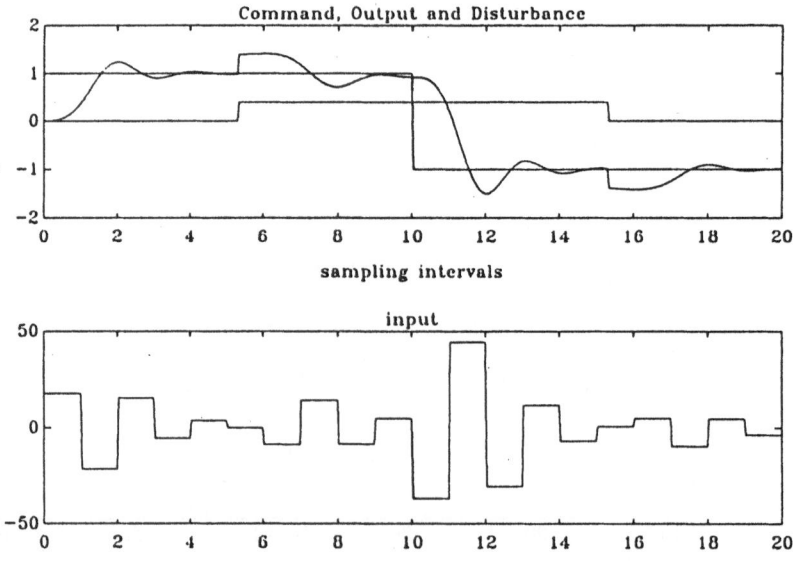

Figure 4: Sampling interval 3 times longer, no penalty on input.

in Section 7. Efficient and numerically reliable algorithm for the computation of the control law is ellaborated in Section 8. The Section 9 is concluding and gives the author's recommentation how to apply the results.

2   Formulation of the modified LQG control problem

When formulating a control problem it is necessary to define : 1) the system model considered, 2) the admissible control strategies among which the optimal one is to be chosen, and 3) the criterion. The system model will be introduced and discussed in Section 3. Here the attention will be paid to the two remainig parts of the problem formulation.

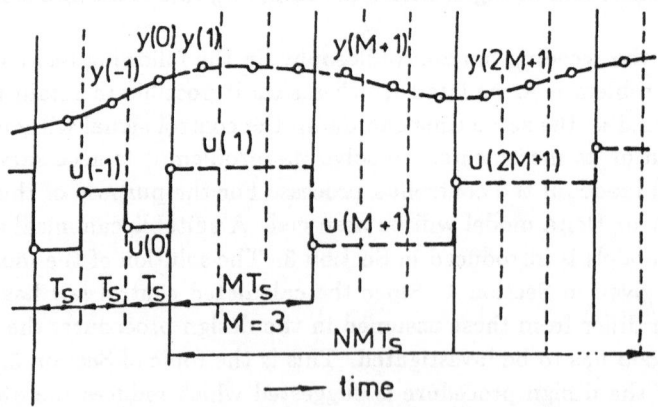

Figure 5: Receding control horizon with piece–wise constant input.

The control strategy will be designed for a receding horizon, finite or infinite, but only those control strategies are taken into the competition which produce the control input which is piece–wise constant for periods longer than the sampling period the control loop operates with. Hence, a fake sampling interval, consisting of M regular sampling intervals, is introduced as shown in Fig.5. However, M samples of the system output are measured within this fake sampling interval.

Let the receding horizon, the control strategy is to be design for, be N fake sampling intervals, ie. NM regular sampling intervals. To simplify the writing we make use of time indexing so that $u(1)$ is the first input which is to be determined on the basis of the input–output data $\mathcal{D}_0 = \{\ldots, u(-1), y(-1), u(0), y(0)\}$ observed on the system up to and including the time index 0. Note that $y(1)$ is not available at the moment when $u(1)$ is to be determined. The time interval between the samples $y(0)$ and $u(1)$ is required to perform the computation. The speed of the computer determines this sample shift.

Following the receding horizon philosophy, we are interested mainly in the control law generating $u(1)$. This is, actually, the only control law which will be really applied.

It is also necessary to define the availability of the command signal w (the output reference, the setpoint sequence). In this paper the case of a positional servo will be considered. It is assumed that the command signal is given up to and including the time index 1, $W_1 = \{\ldots, w(0), w(1)\}$. Note that $w(1)$ says the controller what the next output $y(1)$ should be after the input $u(1)$ is applied. Hence the quadratic criterion, we shall consider, is

$$J = \mathrm{E}\left\{\sum_{j=0}^{N-1}\left[\sum_{i=1}^{M} q_r\, r(jM+i)^2 + q_u\,\Delta u(jM+1)^2\right] + x'(NM)P_N x(NM)\ \Bigg|\ \mathcal{D}_0, W_1\right\} \tag{2}$$

where

$$r(k) = y(k) - w(k) \tag{3}$$

is the control error, $x(NM)$ is a suitably chosen last state and

$$q_r > 0, \qquad q_u \geq 0 \qquad P_N \gg 0$$

The criterion has to be minimized under the restriction

$$\Delta u(jM+i) = 0 \qquad \text{for} \quad i \neq 1 \tag{4}$$

The last term in the criterion (3) is important for stability reasons as it will be explained later on.

To be able to solve the LQG control problem, modified in the above way, it is necessary to define not only the system model, which will be introduced in the next Section, but also the stochastic model describing the expected evolution of the comand signal within the receding horizon. For the case of positional servo it is suitable to consider the future command signal as the generalized random walk

$$w(k) = w(k-1) + e_w(k) \tag{5}$$

where $e_w(k)$ is discrete white noise with arbitrarily time varying variance $Ew^2(k) = \varrho_w(k)$, which is not correlated with $y(k)$ and with the previous input–output data $\mathcal{D}_{k-1}$.

Note that the model (5) covers also the case of regulation with fixed setpoint when $\varrho_w(k) = 0$.

3   Process model

The input–output system model is assumed, for the control design purposes, in the incremental ARMA form

$$\Delta y(k) + \sum_{i=1}^{n} a_i \Delta y(k-i) = \sum_{i=0}^{n} b_i \Delta u(k-d-i) + e(t) + \sum_{i=1}^{n} c_i e(k-i) \tag{6}$$

or in the Delta form

$$\Delta^{n+1} y(k) + \sum_{i=1}^{n} a_i \Delta^{i+1} y(k-i) = \sum_{i=0}^{n} b_i \Delta^{i+1} u(k-i) + \sum_{i=0}^{n} c_i \Delta^i e(k-i) \tag{7}$$

where Δ denotes the backward difference, eg.

$$\Delta y(k) = y(k) - y(k-1) \qquad \Delta^i y(k) = \Delta^{i-1} y(k) - \Delta^{i-1} y(k-1)$$

end $e(k)$ is a discrete white noise with constant variance ϱ.

Both the models (6) and (7) can discribe the same input-output relation but with different parameters, of course. We shall keep the same notation for the different parameters because, as shown in detail in [8], they can be given the common canonical observable state representation

$$y(k) = y(k-1) + h'_s\, s(k-1) + b_0\, \Delta u(k) + e(t) \tag{8}$$
$$s(k) = F_s\, s(k-1) + G_s\, \Delta u(k) + \varsigma(k) \tag{9}$$

If the following parameterless matrices

$$h_s = \begin{bmatrix} 1 \\ 0 \\ 0 \\ \vdots \\ 0 \end{bmatrix} \qquad H_s = \begin{bmatrix} 0 & 1 & 0 & \cdots & 0 \\ 0 & 0 & 1 & \cdots & 0 \\ \vdots & \vdots & \vdots & & \vdots \\ 0 & 0 & 0 & \cdots & 1 \\ 0 & 0 & 0 & \cdots & 0 \end{bmatrix} \tag{10}$$

and the following parametr vectors are introduced

$$\bar{a} = \begin{bmatrix} a_1 \\ a_2 \\ \vdots \\ a_n \end{bmatrix} \qquad \bar{b} = \begin{bmatrix} b_1 \\ b_2 \\ \vdots \\ b_n \end{bmatrix} \qquad \bar{c} = \begin{bmatrix} c_1 \\ c_2 \\ \vdots \\ c_n \end{bmatrix} \tag{11}$$

then the matrix coefficients of the state equation (9) are

$$F_s = H_s - \bar{a} h'_s + \mu I \qquad G_s = \bar{b} - \bar{a} b_0 \qquad \varsigma(k) = [\bar{c} - \bar{a}] e(k) \tag{12}$$

where $\mu = 0$ for ARMA case and $\mu = 1$ for Delta case.

4 Solution of the modified LQG control problem

To be able to solve the control problem formulated in Section 2 in a compact and standard way it is advantageous to incorporate into the model (8) and (9) the control error (3). For this purpose it is suitable to introduce the variable

$$x_1(k) = y(k) - w(k+1) = r(k) - e_w(k+1)$$
$$x_1(k) = x_1(k-1) + h'_s s(k-1) + b_0 \Delta u(k) + e(k) - e_w(k+1)$$

where we make use of (5) and (8).

Now, if the following overall state is introduced

$$x(k) = \begin{bmatrix} x_1(k) \\ s(k) \end{bmatrix}$$

the state space model gets the compact form

$$x(k) = Fx(k-1) + G\Delta u(k) + \xi(k) \tag{13}$$

$$r(k) = h'x(k) + e_w(k+1) \tag{14}$$

where

$$h' = \begin{bmatrix} 1 & 0 & \cdots & 0 \end{bmatrix} = \begin{bmatrix} h'_s & 0 \end{bmatrix}$$

$$F = \begin{bmatrix} 1 & h'_s \\ 0 & F_s \end{bmatrix} \qquad G = \begin{bmatrix} b_0 \\ G_s \end{bmatrix} \qquad \xi(k) = \begin{bmatrix} e(k) - e_w(k+1) \\ [\bar{c} - \bar{a}]\, e(k) \end{bmatrix} \tag{15}$$

As $e_w(k+1)$ is not correlated with the data on which the controller generating $u(k)$ is allowed to operate, the criterion (2) can be replaced by the criterion

$$J = E\left\{ \sum_{j=0}^{N-1} \left[\sum_{i=1}^{M} x'(jM+i) h q_r h' x(jM+i) + q_u\, \Delta u(jM+1)^2 \right] + \right.$$

$$\left. + x'(NM) P_N x(NM) \,\Big|\, \mathcal{D}_0, W_{(1)} \right\} \tag{16}$$

It is important to note that for arbitrary but constant levels of the input and of the output the steady mean value of the state $s(t)$, which is the main component of $x(t)$, is zero. To drive $s(t)$ towards zero means to quiet down the transient process. It is the role of the last term in the criterion (16) to ensure that, at the end of the control horizon the control error be as small as possible and that the system be close to a steady state. (This holds only for incremental models. In case of positional models, it is the increment of the final state $s(NM)$ which has to be penalized [2]. To drive the state of a positional model to zero and at the same time to drive $y(NM)$ to $w(NM) \neq 0$ would be contradictory.)

Now everything is prepared for application of dynamic programing which yields the following

Result 1 (LQG control synthesis for piece-wise constant input)
Let the admissible input within the receding horizon of NM sampling intervals be re-stricted to be constant for each M sampling periods. Then the control law minimizing the criterion (16) and operating in positional servo mode on the system described by the incremental ARMA/Delta model (8,9) or (13,14) is

$$\Delta u(t+1) = -L_1 \hat{x}(t|t) = L_r(w(t+1) - y(t)) - L_s \hat{s}(t|t) \qquad (17)$$

$$L_1 = [L_r, \ L_s] = (G'P_1G + q_u)^{-1}G'P_1F \qquad (18)$$

where $\hat{x}(t|t)$ or $\hat{s}(t|t)$ are the current estimates of the state $x(t)$ or $s(t)$, respectively, and P_1 is the nonnegative definite square matrix of dimension $n+1$ which is obtaind by the $N-1$ steps of the following recurcion for $j = N, N-1, \cdots, 2$.

$$\bar{P}_{j-1} = F'P_jF - F'P_jG(G'P_jG + q_u)^{-1}G'P_jF \qquad (19)$$

$$P_{j-1} = (F^{M-1})'\bar{P}_{j-1}F^{M-1} + Q_M \qquad (20)$$

where

$$Q_M = \sum_{i=0}^{M-1} (F^i)'h \ q_r h' \ F^i \qquad (21)$$

Elimination of $\bar{P}_{j-1}$ from (19) and (20) leads to the following Riccati difference equation which will be of use later on.

$$P_{j-1} = (F^M)'P_jF^M - (F^M)'P_jG(G'P_jG + q_u)^{-1}G'P_jF^M + Q_M \qquad (22)$$

Note that $y(t)$ and $w(t+1)$ are available when $u(t+1)$ is generated and therefore

$$\hat{x}(t|t) = \begin{bmatrix} y(t) - w(t+1) \\ \hat{s}(t|t) \end{bmatrix} \qquad (23)$$

The state $s(t)$ can be estimated in real time using the following Result 2 taken from [3]

Result 2 (State estimation)
For the state model (8, 9) with the parameters (11, 12) the real-time estimation of the state $s(t)$ can be reduced to the following linear difference equation

$$\hat{s}(t|t) = C(t)\hat{s}(t-1|t-1) - G_y(t) \ \Delta y(t) + G_u(t) \ \Delta u(t) \qquad (24)$$

where

$$G_y = [\bar{a} - \tilde{c}(t)] \qquad\qquad G_u = [\bar{b} - \tilde{c}(t)b_0] \qquad (25)$$

$$\tilde{c}'(t) = [\tilde{c}_1(t), \tilde{c}_2(t), \cdots, \tilde{c}_n(t)] \qquad (26)$$

$$C(t) = H_s - \tilde{c}(t)h_s = \begin{bmatrix} -\tilde{c}_1(t) & 1 & 0 & \cdots & 0 \\ -\tilde{c}_2(t) & 0 & 1 & \cdots & 0 \\ \vdots & \vdots & & \vdots & \\ -\tilde{c}_{n-1}(t) & 0 & 0 & \cdots & 1 \\ -\tilde{c}_n(t) & 0 & 0 & \cdots & 0 \end{bmatrix} \qquad (27)$$

The time varying parameters $\tilde{c}(t)$ are generated by the following algebraic recursion

$$\tilde{c}(t) = [H_s R_s(t-1)h'_s + \bar{c}]/(1 + h_s \ R_s(t-1)h'_s) \qquad (28)$$

$$R_s(t) = C(t) \ R_s(t-1) \ C'(t) + [\tilde{c}(t) - \bar{c}][\tilde{c}(t) - \bar{c}]' \qquad (29)$$

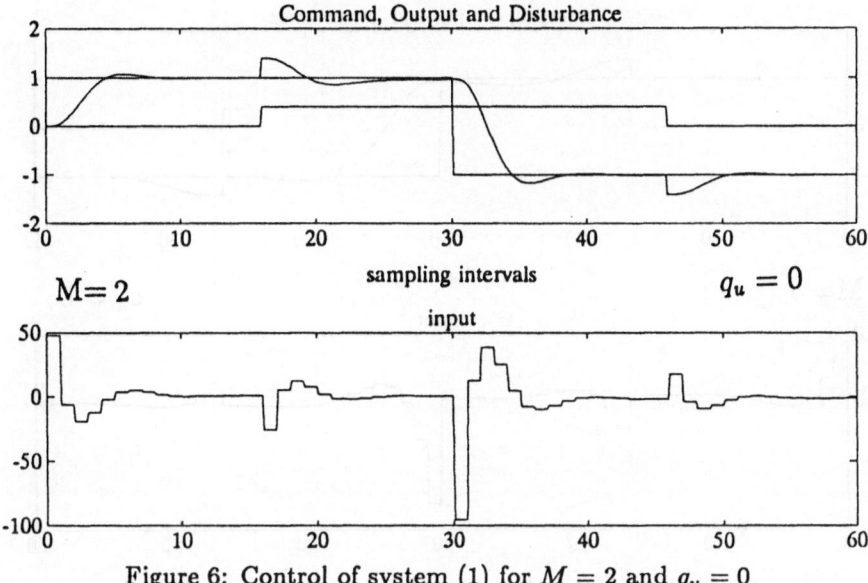

Figure 6: Control of system (1) for $M = 2$ and $q_u = 0$

The recursion (28, 29) replaces the Riccati equation of the standard Kalman filter and for $R_s(0) > 0$ works like sequential reflection of the polynomial c associated with the moving average part of the ARMA/Delta model. In this way the stability of the filter (24) is guaranteed also for the case when the c–polynomial has unstable roots or roots lying at the stability boundery. This means that the incremental models considered in this paper can cover also the positional models if the factor $(1 - z^{-1})$ is incorporated into the c–polynomial. (See [2] for more details.)

The practical effect of the choice $M > 1$ on the control of the system (1) is demonstrated for $M = 2$ in Fig.6 and for $M = 3$ in Fig.7. No penalty on the input increments was applied in both these cases and the receding horizon was chosen sufficiently long so that the effects of structuring the input and of the choice of P_N were not mixed.

5   Stability

As the control law determined according to Result 1 for finite receding horizon and for a different sampling rate than actually applied, the asymptotic stability of the closed control loop has to be checked. The technique of "Fake algebraic Riccati equation" [4] can serve the purpose.

Consider the relation (19) for $j = 1$ and rewrite it in the following way

$$P_1 = F'P_1F - F'P_1G(G'P_1G + q_u)^{-1}G'P_1F + Q_F \tag{30}$$

where

$$Q_F = P_1 - \bar{P}_0 \tag{31}$$

The equation (30) is the algebraic Riccati equation which corresponds to our control law (18) and which could be obtained as the stationary solution for $M = 1$ and $N \rightarrow \infty$ if

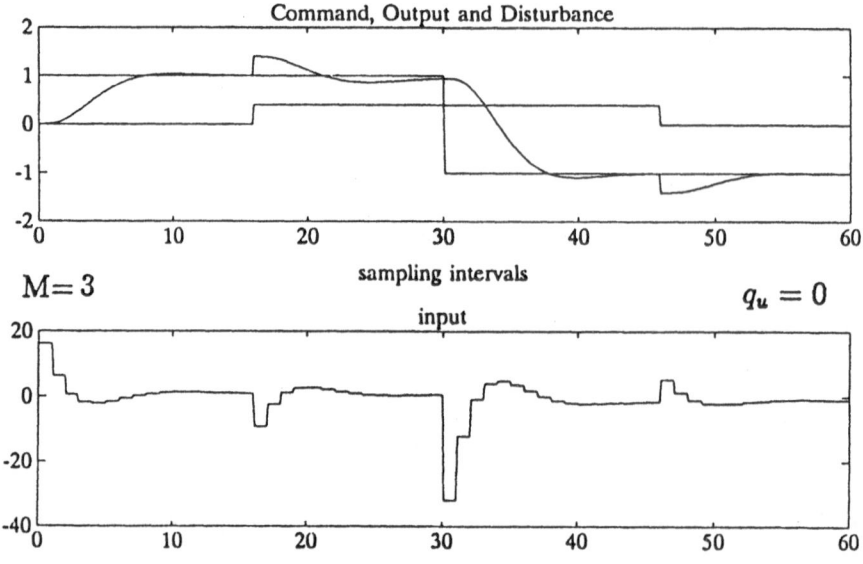

M= 3 $\qquad q_u = 0$

Figure 7: Control of system (1) for $M = 3$ and $q_u = 0$

the state x were weighted by Q_F (31) instaed of $h\ q_r h'$. From the theory of the algebraic Riccati equation [5] the following Result 3 is immediately obtained.

Result 3 *The control loop with the control law (18) and the system (8,9) is stable if the system is stabilizable (the polynomials related to the a- and b- parts of the input–output model (6) or (7) have no unstable common factors), if $q_r > 0$ and if*

$$P_1 \geq \bar{P}_0 \tag{32}$$

Making use of (20) for $j = 1$, the matrix (31) can be expressed in the following way

$$Q_F = P_1 - \bar{P}_0 =$$

$$= P_1 - \bar{P}_0 + Q_M + Q_L \tag{33}$$

where Q_L satisfies the Lyapunov equation

$$Q_L = \bar{P}_0 - (F^{M-1})' \bar{P}_0 F^{M-1} \tag{34}$$

If the system is stable, $Q_L \geq 0$. By definition $Q_M \geq 0$, and to gaurantee the stability it is sufficient to chose P_N sufficiently large in order to force the difference Riccati equation (22) to monotonicity so that $(P_1 - P_0) \geq 0$.

6 Reduction of overshoots

It is typical for the LQG optimal control, especially when applied with fast sampling rates, that the step response of the closed control loop exhibits overshoots which are not well seen by practicing engineers. This, sometimes unfavorable, peculiarity stems

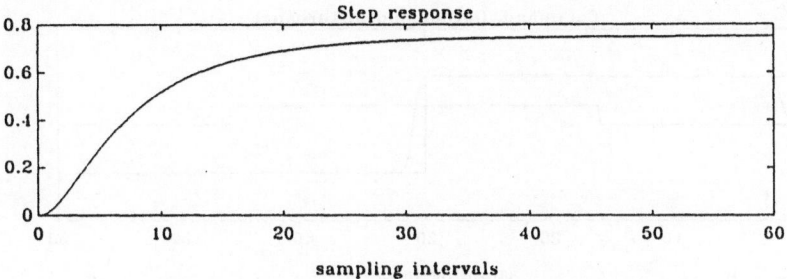

Figure 8: The step response of the system (36)

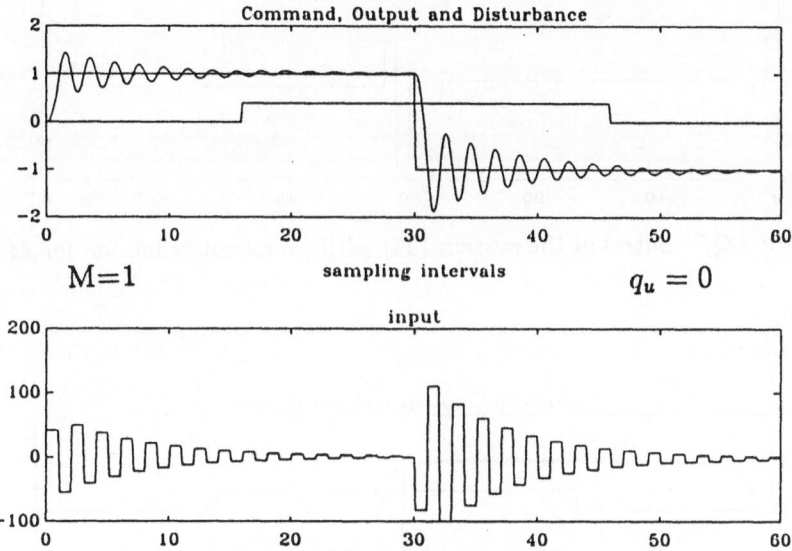

Figure 9: Minimum variance control of system (36)

from the fact that the relatively large and physically unremovable control errors at the beginning of the transient process, when squared, contribute to the minimal value of the quadratic criterion so that even very small and practically insignifacant improvement of this initial part lead to overshoots in the future. The remedy is simple : *Do not penalize the control errors within the first fake sampling interval of the receding horizon !* This means to set $q_r = 0$ in the last step of the recursion (19,20) for $j = 2$, so that

$$P_1^* = (F^{M-1})' \bar{P}_1 F^{M-1}, \qquad L_1^* = (G'P_1^*G + q_u)^{-1} G'P_1^* F \qquad (35)$$

The fact that the practical effect of this simple modification of the LQG control design can be very significant is demonstrated on the example

$$y = \frac{(s + 0.75)}{(s + 2)(s + 1)(s + 0.5)} \, u \qquad\qquad T_s = 0.25 \text{ sec} \qquad (36)$$

taken from [6]. The step response of the system (36) is shown in Fig.8 and its standard minimum variance control is registered in Fig.9. Note that the control, when seen only

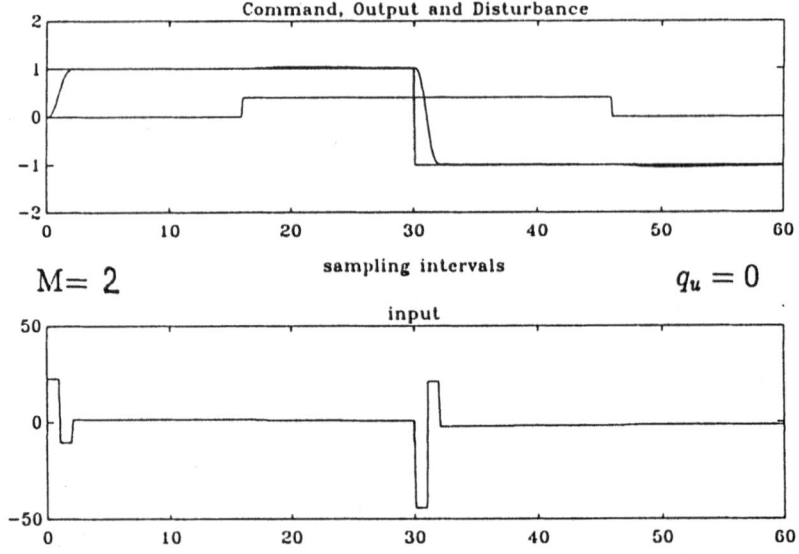

Figure 10: LQG control of the system (36) with overshoot reduction for $M = 2$

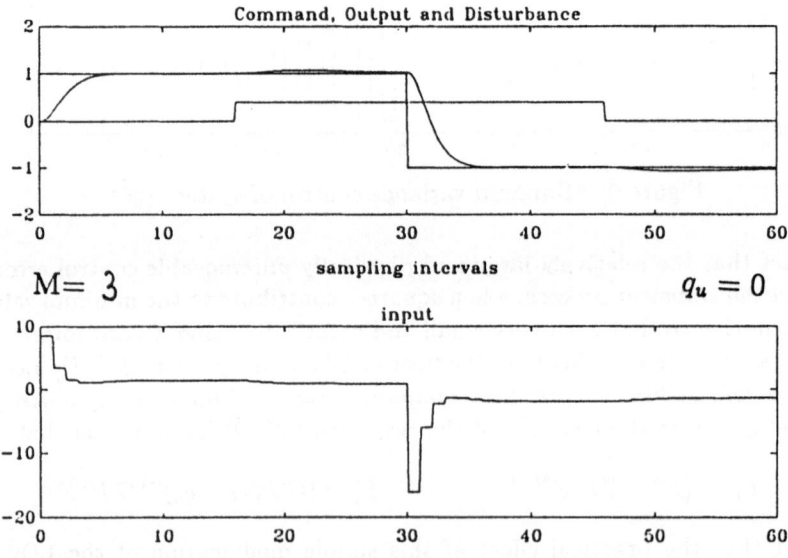

Figure 11: LQG control of the system (36) with overshoot reduction for $M = 3$

through the samples of the output, is theoretically perfect. The controller is not aware of the intersample riples. The plots in Fig.10 and Fig.11 show the control designed for $M = 2$ and $M = 3$, respectively. In both these cases the overshoot reduction was applied and no pealty was set on the input changes.

7 Robustness

It is natural to ask how the modification suggested in previous Sections affect the natural robustness of the LQG control with respect to modelling errors. The Nyquist plot of the open loop transfer function for $z = e^{j\omega T_\bullet}$ can help to answer this question [7]. For this purpose let us find the overall transfer function of the controller.

After the stationarity of the state estimator (24) is reached, ie. for $\lim_{t\to\infty} \tilde{c}(t) = \tilde{c}$ the dynamic relation between the state estimate $\hat{s}(t|t)$ and its inputs, $\Delta y(t)$ and $\Delta u(t)$, can be expressed using the forward shift operator z

$$\hat{s}(t|t) = z(zI - C)^{-1}[G_u\Delta u(t) + G_y\Delta y(t)] \tag{37}$$

From (17) we also have

$$\Delta u(t) = L_r w(t) - z^{-1}L_r y(t) - z^{-1}L_s \hat{s}(t|t) \tag{38}$$

This, together with the relation

$$L_s(zI - C)^{-1}G = \frac{\det(zI - C + GL_s)}{\det(zI - C)} - 1 \tag{39}$$

where L_s is a row vector and G is a column vector, gives the following

Result 4 (The transfer function of the controller)
The transfer function between $y(t)$ and $u(t)$ for the LQG controller, when designed for an incremental ARMA model of the controlled system, is

$$\frac{s(z)}{r(z)} = -\frac{(z - 1)(g(z) - \tilde{c}(z)) + L_r\tilde{c}(z)}{(z - 1)f(z)} \tag{40}$$

where

$$\tilde{c}(z) = \det(zI - C), \qquad f(z) = \det(zI - C + G_y L_s), \qquad g(z) = \det(zI - C + G_u L_s) \tag{41}$$

If such a controller is connected with a system, the transfer function of which is $b^*(z)/a^*(z)$, the function, sometimes called the return difference [7], is

$$1 + \frac{s(z)b^*(z)}{r(z)a^*(z)} = \frac{r(z)a^*(z) + s(z)b^*(z)}{r(z)a^*(z)} \tag{42}$$

The poles of the return difference coincide with the poles of the open control loop while its zeros coincide with the poles of the closed control loop. When the return difference is plotted for $z = e^{j\omega}$ and for ω running from 0 to π, then, according to the

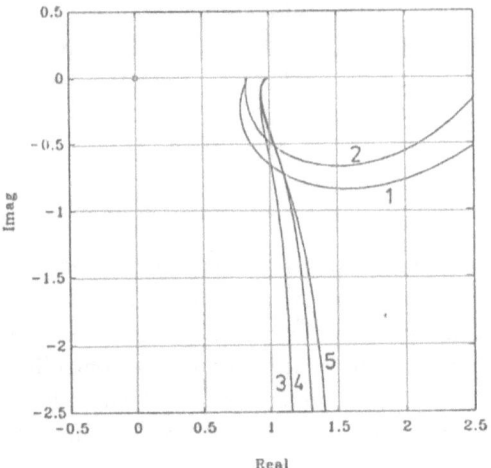

Figure 12: Stability plot for example (1), $1--N=1, 2--3T_s, 3--q_u>0, 4--N=2$ with OR, $5---N=2$

principle of arguments from the theory of functions of complex variables, the number of encirclements in the positive direction around the point $(0,0)$ is equal to the difference between number of unstable poles of the closed control loop and the number of unstable poles of the open loop. The shortest distance between the plotted curve and the critical point $(0,0)$ can be taken as a reasonable measure of the reserve in stability and thus also of robustness.

The return difference plots for control of the systems (1) and (36), for the nominal case $a^*(z)=a(z)$ and $b^*(z)=b(z)$, are given in Fig.12 and in Fig.13. The comparizon clearly shows that the sugested modifications of the LQG control design increase the robustness significantly. More general theoretical results are not available yet.

8   Numerical algorithms

In this Section an algorithm will be derived which makes it possible to solve the control synthesis according to Result 1 in a numerically save and efficient way. First, let us recall some facts, introduce a suitable notation, and introduce the elementary algorithm of dyadic reduction, which forms the numerical kernel of the procedure.

Any nonnegative definite matrix P can be expressed, in many ways, in the following factorized form, which can be understood as a sum of weighted dyads

$$P = M'D\,M = \sum_{i=1}^{\nu} m_i' D_i m_i \tag{43}$$

where M is a rectangular matrix, D is a diagnal matrix of compatible dimension, m_i is the i-the row of the factor M and the weight $D_i \geq 0$ is the corresponding diagonal element of D. Then the quadratic form $x'Px$ can be expressed as a sum of weigthed

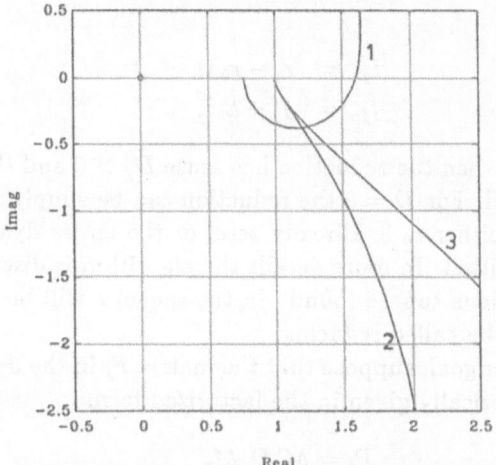

Figure 13: Stability plot for example (36); $1--N = 1, 2--N = 2$ with OR, $3--N = 3$ with OR

squares

$$x'Px = \sum_{i=1}^{\nu} D_i(m_i x)^2 \tag{44}$$

For our purposes it is suitable to introduce the following notation for nonnegative definite quadratic forms.

$$x'Px = \|D; Mx\|^2 \tag{45}$$

where D should be understood as a vector of weights assigned to corresponding rows. As no confusion can occur, the same notation D will be used both for the column vector of nonnegative weights and for the diagonal matrix. The factorization (43) is not unique and, for numerically given D and M, can be modified in various ways without changing the value of the quadratic form for arbitrary x. The algoritm of dyadic reduction is a suitable numerical tool for this purpose.

Consider the sum of two weighted dyads $f'D_f f + r'D_r r$ where

$$
\begin{aligned}
f &= [1, f_1, f_2, \cdots, f_n] \\
r &= [r_0, r_1, r_2, \cdots, r_n]
\end{aligned}
$$

Note that $f_0 = 1$. The algorithm of dyadic reduction modifies this sum of dyads

$$f'D_f f + r'D_r r = \tilde{f}'\tilde{D}_f \tilde{f} + \tilde{r}'\tilde{D}_r \tilde{r} \tag{46}$$

in such a way that

$$\tilde{f}_0 = f_0 = 1, \qquad \tilde{r}_0 = 0 \tag{47}$$

By this modification the dimension of the nonzero part of r is reduced by one, hence the name. It is easy to prove that the following algorithm performs the task.

$$
\begin{aligned}
\tilde{D}_f &= D_f + D_r r_0^2 \tag{48} \\
\tilde{D}_r &= (D_f/\tilde{D}_f)D_r \tag{49} \\
k_r &= (D_r/\tilde{D}_f)r_0 \tag{50}
\end{aligned}
$$

for $j = 1, 2, \cdots, n$

$$\tilde{r}_j \;\; = \;\; r_j - r_0 f_j \tag{51}$$
$$\tilde{f}_j \;\; = \;\; f_j + k_r \tilde{r}_j \tag{52}$$

Note that in all cases when the reduction has sense $\tilde{D}_f > 0$ and the divisions in (49) and (50) can be performed. For $\tilde{D} = 0$ the reduction can be simply skipped, as in such a case, besides $D_f = 0$, either r_0 is allready zero, or the entire dyad $r'D_r r$ has a zero weight D_r and can be omitted. In more details the algorithm is discussed in [2] and [8] where also other applications can be found. In the sequel r will be called the reduced row, while the row f will be called reducing.

To proceed towards our goal, suppose that the matrix P_j in the dynamic programing recursion (19, 20) is numerically given in the factorized form

$$P_j = M_z' D_z M_z \tag{53}$$

where M_z is a lower triangular matrix with ones on the main diagonal (monic LT-matrix) and D_z is diagonal. Then the minimazation step of dynamic programing, which produced the relation (19), reads

$$\min_{\Delta u(j)} \mathrm{E}\left[\hat{x}'(j|j) P_j \hat{x}(j|j) + q_u \Delta u(j) \;\middle|\; D_{j-1}, W_{j-1} \right] =$$

$$= \min_{\Delta u(j)} \left[\| D_z; M_z(F\hat{x}(j-1|j-1) + G\Delta u(j)) \|^2 + \| q_u; \Delta u(j) \|^2 \right] + \beta(j) \tag{54}$$

where $\beta(j)$ are the covariance terms which cannot be influenced and therefore do not enter the minimization procedure. Omitting the time indexing for simplicity, the term, which is to be minimized, can be expressed as follows.

$$\| D_z; M_z(F\hat{x} + G\Delta u) \|^2 + \| q_u; \Delta u \|^2 = \left\| \begin{bmatrix} q_u \\ D_z \end{bmatrix}; \begin{bmatrix} 1 & 0 \\ M_z G & M_z F \end{bmatrix} \begin{bmatrix} \Delta u \\ \hat{x} \end{bmatrix} \right\|^2 \tag{55}$$

Considering the special structure of matrices F and G, introduced by (15)

$$F = \begin{bmatrix} 1 & h_s' \\ 0 & -\bar{a} + H_s + \mu I \end{bmatrix} \qquad G = \begin{bmatrix} b_0 \\ \bar{b} - \bar{a} b_0 \end{bmatrix}$$

and considering also the corresponding partitioning of the monic LT–matrix M_z

$$M_z = \begin{bmatrix} 1 & 0 \\ m_r & M_s \end{bmatrix}$$

it is seen that, in order to determine the matrix products $M_z G$ and $M_z F$ in (55), the significant part of computation is

$$m_a = m_r - M_s \bar{a} \qquad\qquad m_u = m_a b_0 + M_s \bar{b} \tag{56}$$

where M_s is a monic LT-matrix. After this calculations, the quadratic form (55), which is to be minimized with respect to Δu, gets the form depicted in Fig.14. To minimize

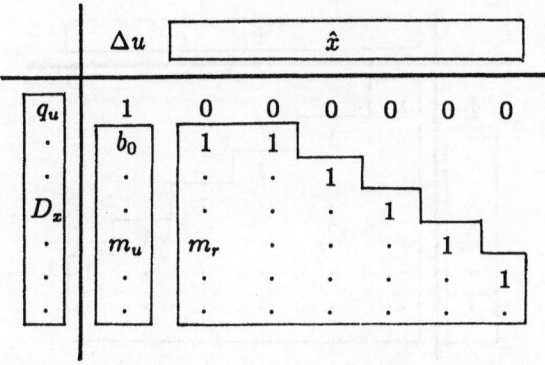

Figure 14: Scheme of the quadratic form (55)

such a quadratic form, represented by q_u, b_0 is inherent in graph. In principle, the point w_0 is included, but indeed Δu, and its values, the vector and... underlying column as prior. As shown in Figure 14, we have that neither the series in the data, matrix of the column... into the mass denoted $D_x^T x$. If so described, if the reduction is performed downward to... than the more along the first row, or the numerical area in the scheme, available at the still side (Fig. 2) becomes m_u, elements, program, as indicated in Fig. 15. After this modification, in the scheme of the quadratic form, which is found in the cells of the q_u and $\bar{q}_u$ as presented in the first row. Here, the quadratic form, represented by the down...

$$\ldots \ldots \ldots \quad [56]$$

where .... for each mapping as in............

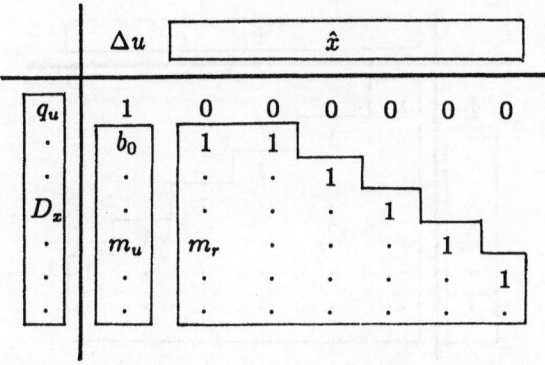

Figure 15: Application of dyadic reduction for minimization of the quadratic form (55)

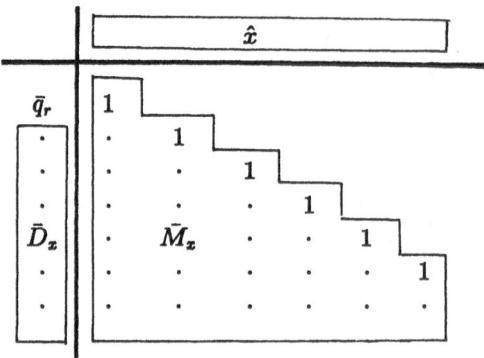

Figure 16: Scheme of the quadratic form (60)

such a quadratic form with respect to Δu it is sufficient to apply the dyadic reduction so that the one, lying under Δu, is used to reduce the rest of the underlying column to zeros, as shown in Fig.14. Note that neither the zeros in the empty space of the scheme nor the ones indicated in Fig.15 are destroyed if the reduction is performed downwards. At the same time the first row of the numerical area in the scheme, originally filled fith zeros (Fig.14), becomes, in general, nonzero as indicated in Fig.15. After this modification the only square of the quadratic form, which depends on the choise of Δu, is produced by this first row. Hence, the qudratic form is minimized by the choise

$$\tilde{q}_u(\Delta u - L\hat{x})^2 = 0 \tag{57}$$

where $L = L_j$ is the optimal control law for this j–th step. The minimum of the quadratic form then is

$$\min_{\Delta u}\left\|\begin{bmatrix} q_u \\ D_x \end{bmatrix}; \begin{bmatrix} 1 & 0 \\ M_x G & M_x F \end{bmatrix}\begin{bmatrix} \Delta u \\ \hat{x} \end{bmatrix}\right\|^2 = \left\|\bar{D}_x; \bar{M}_x\hat{x}\right\|^2 \tag{58}$$

Now, let us consider the regular case for $M = 1$ when the second part of the recursion (20) has the form

$$P_{j-1} = \bar{P}_{j-1} + hq_r h' \tag{59}$$

and the corresponding quadratic form is

$$\hat{x}'(j-1|j-1)P_{j-1}\hat{x}(j-1|j-1) = \left\|\begin{bmatrix} q_r \\ \bar{D}_x \end{bmatrix}; \begin{bmatrix} h' \\ \bar{M}_x \end{bmatrix}\hat{x}\right\|^2 \tag{60}$$

The "numerical shape" of this quadratic form shows the scheme Fig.16. To conclude the recursion numerically it is now sufficient to reduce the last row. Again the dyadic reduction can be employed to perform the task. However, not to destroy the desired structure the last row must be reduced from right to left using, as reducing, the rows with ones lying above the zeroed entries. The result is depicted in Fig.17 and the propagated matrix P_{j-1} is obtained in the factorized form

$$P_{j-1} = \tilde{M}'_x \tilde{D}_x \tilde{M}_x \tag{61}$$

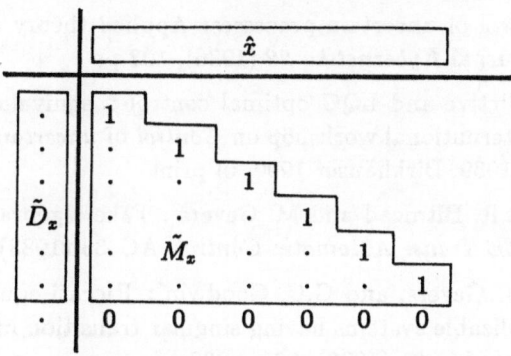

Figure 17: Reduction of the last row

where $\tilde{M}_x$ is again a monic LT–matrix like M_x in (53).

The propagation of the matrix P for steps with $\Delta u = 0$ can be performed in the same way except that the minimization, realized by reducing the column under Δu in the scheme Fig.14, is skipped, as this entire column is automatically zeroed by $\Delta u = 0$. Of course, the calculation of m_u can also be skipped.

An algorithm which propagates the parameters $\tilde{c}(t)$ of the state estimator (24) and replaces the recursion (28, 29) can be found in [2] or [8].

9   Conclusions

The paper will be concluded by the author's recomendation how to apply the presented results.

1. Choose a fake sampling interval, as a multiple of the true sampling interval, so that approximately 5 – 10 fake sampling intervals cover the expected step response of the controlled process.

2. Design the control law for the fake sampling rate, however with penalties set on all expected control errors within the receding horizon and no penalties on the process input.

3. To reduce overshoots omit penalties on the control errors within the first fake sampling interval.

4. Apply the control law with full sampling rate.

5. Use factorization algorithms to compute both the control law and the state estimator.

References

[1] Böhm J.: LQ self–tuners with signal level constrains. 7th IFAC/IFORS Symposium on Identification and System Parameter Estimation, York, UK, July 1985, Preprints, 131 – 135

[2] Peterka V.: Control of uncertain processes: Applied theory and algorithms. Supplement to the journal *Kybernetika 22* (1986), 102 pp

[3] Peterka V.: Predictive and LQG optimal control : equivalences, differences and improvements. International workshop on *Control of uncertain systems*, University of Bremen, June 1989, Birkhäuser 1990, in print

[4] Poubelle M.A., R.R. Bitmead and M. Gevers : Fake algebraic Riccati techniques and stability. *IEEE Trans. Automatic Control*, AC–33 (1988), 379 – 381

[5] de Souza C.E., M. Gevers, and G.C. Goodwin : Riccati equations in optimal filtering of nonstabilizable systems having singular transition matrices. *IEEE Trans. Automatic Control*, AC–31 (1986), 831 – 838

[6] Peterka V.: Combination of finite settling time and minimum integral of squared error in digital control systems. *Proc. 2nd World Congress IFAC*, Basel 1963, Butterworth, London 1964, 196-203

[7] Åström K.J. and B. Wittenmark : *Computer Controlled Systems, Theory and Design*. Prentice–Hall, 1984

[8] Peterka V.: Algorithms for LQG self-tuning control based on input–output Delta models. Proc. 2nd IFAC Workshop on *Adaptive Systems in Control and Signal Processing*, Lund, Sweden 1986, pp.13-18

Adaptation and Robustness

K J Hunt and P J Gawthrop

Control Group, Department of Mechanical Engineering
University of Glasgow, Glasgow G12 8QQ
Scotland

Abstract

Recent work has moved towards a unification of the areas of robust and adaptive
control, two areas which have traditionally received separate attention. In this work
we review the issues in the development of both *adaptive robust control* and *robust
adaptive control* and suggest further steps towards the practical exploitation of the
field.

1 Introduction

The treatment of plant uncertainty is the central issue of control system design. It is well
known that the use of time-invariant feedback can itself reduce the effect of uncertainty.
This fact was emphasised by Horowitz (Horowitz, 1963). However, the development of
control system design algorithms in which the effect of uncertainty is central has since
been tackled in different ways.

Broadly, there are two approaches to the treatment of uncertainty in control system
design which in the past have received separate attention. In *robust control* (Francis,
1987)(Lunze, 1988) the plant model used as the basis of a control design includes not only
a nominal model, but also a measure of the uncertainty associated with that model. In
adaptive control (Astrom and Wittenmark, 1989), on the other hand, the design normally
proceeds with recourse to the Certainty Equivalence Principle; the plant model, which is
normally obtained from a system identification scheme, is taken as a true representation
of the system and no explicit account is taken of the uncertainty associated with the
estimated model.

An interesting exception to this classification is *dual control* (Feldbaum, 1965). This
is an adaptive control method based on stochastic multi-step criterion optimisation. The
key feature of dual control is that the system will introduce perturbation signals to try
and improve estimation when the estimated model is uncertain. In the special case where
the criterion is minimised over only one step *cautious control* is obtained. In cautious
control the controller gain is reduced in proportion to any measured increase in estimator
uncertainty. Unfortunately, due to the computational intractability of the general dual
control problem this area is mainly of only theoretical interest.

It is the dichotomy in the differing approaches to system modelling which has resulted in the historically separate development of robust and adaptive control. Recently, however, there has been a significant movement towards the unification of robust and adaptive control methods (Goodwin et al., 1985; Gawthrop, 1985a; Kosut, 1988; Goodwin et al., 1989).

Robust control proceeds from a nominal system model and a description of the uncertainty associated with that model. The implied bounds on closed-loop uncertainty and achievable performance means that a given design for a specific system may result in the closed-loop specifications not being met. In this instance the closed-loop specifications may have to be relaxed. A further difficulty in robust control is the requirement to address all plants within a given model uncertainty set. This may result in a design which is overly conservative for the actual plant uncertainty likely to be encounted. One way of dealing with this issue using a fixed robust design is to explicitly model the plant uncertainty using structured perturbations (Doyle, 1982).

In adaptive control, on the other hand, a plant model is estimated and the control synthesis proceeds as if this model were a true representation of the plant. This procedure may lead to difficulties when the estimated model is inaccurate and, more seriously, when the *structure* of the plant model is incorrect. This problem has received much attention in the adaptive control literature, leading to a variety of design guidelines for the application of adaptive control.

However, the two approaches (robust non-adaptive and adaptive) control are not necessarily competitive. In particular they can be combined in at least two ways to give:

- robust adaptive control

- adaptive robust control

Robust adaptive control. The former approach looks at an adaptive controller and attempts to analyse the robustness of the adaptive controller. There is a large literature in this area (Anderson et al., 1986) (Kosut and Johnson, 1984) (Kosut, 1985) (Kosut and Friedlander, 1985) (Gawthrop, 1978) (Gawthrop, 1980) (Gawthrop, 1981) (Gawthrop and Lim, 1982) (Gawthrop, 1987b) (Gawthrop, 1985b) (Gawthrop, 1987c) (Gawthrop, 1987a) (Gawthrop and Kharbouch, 1988) . More recently, attention has turned to using this analysis as an aid to *synthesising* robust adaptive controllers (Gawthrop, 1985a) (Gawthrop, 1988) based on Horowitz's quantitative feedback theory (QFT) (Horowitz, 1982). This approach to QFT-based adaptive control is described in Section 4.2. It must be emphasised that QFT is used to analyse the robustness of the adaptive algorithms and *not* for designing the underlying controller as discussed in the next paragraphs.

Adaptive robust control. The latter approach is based on the idea of using adaptation in conjunction with a robust design method. The role of adaptation in adaptive robust control is to reduce, by estimation, the plant uncertainty. In this way the extent of uncertainty which the robust design must address can be reduced.

The differentiating feature, and the implied requirement, of adaptive robust control is the need to explicitly quantify the uncertainty in the estimated plant model (Kosut, 1988; Goodwin et al., 1989). Moreover, this uncertainty must be in a form compatible with the uncertainty representation expected by the robust control design scheme to be used.

A further possibility which can reduce the model uncertainty is to estimate multiple plant models concurrently (Peterka, 1989). Each model estimated differs in assumed order and time delay. The estimation can be arranged in such a way that a confidence measure is associated with each model. The model with highest confidence is chosen as the basis for control design.

These estimator issues (uncertainty quantification and multiple models) are discussed in Section 3.

The key issue in the selection of an appropriate control design method is to match the plant description, together with the quantitative uncertainty measure, produced by the estimator to a control design utilising a plant description in this form. Potential robust control designs which meet this criterion are discussed in Section 4.1.

This argument is based upon the traditional *indirect* approach to adaptive control; perform a model estimation step followed by a controller synthesis whenever controller adaptation is required. Another approach is to begin with a nominal plant model together with its *a priori* uncertainty description and to perform an off-line robust design. Following this the parameters of the robust controller are adapted on-line using some scheme which takes account of reduced uncertainty. This has been done in the context of Horowitz's quantitative feedback theory (QFT) (Yaniv et al., 1986). In this way the need to recompute the robust controller is avoided.

2 Modelling Uncertainty

To perform a robust design requires a specification of the type of plant uncertainty together with a description of the range of uncertainty. There are many ways of mathematically representing uncertainty; see, for example the book by Lunze (Lunze, 1988). Some examples appear in this section.

Assume that the plant input-output relation is described by

$$y = G_T u \tag{1}$$

where y and u are the measured output and the control input, respectively. G_T is the true plant transfer function.

Unstructured plant perturbation

The true plant is decomposed into a parametric nominal part G_0 together with an unstructured perturbation G_Δ^g as follows:

$$G_T = G_0 + G_\Delta^g \tag{2}$$

The perturbation G_Δ^g represents the global uncertainty associated with the plant. This global uncertainty includes parameter inaccuracies, parameter time variations and plant undermodelling.

A frequency-dependent upper bound $\delta^g(\omega)$ on the global uncertainty can be defined by:

$$|G_\Delta^g(\omega)| \leq \delta^g(\omega) \quad \forall \omega \tag{3}$$

This rather crude measure can lead to conservative design.

Structured plant perturbation

There are many ways of defining a *structured* plant perturbation. One of these is the QFT *template* (Horowitz, 1963)(Horowitz, 1982) where the bound in inequality (3) is replaced by set inclusion:

$$G_\Delta^g(\omega) \in T(\omega) \quad \forall \omega \tag{4}$$

and $T(\omega)$ is a region of the complex plane (usually represented on a Nichols chart). This representation can often be related directly to the actual plant characteristics.

Robust control

Robust control design proceeds from a *constant* nominal model $\bar{G}_0$. Given the upper bound on the global uncertainty, the robust controller must stabilise *all* plants G where $G(\omega)$ belongs to the set

$$S^g(\omega) = \{\bar{G}_0(\omega) + G_\Delta(\omega) : |G_\Delta(\omega)| \leq \delta^g(\omega)\} \tag{5}$$

Adaptive control

In adaptive control an estimate $\hat{G}$ of the nominal plant model is used. We associate with this estimate an *a posteriori* uncertainty G_Δ^l defined in relation to the true plant as

$$G_T = \hat{G} + G_\Delta^l \tag{6}$$

Assuming that an upper bound $\delta^l(\omega)$ on the *a posteriori* uncertainty is available, we have

$$|G_\Delta^l(\omega)| \leq \delta^l(\omega) \quad \forall \omega \tag{7}$$

Thus, the adaptive controller must stabilise all plants G where $G(\omega)$ belongs to the set

$$S^l(\omega) = \{\hat{G}(\omega) + G_\Delta(\omega) : |G_\Delta(\omega)| \leq \delta^l(\omega)\} \tag{8}$$

If the model *structure* is incorrect, then the parameter estimator cannot reduce G_Δ^l to zero.

3 Parameter Estimation and Adaptive Control

3.1 Adaptive robust control

In adaptive robust control we begin with a robust control design method having a fixed nominal model together with its global uncertainty and associated upper bound $\delta^g(\omega)$. The robust design is then combined in a standard indirect adaptive control scheme with a parameter estimation algorithm. The differentiating feature of this scheme is that the estimation produces not only an estimate $\hat{G}$ of the plant, but also delivers a quantitative measure of the upper bound $\delta^l(\omega)$ on the *a posteriori* uncertainty.

The adaptive robust controller incorporates a decision scheme which will utilise the minimal model uncertainty i.e. the estimated plant model is used whenever

$$\delta^l(\omega) < \delta^g(\omega) \tag{9}$$

and the constant nominal model associated with the fixed robust design is used otherwise.

Such a scheme is clearly superior to a fixed robust design since the controller must only stabilise all systems G such that $G(\omega)$ lies in the intersection of the sets (5) and (8) i.e.

$$G(\omega) \in S^g(\omega) \cap S^l(\omega) \tag{10}$$

This use of uncertainty set quantification as the basis of adaptive robust control has been developed by Kosut (Kosut, 1988) and Goodwin *et al* (Goodwin et al., 1989).

3.2 Multiple Models

In the implementation of a parameter estimation scheme a number of design variables must be selected *a priori*. Key design variables relating to the model structure are the order n of the estimated model and the model time delay d. In stochastic models it is also often desirable to pre-select the moving average part of the model (the c-parameters).

A method for reducing the uncertainty in the estimated model is to consider more than one prior choice of the design triple $\{n, d, c\}$. This involves running a number of estimators in parallel, one for each model (Peterka, 1989).

Using the Bayesian approach the confidence in each model can be quantified, and the best model selected. In this approach each prior choice of $\{n, d, c\}$ is considered as an uncertain hypothesis. The probability of each hypothesis, conditioned on the observed data, is determined in parallel in real-time.

Formally, assume that a finite number N of choices of the design triple $\{n, d, c\}$ has been selected. Each choice is considered as an uncertain hypothesis H:

$$H_j : \{n, d, c\} = \{n_j, d_j, c_j\} \quad j = 1, 2, \ldots N \tag{11}$$

These uncertain hypotheses are determined using the observed data. Denote the set of plant input-output data up to the current time t by $\mathcal{D}_t$. The model estimate resulting in the highest confidence can be determined by comparing the probabilities of each hypothesis H i.e. the following probabilities must be calculated:

$$p(H_j|\mathcal{D}_t) = p[\{n, d, c\} = \{n_j, d_j, c_j\}|\mathcal{D}_t] \quad j = 1, 2, \ldots N \tag{12}$$

An analytical formula for these probabilities can be determined using Bayesian analysis. Using this technique in conjunction with the adaptive robust control scheme outlined in the preceding section is a further step towards reducing the uncertainty associated with model estimation.

4 Robust Control Design

The preceding section looked at desirable properties of adaptive robust control from the parameter estimator point of view. Estimation was viewed as a mechanism for reducing the plant uncertainty. We now look at candidate robust control design algorithms to be used in conjunction with these estimators.

4.1   Analytical synthesis

In discrete time methods parameter estimation usually produces a parametric description
of the plant in transfer-function form:

$$G(q^{-1}) = \frac{B(q^{-1})}{A(q^{-1})} = \frac{b_1 q^{-1} + \ldots + b_m q^{-m}}{1 + a_1 q^{-1} + \ldots + a_n q^{-n}} \tag{13}$$

where q^{-1} is the unit delay operator.

It is also possible to identify a *continuous-time* model directly (Unbehauen and Rao,
1987)(Gawthrop, 1987a) in the form:

$$G(s) = \frac{B(s)}{A(s)} = \frac{b_0 s^m + \ldots + b_m}{s^n + a_1 s^{n-1} + \ldots + a_n} \tag{14}$$

The key issue in the selection of an appropriate control design method is to match
this plant description, together with the quantitative uncertainty measure, to a control
design utilising a plant description in this form.

The mainstream of robust control research, however, is based upon a state-space plant
description and is not readily amenable to on-line analytical synthesis. An exception to
this, and a possible candidate for use in an adaptive robust control algorithm, is H_∞ design
based on polynomial optimisation (Kwakernaak, 1986; Grimble, 1986). The *polynomial
equation approach* to LQ synthesis was developed by Kučera (Kucera, 1979) and has
provided a useful basis for the development of LQ self-tuning control (Grimble, 1984; Hunt,
1989). The polynomial approach was then used for H_∞ optimisation in the previously
cited works. Given the close match between the estimator plant model structure and that
used in polynomial optimisation, this approach has potential for adaptive robust control.

4.2   QFT and adaptive control

The QFT (Quantitative Feedback Theory) approach has been developed by Horowitz
and co-workers. This provides an alternative to analytical synthesis methods for robust
control. QFT, which is based upon a straightforward application of linear time-invariant
feedback, is so called because it gives a *quantitative* assessment of the benefits and costs
of the feedback.

QFT is based upon a plant model of the form

$$y = Gu + d \tag{15}$$

where y, u and d are the measured plant output, the control input and the measurement
noise, respectively. Plant uncertainty is characterised by bounds on the frequency response
as in Equation (4).

For setpoint tracking applications QFT utilises a two degrees of freedom control structure described by

$$u = H_1 w - H_2 y \tag{16}$$

where w is a reference, or command, signal and H_1 and H_2 are a reference feedforward
compensator and a feedback compensator, respectively.

Selection of H_1 and H_2 depends upon the well known trade-off between sensitivity reduction and sensor noise amplification. QFT proceeds by finding H_2 with the minimum possible loop gain to satisfy sensitivity requirements and then selects an H_1 to achieve the desired setpoint response. This minimum loop gain may, however, give a feedback transfer-function H_2 which produces excessive sensor noise amplification. In this case the sensitivity specifications must be relaxed. It is this inflexible trade-off between the sensor noise and sensitivity requirements that the QFTS algorithm seeks to address.

The QFTS algorithm (Gawthrop, 1985a; Gawthrop, 1988) uses QFT to design a self-tuning controller which can control a range of uncertain plants.

The QFTS algorithm is based upon a continuous time version (Gawthrop, 1987a) of the self-tuning GMV (generalised minimum variance) controller (Clarke and Gawthrop, 1975; Clarke and Gawthrop, 1979) based on parametric identified models of the form of Equation (14).

This continuous time self-tuning controller leads to an artificial notional feedback system having a number of design polynomials. The robustness of this self-tuning controller has been analysed and is characterised by a transfer function depending on the plant and the design polynomials.

The key idea behind the QFTS algorithm is first to select the self-tuner's design polynomials to correspond to a specific QFT design. Due to the fact that the feedback loop of the self-tuner is a notional artifact, the sensitivity properties of the QFT design are destroyed. However, in the self-tuning implementation the sensitivity properties are recovered asymptotically through estimation while the low sensor noise amplification property is retained. Some example appear in (Gawthrop, 1988).

5   Conclusions

The paper has surveyed a cross section of ongoing research committed to combining desirable properties from robust (non-adaptive) and adaptive control. We conclude by mentioning some potential future directions of this work.

Controller performance can be improved by reducing uncertainty at the outset. Thus, if possible, the controller (adaptive or non-adaptive) should be designed to capture all available information about the plant and its environment. Using a 'black-box' approach to controller design is *not* the way forward.

For this reason, we believe that a key direction of future research is the design of controllers which can incorporate prior plant knowledge. As such prior knowledge is encapsulated in the form of non-linear differential equations, linear discrete-time representations (as Equation (13)) are unlikely to have a future in this line of research. We believe, therefore, that a *model-based* approach is necessary if we are truly to incorporate prior knowledge in our designs: each controller should contain a complete model of the plant together with uncertainties specified in terms of the physical plant components. This leads to challenging research to characterise uncertainty, based on physical parameters embedded in a partially known (and probably intrinsically non-linear) physical system model, and to develop corresponding design techniques for adaptive and non-adaptive control.

To summarise, there is a need for research into modelling, control and parameter estimation in the context of physically-based *partially known systems*. Such methods, though generic in concept, must be able to encapsulate domain and plant specific information.

References

Anderson, B., Bitmead, R., Johnson, C., Kokotovic, P., Kosut, R., Mareels, I., Praly, L., and Reidle, B. (1986). *Stability of Adaptive Systems: Passivity and Averaging Analysis*. MIT Press.

Astrom, K. and Wittenmark, B. (1989). *Adaptive Control*. Addison-Wesley.

Clarke, D. and Gawthrop, P. (1975). Self-tuning controller. *Proceedings IEE*, 122(9):929–934.

Clarke, D. and Gawthrop, P. (1979). Self-tuning control. *Proceedings IEE*, 126(6):633–640.

Doyle, J. C. (1982). Analysis of feedback systems with structured uncertainties. *Proc. IEE, Pt. D*, 129:242–250.

Feldbaum, A. A. (1965). *Optimal Control Theory*. Academic Press, New York.

Francis, B. (1987). *A Course in H-infinity Control Theory*. Springer.

Gawthrop, P. (1978). On the stability and convergence of self-tuning controllers. In *Proceedings of the IMA conference on "The Analysis and Optimisation of Stochastic Systems"*. Academic Press.

Gawthrop, P. (1980). On the stability and convergence of a self-tuning controller. *Int. J. Control*, 31(5):973–998.

Gawthrop, P. (1981). Some properties of discrete adaptive controllers. In Harris, C. and Billings, S., editors, *Self-tuning and adaptive control — theory and applications*. Peter Peregrinus.

Gawthrop, P. (1985a). Comparative robustness of non-adaptive and adaptive control. In *Proceedings of the IEE conference "Control '85", Cambridge, U.K.*

Gawthrop, P. (1985b). Robust self-tuning control of n-input n-output systems. In *Preprints of the 7th IFAC/IFORS Symposium on Identification and System Parameter Estimation, York, U.K.*

Gawthrop, P. (1987a). *Continuous-time Self-tuning Control. Vol 1: Design*. Research Studies Press, Engineering control series., Lechworth, England.

Gawthrop, P. (1987b). Input-output analysis of self-tuning controllers. In Singh, editor, *Encyclopedia of Systems and Control*. Pergamon.

Gawthrop, P. (1987c). Robust stability of a continuous-time self-tuning controller. *International journal of Adaptive Control and Signal Processing*, 1(1):31–48.

Gawthrop, P. (1988). Quantitative feedback theory and self-tuning control. In *Proceedings of the IEE conference "Control '88", Oxford, U.K.*

Gawthrop, P. and Kharbouch, M. (1988). Two-loop self-tuning cascade control. *Proc. IEE Pt.D*, 135(4):232–238.

Gawthrop, P. and Lim, K. (1982). On the robustness of self-tuning controllers. *Proc. IEE*, 129 ptD:21–29.

Goodwin, G. C., Hill, D. J., and Palaniswami, M. (1985). Towards an adaptive robust controller. In *Proc. IFAC Symposium on Identification and System Parameter Estimation, York, England.*

Goodwin, G. C., Mayne, D. Q., and Salgado, M. E. (1989). Uncertainty, information and estimation. In *Proc. IFAC Symposium on Adaptive Systems in Control and Signal Processing, Glasgow, Scotland.*

Grimble, M. (1984). Implicit and explicit LQG self-tuning regulators. *Automatica*, 20:661–670.

Grimble, M. J. (1986). Optimal H_∞ robustness and the relationship to LQG design problems. *Int. J. Control*, 43:351–372.

Horowitz, I. (1963). *Synthesis of Feedback Systems.* Academic Press.

Horowitz, I. (1982). Quantitative feedback theory. *Proc. IEE*, 129 Pt. D.(6):215–226.

Hunt, K. (1989). *Stochastic Optimal Control Theory with Application in Self-tuning Control.* Springer-Verlag: Berlin.

Kosut, R. (1985). Methods of averaging for adaptive systems. In *Preprints of Yale workshop on Applications on Adaptive Control.*

Kosut, R. and Friedlander, B. (1985). Robust adaptive control: Conditions for global stability. *IEEE Transactions on Automatic Control*, AC-30(7):610–624.

Kosut, R. and Johnson, C. (1984). An input-output view of robustness in adaptive control. *Automatica*, 20(5):569–582.

Kosut, R. L. (1988). Adaptive robust control via transfer function uncertainty estimation. In *Proc. American Control Conference, Atlanta, USA.*

Kucera, V. (1979). *Discrete Linear Control: The Polynomial Equation Approach.* Wiley, Prague.

Kwakernaak, H. (1986). A polynomial approach to minimax frequency domain optimization of multivariable feedback systems. *Int. J. Control*, 44:117–156.

Lunze, J. (1988). *Robust Multivariable Feedback Control.* Prentice Hall.

Peterka, V. (1989). Self-tuning control with alternative sets of uncertain process models. In *Proc. IFAC Symposium on Adaptive Systems in Control and Signal Processing, Glasgow, Scotland.*

Unbehauen, H. and Rao, G. (1987). *Identification of Continuous Systems.* North-Holland, Amsterdam.

Yaniv, O., Gutman, P., and Neumann, L. (1986). An algorithm for adaptation of a robust controller to reduced plant uncertainty. In *Preprints of the 2nd IFAC workshop on Adaptive Systems in Control and Signal Processing, Lund, Sweden.*

Digital Controllers for ŠKODA Rolling Mills

Pavel Ettler and František Jirkovský

ŠKODA Works 30-55
316 00 Plzeň, Czechoslovakia

1 Introduction

Despite of the fact that much was written and spoken about adaptive control, industrial implementations of the control which can be considered from a theoretical point of view as **really** adaptive are far not common.

Nowadays when technical limits like a processor speed, an amount of computer memory and dynamic properties of sensors and actuators cannot be in most cases considered as causes of such slow utilization of the adaptive control, these causes must be sought elsewhere. According to our experience implementation of a really adaptive control is prevented by:

1. the control aim can be almost reached by simple controllers

2. considered type of a technological process is already satisfactory controlled by conventional controllers

3. *ad hoc* approaches are seemed by control engineers to be more straightforward, less expensive and less time-consuming

4. it exists some kind of a *communication barrier* between academic people and control engineers and different motivations for both theses groups

Our relatively small group of people interesting in implementation of modern approaches of automatic control has a couple of good results in this field [7]. This would have been impossible without close contacts with people from theoretical institutes – in our case from The Institute of Automation an Information Theory of The Czechoslovak Academy of Sciences (ÚTIA) and Department of Technical Cybernetics of the Technical University in Plzeň. This co-operation began about ten years ago and was originally motivated by the problem of the rolled strip thickness control. Nowadays the number of our control aims has grown – and with it the need for an unified theoretical approach and for corresponding control software (if it even can be unified) – has increased as well.

2 Our control problems

The following sections depict control problems with which we have or will have something to do. We tried to use the Bayesian stochastic approach by their solution [13,14].

2.1 Strip thickness control

During rolling of a metal strip – both cold and hot – one of the main quality index is the variation of the outgoing strip thickness from its nominal value. Due to the changing thickness and non-homogenity of the ingoing strip, changing temperatures of the mill and the strip itself and other influences, from which only some can be measured, the problem in common leads to the adaptive control. We succeeded in finding the mathematical model, parameters of which estimated in real-time are used for evaluation of the coefficients of the *cautious* controller [5] . This is an application of the well-known separation principle. Model parameters are identified by the algorithm called the directional forgetting which ensures the numerical stability by the LDL^T-like matrix factorization [11] .

This approach brought good results on reversive twenty–roll cold rolling mills. The first of them in Kovohutě Rokycany is equipped with such a controller since 1984 [6].

These implementations proved advantages of the stochastic approach to control – even if it took a relatively long time of study and experiments. In the first stage, when we were looking for an appropriate mathematical model based on measured data, we combined the *ad hoc* methods with theoretical approaches of ÚTIA [8,9] , which were solved algorithmically at that time but were not implemented in software suitable for a straightforward industrial application . We felt the need for an **integrated reliable** software support even if we have not been perfectly convinced of a possibility of its **final** realization.

2.2 Flatness control

Flatness or profile control is another interesting control task concerning rolling, especially rolling of a wide, thin strip. This problem can be theoretically solved by the same approach mentioned in the preceding section. But unlike the thickness control the necessary sensors of which make now no technical problems at all, the cornerstone of the flatness control seems to be the flatness or profile sensor.

Our people spent some time by research in this field, both on profile sensor development and its data processing, especially during 1977–1980 on the Innocenti-Sendzimir twenty-roll rolling mill in Kovohutě Povrly. The Bayesian approach promoted by ÚTIA was tested for this essentially multidimensional case. Addmissible results were achieved as regards visualization of the strip profile but this work did not lead to the closed loop control mainly because of the sensor.

Only very few world firms have mastered the really reliable implementation of the flatness meter and it has seemed to be more economic to buy such a good working sensor from its producer – together with the flatness control of course – then to go on in developing the own one.

2.3   Position control

Control of the roll positioning system has been traditionally realized by a manually adjustable continuous (analogous) controller. The first attempt to make it digitally was done on the four-high cold rolling mill in Kovohutě Rokycany but in that case it concerned the comparatively slow electromechanical screwdown system. Nevertheless this digital servoregulator together with an adaptive controller, similar to the above mentioned one, works satisfactory since 1983 [3] .

Current task is to control digitally much faster electrohydraulic roll positioning system. The first rolling mill – the reversive two–high hot strip rolling mill – equipped with such a servoregulator, which is one part of a multilevel digital controller depicted in another section, is just in commissioning.

An interesting fact seems to be (according to us) that in this case we did not decide to utilize the same stochastic approach to control as in another cases. Here we were probably influenced by some of the causes enumerated in the beginning. After some time of looking for satisfactory mathematical model which corresponds to the reality in all cases, we decided for a digitally realized PID controller, even if a modified one. During tuning the controller's coefficients we tested several *signal oriented* algorithms [12,1] but they did not bring remarkably better quality of regulation in comparison when the coefficients were tuned manually.

2.4   Roll eccentricity compensation

Roll eccentricity and ovality did not caused serious problems to us in the past , while till the last year we were involved in control of the twenty-roll mills with very small roll diameters. This task has appeared for us as we are now interested in the thickness controll on two–high hot– and especially four–high cold rolling mills with considerably greater rolls. We have an advantage of a possibility to let us inspire by various solutions of other firms [4,10,15] . Here also can be fruitful to utilize the Bayesian approach to testing of statistical hypotheses, at least for the detection of the periodic variations of the thickness imprinted into the outgoing strip. Mathematical models belonging to simple hypotheses differ in their periodical components. Interval of used frequencies can be easily determined from the interval of admissible roll diameters. As this interval is relatively small the dominant frequency can be found with significantly less computing effort and with finer resolution then when we use the Fast Fourier Transform algorithm [2] . The first results of the solution of this subproblem which are based on both simulated and measured data seem to be promising .

The Bayesian testing of statistical hypotheses can also be used as a tool for the fault diagnostics in industrial systems. This would probable be one of our area of interest in the future. The adaptive filtering is for us another interesting topic.

3   Unified realization of technical solution

Although our first adaptive thickness controller was implemented in a minicomputer, our hardware is now based on microcomputers with sixteen bits processors. Today's technical solution uses several multiprocessor industrial microcomputers connected together via the BITBUS network.

On the two–high hot rolling mill in Kovohutě Rokycany this system contents one two-processor microcomputer equipped with input/output boards which realizes the servoregulator, the controller of the electromechanical screwdown system, mill stretch compensator and logical functions. The second microcomputer connected with the first one via the BITBUS network communicates with operator and with the microcomputer-based isotope thickness meter. Both microcomputers are connected via serial links with a personal computer. This serves as a program development system and as a tool for data storing and analysis.

Software solution is based on the iRMX 86 real-time operating system on both industrial and personal types of computers.

Similar technical solution is used for other rolling mills but for example in the case of the four–high cold rolling mill in Frýdek–Místek it consists of four industrial microcomputers. The BITBUS network serves as a data highway again.

4   Conclusions

Industrial utilization of the adaptive control or its elements is determined by a long-time conception and by co-operation with academic people. Our group of control engineers fullfils these requirements and has gained considerable experience in this field during its about ten years of existence.

The Bayesian approach applied to solutions of control problems concerning rolling mills brought in same cases good results in the other cases seems to be promising and must be still elaborated.

References

[1] ÅSTRÖM K.J., HÄGGLUND T.: Automatic tuning of simple regulators with specifications on phase and amplitude margins. In: Automatica Vol. 20, No.5, 1984, pp.645–651.

[2] BRIGHAM E.O.: The Fast Fourier Transform. Prentice Hall, New Jersey 1974

[3] CENDELÍN J, JIRKOVSKÝ F.: Microcomputer control system for automatic thickness control. In: Preprints of the conference AUTOS'84, Plzeň 1984. In Czech.

[4] EDWARDS W.J. and others: Roll eccentricity control for strip rolling mills. In: Preprints of IFAC World Congress 1987, pp.200–211.

[5] ETTLER,P.: An adaptive controller for ŠKODA twenty–roll cold rolling mills. In: Proceedings of 2nd IFAC Workshop on Adaptive Systems in Control and Signal Processing, Lund Sweden, Lund Institute of Technology, 1986, pp.277–280.

[6] JIRKOVSKÝ,F. ,ETTLER,P.: ŠKODA cold rolling mills (Microsystem 85 Reports), The Euromicro Journal, Vol. 17, Nr. 4, April 1986, Elsevier Science Publishers B.V. (North–Holland)

[7] JIRKOVSKÝ F. ,ETTLER P. ,JANEČEK E.: Trends in application of adaptive controllers for rolling mills. In Automatizace 1989/5, pp. 139–143. In Czech.

[8] KÁRNÝ M.: Algorithms for determining the model structure of a controlled system. In Kybermetika 1983/2, Academia Praha.

[9] KÁRNÝ M. ,KULHAVÝ R.: Structure determination of regression-type models for adaptive prediction and control. In: Bayesian analysis of time series and dynamic models (J.C.Spall, ed.). Marcel Dekker, New York 1988, pp.313–345.

[10] KITAMURA A. and others: Recursive identification technique for roll eccentricity control. In: Preprints of IFAC World Congress 1987, pp.126–131.

[11] KULHAVÝ R.: Tracking of slowly varying parameters by directional forgetting. Preprints of IFAC World Congress, 1984, Budapest, Vol.X, pp.178–183.

[12] MARŠÍK J.: A new conception of digital adaptive PSD control. In Problems of Control and Information Theory, 12, Budapest 1983, pp.267–279.

[13] PETERKA V.: Bayesian approach to system identification. In: Trends and Progress in System Identification (P.Eykhoff, ed.). IFAC Series for Graduates, Research Workers and Practicing Engineers 1, Pergamon Press, 1981, pp.239–303.

[14] PETERKA V. and others: Algorithms for adaptive microprocessor control of industrial processes. ÚTIA ČSAV, Praha 1982. In Czech.

[15] SIEMENS: New method compensates backup roll eccentricity, affording closer tolerances. In: Ideas for steel. A19100-E283-B122-X-7600

AN EXPERT SYSTEM AS A CONTROLLER
AND ITS CLOSED LOOP BEHAVIOUR

O. Kříž, R. Jiroušek
Institute of Information Theory and Automation
Czechoslovak Academy of Sciences
Prague

In our institute, a new concept of expert systems has been developped [1], [2], [3]. This contribution is an effort to investigate its non–traditional application in the field of automatic control. Therefore, exaplanation of the theoretical foundations of the system INES are deliberately omitted and the interested reader should refer to [2].

INES is a type of an expert system designed for decision–making (diagnosing) under uncertainty. Decision making with certain specific features exists in the field of automatic control, too. Here, parameters of the system's model are identified and serve as a basis for selection of the corresponding parameters of the controller. The goal of the choice is to make the resulting behaviour of the controlled system optimal in the sense of a given criterion. System and controller are supposed to have fixed structure (algebraically expressible).

Classical ES decide what diagnosis should be made i. e. what respective value from the range of the diagnosis variable. The decision for a specific case is the function of concrete values of symptom variables for this specific case. The general information that makes this concrete decision possible is stored in a so called knowledge base. It corresponds to a set of axioms existing for the branch of science, it is supplied by a person called the expert and the final diagnosis is inferred on its basis solely.

When dealing with the issue of potential application of ES in automatic control, it is necessary to determine the mutual correspondence of basic notions in both conceptual frames. From different possible approaches we have chosen the following one:

The basic setting is a system and a controller connected in a closed loop. Noise that perturbs the output of the system belongs to certain family of distributions e. g. gaussian. The diagnosis variable is the output r_T of the controller. Symptom variables $s_{T-4}, s_{T-3}, \ldots, s_{T-1}, r_{T-4}, \ldots, r_{T-1}$ are computed from the outputs of the system $s(t-1), s(t-2), \ldots$ measured at previous time points as well from the control in the past $r(t-1)$, $r(t-2), \ldots$. The control law that is to be minimized is the average of squares of differences between the prescribed $s_0(t)$ and actual $s(t)$ value of the system output s at successive discrete time points $1, 2, \ldots, t,\ t \in N$

$$\lim_{n \to \infty} \frac{1}{n} \sum_{t=1}^{n} [s(t) - s_0(t)]^2 \longrightarrow \min$$

The expert as a source of knowledge is here substituted by an "optimal" controller. What is the "optimal"?

In this study with the given system, noise and criterion the structure and parameters of the best controller can be determined from the theory and therefore this controller will be considered as an ideal whose behaviour should be imitated.

In real–world situations we observe the closed–loop behaviour of the system and a concrete controller (e. g. PID). If we like the control process (either objectively: reasonable dispersion, not exceed boundaries or sub-jectively by mere observing the graph of system's history) we declare the controller to be the "optimal" and try to imitate its behaviour. No specific information about the real structure of the controller is needed. We just simply add the record of its closed loop performance to the input informa-tion that will be used for generating of the knowledge base. As a special case and that is the main motivation of this study a human operator can be con-sidered. Such people are said often to have better results than any artificial controller. They know how to control but it is practically impossible to get this knowledge out of them in a formalized way. With this approach, taking their behaviour as the input information will suffice. Having data about the system with different controllers which all are considered to be "optimal", all this data should be used for subsequent processing. Input information is one or several data sets with measured variables at different points at the system. Each record corresponds to certain time point, measurements are taken equidistantly.

In classical ES, rules of type

IF   A   THEN   B   (w)

are considered as basic pieces of information. Expert system INES [2] used for the purposes of this study interprets this rule as follows:

The propositions A and B describe the closed loop behaviour. A corresponds to symptom variables e.g. $A \equiv s_{T-3}(t) \in a_1 \, \& r_{T-1}(t) \in a_2, \quad a_1, a_2 \in R$. B is a hypothesis about potential control r_T at time t e.g. $B \equiv r_T(t) \in b$, $b \in R$. If the state of the system and of the controller in the past satisfies A then the optimal controller would behave according to B with weight w equal to conditional probability $P(B|A)$.

It is more natural for INES and more convenient for the user to supply information not as individual rules but as small–dimensional distributions (or contingency tables) which are marginals (conditional probabilities can be easily transformed to this form) of a certain joint distribution P that is supposed to describe completely the closed loop behaviour of the system with the "optimal" controller.

The data sets with observed behaviour (globally refered to as the training data set) are used then for generation of the marginals (usually $\sim 10 - 500$). Then a part of the system INES called INTEGRATOR integrates those marginals into structures able to generate the approximation $\tilde{P}$ of the theoretical all–describing distribution P.

These structures represent the knowledge base for the expert system – driven controller. It should be emphasized that prior to computing of marginals the ranges of variables must be discretized to several equivalence classes to keep the sizes of structures algorithmically tractable.

With the knowledge base constructed from the training set let us now describe the gist of the decision that takes part in the subsystem of INES called APPLICATOR. The last decisions and system reactions are stored again in shift registers of sufficient length. The APPLICATOR takes all these available "symptoms" fulfilling A and computes conditional probabilities $\tilde{P}(d_i/A)$ for all decisions d_i as $\tilde{P}(d_i, A)/\tilde{P}(A)$.

The length of regression m ($\equiv$ the length of the shift registers storing state s, control r and difference of state Δs) was selected to be 5 (a value supposed not to be exceeded by the order of the system).

The total number of discretized variables is therefore 15: $s_{T-4}, r_{T-4}, \Delta s_{T-4}, \ldots, s_T, r_T, \Delta s_T$ and they correspond to $s(t-4), r(t-4), \Delta s(t-4), \ldots, s(t), r(t), \Delta s(t)$. Symptom variables are $s_{T-4}, r_{T-4}, \ldots, s_{T-1}, r_{T-1}, \Delta s_{T-1}$, diagnosis variables is r_T and $s_T, \Delta s_T$ being result of r_T are not interesting for its determination. The 28 marginal distributions generated from the training set (9995 15–tuples) and used as input to the INTEGRATOR were chosen as all pairs from $\{r_{T-3}, \Delta s_{T-3}, r_{T-2}, \ldots r_T, \Delta s_T\}$ plus diagnosis variable r_T i.e. $(r_{T-3}, \Delta s_{T-3}, r_T), (r_{T-3}, \Delta s_{T-2}, r_T), \ldots, (r_{T-1}, \Delta s_{T-1}, r_T)$.

The system to be controlled is a regression model of the first order, stable, with minimal phase and static gain 95. It is described by the equation

$$s(t) = 0.98\, s(t-1) + r(t) + 0.9\, r(t-1) + e(t)$$

where $e(t)$ is a discrete–time white noise $\sim N(0,1)$. The goal is to keep the output of the system $s(t)$ as close to zero level as possible i.e. $s_0(t) \equiv 0$. The "optimal" controller (O)

$$r(t) = -0.98\, s(t-1) - 0.9\, r(t-1)$$

reduces the system output to actual noise and achieves thus the theoretical dispersion 1.0.

To get other references to compare ES–driven controller (E) with, the "zero" controller (Z)

$$r(t) \equiv 0$$

and the following simple proportional controller (P)

$$r(t) = A_R\,(k + 1 - D_s(s(t-1)))$$

were introduced.

To describe transformation from discrete to analogue values and vice versa 6 conversion functions

$$A_x,\ D_x \qquad x \in \{s, r, \Delta s\}$$

were defined

$$A_x :< 1, 2, \ldots, k > \longrightarrow R \qquad
\begin{aligned}
A_x(i) &= &(x_i - x_{i-1})/2 &\quad \text{if } i \in < 2, k-1 >\\
A_x(1) &= &\ell \cdot x_1 &\\
A_x(k) &= &\ell \cdot x_{k-1} &
\end{aligned}$$

$$D_x : R \longrightarrow < 1, 2, 3, \ldots, k > \qquad D_x(x) = i \quad x \in (x_{i-1}, x_i >$$

where division points $\{x_i\}_0^k$ of the real line R were established from experimental distribution $\overline{P}_x$ to preserve the same probability

$$\overline{P}_x\,(< x_{i-1}, x_i)) = \frac{1}{k}$$

for all equivalence intervals $< x_{i-1}, x_i >$, $i = 1, 2, \ldots, k$.

In our study $k = 5$, $\ell = 1.2$ and division points sequences $\{x_i\}_0^k$,

$x \in \{s, r, \Delta s\}$ were found to be

$$
\begin{aligned}
\{s_i\}_0^5 &= \{-\infty, -0.8, -0.18, 0.18, 0.8, +\infty\} && \text{system} \\
\{r_i\}_0^5 &= \{-\infty, -2.0, -0.5, 0.5, 2.0, +\infty\} && \text{controller} \\
\{\Delta s_i\}_0^5 &= \{-\infty, -1.0, -0.4, 0.4, 1.0, +\infty\} && \text{differences} \\
&&& \Delta s(t) = s(t) - s(t-1)
\end{aligned}
$$

E. g.

$$
\begin{aligned}
s(t) &= -0.318 && \text{is mapped by } D_s \text{ to 2,} \\
r(t) &= -0.318 && \text{is mapped by } D_r \text{ to 3,} \\
A_r \left(D_r(-0.318) \right) && \text{is 0.00.}
\end{aligned}
$$

The decision–making of the ES–driven controller (E) can be described then as

$$
r(t) = A_r \left(\text{argmax}_{i \in <1,2,\ldots,k>} \left\{ \tilde{P}(r_T(t) = i \,|\, r_{T-4}(t) = D_r(r(t-4)) \& \cdots \right. \right.
$$
$$
\left. \left. \cdots \& \Delta s_{T-1}(t) = D_{\Delta s}(\Delta s(t-1)) \right) \right\}
$$

conditional probabilities $\tilde{P}(\cdot|\cdot)$ are established as mentioned above by the module APPLICATOR of INES.

Testing of the total decision quality was performed in two phases:

1. The opened–loop behaviour of ES-driven controller r_E was described in [3]. In this scheme, the system was controlled by optimal controller r_O in closed–loop and its discretized outputs were compared with the decision of the r_E that was supplied all the necessary information from the past. Decision errors

$$
\lim_{t \to \infty} \frac{\text{card}\left(\{k \le t \,|\, r_E(k) \ne D_r(r_O(k))\} \right)}{t}
$$

were about 3 % on the training data set.

2. Closed–loop behaviour of ES–driven controller is visible in the table compared with other controls.

DISPERSIONS: $\frac{1}{N} \sum_{t=1}^{N} s_R^2(t)$, $R \in \{O, E, P, Z\}$, $N(0,1)$

R	\multicolumn{6}{c}{N}
	500	1000	1500	2000	2500	3000
O	1.084	1.0130	1.057	1.057	1.0426	1.0403
E	3.039	2.836	2.691	2.714		
P	6.039	6.333	6.155	6.2005		
Z	34.765	30.126	27.320	27.467	26.411	24.564

Prior to the simulation we believed the results to be better but a more detailed analysis of sudden increases in dispersion revealed that to blame is

1. insufficient discretization k of variables

2. unnecessarily high order of regression m, as the controller was instructed to respect statistical dependencies that, in fact, didn't exist.

Probable remedy would be to increase discretization to $k = 20$ and change the order to $m = 2$.

To estimate "robustness" of the ES–driven controller, the dispersion σ^2 instead of 1.0 (for which knowledge base was constructed) was used at the simulation time.

DISPERSIONS: $\frac{1}{N} \sum_{i=1}^{N} s_R^2(t)$, $N = 1000\ R \in \{O, E, P, Z\},\ N(0, \sigma^2)$

R	σ			
	0.5	1.0	2.0	5.0
O	0.2512	1.006	4.01	25.01
E	1.27	2.836	23.59	272.9
P	5.83	6.333	10.51	46.55
Z	7.53	30.126	120.49	753.17

The results show that for increased σ, the simple proportional controller is more robust. It may be interpreted in such a way that ES can interpolate efficiently but cannot extrapolate. Therefore, the knowledge base should be obtained from more perturbed system and then this information will appear in the knowledge base too.

In any case, we believe that the results justify further research in probabilistic AI–methods for control.

Acknowledgement

The authors wish to express their acknowledgement to their colleagues A. Halousková and M. Kárný who supplied samples of optimal closed–loop behaviour and were helpful with interpretation of results.

References

[1] Perez, A. (1983): Probability approach in integrating partial knowledge for medical decison making (in Czech). Trans. BMI'83 Conference on Biomedical Engineering Mariánské Lázně, 221–226.

[2] Perez, A. and Jiroušek R. (1985): Constructing an intensional expert system INES. In: J. H. van Bemmel, F. Grémy and J. Zvárová (Eds.) Medical Decision–Making: Diagnostic Strategies and Expert Systems. North–Holland, 307–315.

[3] Jiroušek, R., Perez, A., Kříž, O.: Intensional way of knowledge integration for expert systems. In: DIS'88 – Distributed Intelligence Systems, June 27 – July 1, 1988, Varna, Bulgaria, 219–227.

H_∞ ROBUST GENERALIZED MINIMUM VARIANCE SELF-TUNING CONTROLLER

M.J. Grimble
Industrial Control Unit, University of Strathclyde
Marland House, 50 George Street, GLASGOW G1 1QE, UK.

Summary

There are many methods of implementing H_∞ self-tuning controllers but the technique proposed here is very straightforward. To obtain the simple controller calculation algorithm a constraint must be placed on the cost weight selection. However, both error and control signal spectra can be minimized, or equivalently sensitivity and control sensitivity functions. The controller can also be selected so that the closed-loop response of the system is predetermined via the selection of prespecified closed-loop poles.

Acknowledgement

We are grateful for the support of the United Kingdom Science and Engineering Research Council, British Aerospace and the European Economic Community Science Programme.

1. INTRODUCTION

The main task of an ideal adaptive controller is to adapt to systems with time-varying parameters or nonlinear elements. A nonlinear system can often be represented by a linear system with time-varying parameters and hence it is the ability to adapt to time-varying plants which is needed. Currently most of the self-tuning theory is for systems with fixed parameters but steps are being taken to move towards the ultimate objective of adaptation for time varying systems.

In the following an H_∞ optimal control law is proposed which has a degree of robustness to time varying uncertainties and this can be quantified using recent stability lemmas (Chiang, 1988 [1]). The identification algorithms described may be made to track both slow and fast parameter variations.

1.1 *The control law*

The generalised minimum variance (GMV) control law was used to motivate the H_∞ design approach developed below. The GMV control law was introduced by Clarke and Hastings-James [2] and is used as a basis of some of the most successful self-tuning control schemes (see for example Clarke and Gawthrop [3] and Koivo [4]). The control law has the great advantage of simplicity and this was the main incentive for seeking an H_∞ equivalent. A generalized H_∞ (GH_∞) control law was derived by Grimble for both scalar (1987, [5]) and multivariable systems (1989 [6]). A particular version of this control law is obtained in the following for application in self-tuning systems where reasonably fast computations are needed.

The simplifications achieved in the control law employed does of course lead to some restrictions:

(1)   If the system is non-minimum phase and open loop unstable both large and small values for the cost-function weights can result in instability. Thus, restrictions must be placed on the form and magnitude of the cost-function weighting polynomials.

(2)   These restrictions can limit the weighting which can be attached to say the error term in the cost-function so that the error spectrum cannot be penalised with arbitrary weightings.

1.2 *Marine and Aerospace Applications*

The main problem in the design of marine control systems is to design a controller where the vessel model is poorly defined. The H_∞ self-tuner is particularly appropriate for such systems where the adjustment to parameteric uncertainty is via the self-tuning action and the robustness to modelling errors is through the robust design. In aerospace applications flight control systems must cope with very large gain variations and be totally reliable. There is therefore considerable potential for a truly robust adaptive system.

2. SYSTEM MODEL

The closed-loop scalar discrete-time system to be controlled is shown in Fig. 1. The reference $r(t) = W_r(z^{-1})\zeta(t)$ and disturbance $d(t) = W_d(z^{-1})\xi(t)$ subsystems are driven by zero-mean mutually independent, white noise signals $\zeta(t)$ and $\xi(t)$, respectively. The variances $E\{\zeta(t)^2\} = Q_o$ and $E\{\xi(t)^2\} = Q$. The plant $m(t) = W(z^{-1})u(t)$ has the control input $u(t)$. The subsystems can be represented in polynomial form as:

$$[W\ W_r\ W_d] = A^{-1}[B\ E_r\ C] \tag{1}$$

where the arguments of the polynomials (in the unit-delay operator z^{-1}) are omitted for notational simplicity. The plant can also be represented in terms of coprime polynomials as $W = A_o^{-1}B_o$. The greatest common factor of A and B is denoted by U_o, so that $A = U_o A_o$ and $B = U_o B_o$. The system polynomials can be written in the form:

$$A(z^{-1}) = 1 + A_1 z^{-1} + ... + A_{n_a} z^{-n_a}$$

$$B(z^{-1}) = (B_o + B_1 z^{-1} + ... + B_{n_b} z^{-n_b})z^{-k}$$

where $E_r(0) = E_o > 0$ and $C(0) = C_o > 0$. Note that $B(z^{-1})$ can also be written as:

$$B(z^{-1}) = B_k(z^{-1})z^{-k}.$$

2.1 *Stability Theorem*

The stability of the closed-loop system under a time-invariant feedback controller C_o is determined by the following lemma:

Lemma 2.1: The closed-loop system with both plant and controller free of unstable hidden modes is asymptotically stable if, and only if, the controller $C_o = M_o^{-1}N_o$, where M_o and N_o are asymptotically stable rational transfer-functions satisfying:

$$A_o M_o + B_o N_o = 1.$$

Proof: Kucera (1980[8]).

In the following solution of the GMV problem the controller is parmeterized, using the above results, to ensure that internal stability is preserved. That is, the controller is constrained to ensure that unstable pole or zero cancellations, within the plant, cannot occur.

2.2 *Cost-function*

By analogy with the cost-function employed in the GMV control problem [2], [3], let the fictitious output to be minimized have the form:

$$\phi(t) = P_c e(t) + F_c u(t) \tag{2}$$

where $P_c = P_{cd}^{-1} P_{cn}$ and $F_c = F_{cd}^{-1} F_{cn}$ denote appropriate weighting transfer-functions and the cost-function:

$$J = E\{\phi(t)^2\} \tag{3}$$

The signals $e(t)$ and $u(t)$ are given as:

$$e(t) = r(t) - y(t) = (1 - WA_oN_o)(r(t) - d(t)) \tag{4}$$

$$u(t) = A_oN_o(r(t) - d(t)) \tag{5}$$

The spectral density of the signal $r(t) - d(t)$ is denoted by:

$$\Phi_o(z^{-1}) = \Phi_{rr}(z^{-1}) + \Phi_{dd}(z^{-1}) \tag{6}$$

The power spectrum of the signal $\{\phi(t)\}$ is denoted by $\Phi_{\phi\phi}(z^{-1})$.

3. COST FUNCTION MINIMIZATION

The GMV optimal controller, to minimize error and control variances, is defined by the following theorem. Attention turns to the H_∞ optimal control problem in the following section.

Theorem 3.1: *Generalized Minimum Variance (GMV) Controller*

Consider the system described in §2 and assume that the cost-function (3) is to be minimised:

$$J = E\{(P_c e(t) + F_c u(t))^2\}$$

where the dynamical weighting terms are defined as:

$$P_c = P_{cd}^{-1} P_{cn}, \quad P_{cd}(0) = 1 \text{ and } P_{cn}(0) \neq 0 \tag{7}$$

$$F_c = F_{cd}^{-1} F_{cn}, \quad F_{cd}(0) = 1 \text{ and } F_{cn} = F_{nk} z^{-k} = (F_{n_0} + F_{n_1} z^{-1} + ... + F_{nn} z^{-n}) z^{-k} \tag{8}$$

The GMV control law is given as:

$$C_o = (HP_{cd})^{-1} GF_{cd} \tag{9}$$

where H, G are the unique minimal degree solutions with respect to F, of the diophantine equations:

$$P_{cd}AF + z^{-k}G = P_{cn}D_f \tag{10}$$
$$F_{cd}BF - z^{-k}H = F_{cn}D_f \tag{11}$$

subject to the constraint that the weighting elements are selected so that D_c is strictly Hurwitz:

$$D_c \triangleq (P_{cn}F_{cd}B_o - F_{cn}P_{cd}A_o)z^k \tag{12}$$

The definition of the noise sources also ensures that the polynomial D_f is strictly Hurwitz and satisfies:

$$D_f D_f^* = CQC^* + E_r Q_o E_r^* \tag{13}$$

Proof: The proof is given in Grimble (1986[7]). Note that the definition of the weighting function $F_{cn} = F_{nk}z^{-k}$ ensures that the spectral factor D_c is a function of z^{-1} (there are no terms in positive powers of z).

•

The following lemmas (Grimble, 1986[7]) explain some of the main features of the GMV controller.

Lemma 3.1: The minimal-degree solution of the diophantine equations (10) and (11) is unique.

•

Lemma 3.2: The closed-loop system is guaranteed to be internally stable and the degree of stability can be assessed from the characteristic polynomial:

$$\rho_c(z^{-1}) = D_c(z^{-1})D_f(z^{-1})U_0(z^{-1}) \tag{14}$$

where U_0 is the greatest common divisor of A and B.

•

Lemma 3.3: The minimum value for the variance of $\varphi(t)$ and the cost associated with the individual terms in the cost-function are given as:

$$J_{min} = \frac{1}{2\pi j} \oint_{|z|=1} (F^*F) \frac{dz}{z} \tag{15}$$

$$J_e \triangleq \{(P_c e^0(t))^2\} = \frac{1}{2\pi j} \oint_{|z|=1} (\frac{P_{cn} H}{D_c U_0})^* (\frac{P_{cn} H}{D_c U_0}) \frac{dz}{z}$$

$$J_u \triangleq E\{(F_c u^0(t))^2\} = \frac{1}{2\pi j} \oint_{|z|=1} (\frac{F_{cn} G}{D_c U_0})^* (\frac{F_{cn} G}{D_c U_0}) \frac{dz}{z}.$$

Lemma 3.4: The minimum variance output is given as:

$$\varphi^0(t) = F\frac{A}{D_f} (r-d) \tag{16}$$

•

4. H_∞ COST FUNCTION MINIMIZATION

The Generalized H_∞ (GH_∞) controller which is the basis of the self-tuning algorithm is derived in this section. The derivation follows a similar approach to that referred to as LQG embedding (Grimble, 1986 [10]).

4.1 *Auxiliary problem*

The following lemma, employed by Kwakernaak (1985[9]) in the continuous-time minimax optimisation problem, is the key to linking the H_∞ and GMV problems.

Lemma 4.1: Consider the auxiliary problem of minimising:

$$J = \frac{1}{2\pi j} \oint_{|z|=1} \{X(z^{-1})\Sigma(z^{-1})\} \frac{dz}{z} \tag{17}$$

Suppose that for some real rational $\Sigma(z^{-1}) = \Sigma^*(z^{-1}) > 0$ the criterion J is minimised by a function $X(z^{-1}) = X^*(z^{-1})$ for which $X(z^{-1}) = \lambda^2$ (a real constant on $|z|=1$. Then this function also minimises $\sup_{|z|=1} (X(z^{-1}))$. •

4.2 *Application of the lemma*

To invoke Kwakernaak's lemma, note that at the minimum of the cost-function, the integrand is given by (15) as F^*F which must satisfy $F^*F = \lambda^2\Sigma$. Write $\Sigma = B_\sigma^* B_\sigma$, where B_σ is Hurwitz.

Let the cost-function weighting terms be redefined to include the F operator:

$$P_c = P_{cd}^{-1}P_{cn}B_\sigma \quad \text{and} \quad F_c = F_{cd}^{-1}F_{cn}B_\sigma. \tag{18}$$

The diophantine equations (10) and (11) become:

$$P_{cd}AF + z^{-k}G = P_{cn}B_\sigma D_f \tag{19}$$

$$F_{cd}BF - z^{-k}H = F_{cn}B_\sigma D_f \tag{20}$$

and

$$D_c = B_\sigma (P_{cn}F_{cd}B_o - F_{cn}P_{cd}A_o)z^k \qquad (21)$$

must be strictly Hurwitz.

Let F_s be strictly Hurwitz and satisfy: $F_s F_s^* = FF^*$ then from the above results $F_s F_s^* = \lambda^2 B_\sigma B_\sigma^* \Rightarrow B_\sigma = \lambda^{-1}F_s$ where $\lambda \geq 0$. The equations therefore become:

$$P_{cd}AF + z^{-k}G = P_{cn}\lambda^{-1}F_s D_f \qquad (22)$$

$$F_{cd}BF - z^{-k}H = F_{cn}\lambda^{-1}F_s D_f \qquad (23)$$

The solution (G.H,F) of the diophantine equations (22) and (23), can be shown to be such that the F polynomial is non-minimum phase. To prove this result assume that this is not the case and write $F = F^+F^-$, where F^+ is Hurwitz. The diophantine equations (22) and (23) may now be simplifed by noting that $G = G_o F^+$ and $H = H_o F^+$.
After division:

$$P_{cd}AF^- + z^{-k}G_o = P_{cn}\lambda^{-1}F_s^- D_f \qquad (24)$$

$$F_{cd}BF^- - z^{-k}H_o = F_{cn}\lambda^{-1}F_s^- D_f \qquad (25)$$

where F_s^- is strictly-Hurwitz and satisfies:

$$F_s^- F_s^{-*} = F^- F^{-*}. \qquad (26)$$

A further justification of the above assertion is now given. Recall that a minimum-degree solution of (24) and (25) is required, with respect to the F polynomial, and note that the solution (G_o, H_o, F^-) satisfies:

$$\deg\{F^-\} \leq \deg\{F\}.$$

If (G,H,F) represents a minimum-degree solution of (24), (25) then clearly F^+ must be a scalar. This scalar can be set arbitrarily at the value unity and the desired solution F can be taken to be equal to F^-.

4.3 *Implied equation*

Adding equation (22) x BF_{cd} and equation (23) x $(-AP_{cd})$ gives:

$$z^{-k}(G_oBF_{cd} + H_oAP_{cd}) = (P_{cn}BF_{cd} - F_{cn}AP_{cd})\lambda^{-1}F_s^-D_f$$

or

$$(G_oF_{cd}^-B + H_oP_{cd}A) = L_kF_s^-D_f\lambda^{-1} \tag{27}$$

where

$$L_k \triangleq (P_{cn}B_kF_{cd}^- - F_{nk}AP_{cd}). \tag{28}$$

Controller:

The controller follows from (9) substituting for G and H, and cancelling to obtain:

$$C_o = (H_oP_{cd})^{-1}(G_oF_{cd}) \tag{29}$$

4.4 *Physical interpretation of results*

Before summarizing the above results and leaving this section, note that the H_∞ controller is the same as the GMV controller but for the special weightings: $P_{cn}B_\sigma$ and $F_{cn}B_\sigma$ where $B_\sigma = \lambda^{-1}F_s$. This choice of weightings ensures that the function being minimized $X(z^{-1}) = \Phi_{\varphi\varphi}(z^{-1}) = \lambda^2$. That is, the optimal solution is equalizing (constant over all frequencies) and has a magnitude of λ^2.

4.5 *Relationship between F^- and F_s^-*

It may easily be shown (Grimble, 1986[10]) that if F^- is written as:

$$F^- = f_o + f_1z^{-1} + ... + f_{n_f}z^{-n_f} \tag{30}$$

then

$$F_s^- = f_{n_f} + ... + f_1 z^{-n_f+1} + f_0 z^{-n_f} \tag{31}$$

where F^- is strictly non-Hurwitz and F_s^- is strictly Hurwitz. Note that the superscript on F^- and F_s^- will now be dropped for notational convenience.

Theorem 4.1 : Generalized H_∞ optimal controller

Consider the system described in §2 and shown in Fig. 1. Assume that the cost-function $J_\infty = \sup_{z=1} (X(z^{-1})$ is to be minimized where,

$$X(z^{-1}) = \Phi_{\varphi\varphi}(z^{-1}) \tag{32}$$

where the signal

$$\varphi(t) \triangleq P_c e(t) + F_c u(t). \tag{33}$$

Define the function $L_k = P_{cn} B_k F_{cd} - F_{nk} A P_{cd}$ and assume that the cost weightings are chosen so that L_k is strictly Hurwitz. The H_∞ optimal controller can be calculated using:

$$C_0 = (H_0 P_{cd})^{-1}(G_0 F_{cd}) \tag{34}$$

The polynomials (G_0, H_0) are given by the minimal-degree solution (G_0, H_0, F), with respect to F, of the equations:

$$P_{cd} A F + z^{-k} G_0 = P_{cn} F_s D_f \lambda^{-1} \tag{35}$$

$$F_{cd} B F - z^{-k} H_0 = F_{cn} F_s D_f \lambda^{-1} \tag{36}$$

where F_s is strictly Hurwitz and satisfies: $F_s F_s^* = F F^*$. $\tag{37}$

The stability of the system is determined by the implied equation:

$$G_o F_{cd} B + H_o P_{cd} A = L_k F_s D_f \lambda^{-1} \qquad (38)$$

The minimum cost: $J_{min} = \lambda^2$.

Proof: By collecting together previous results. •

4.6 *Remarks*

The control law has some interesting properties:

(a) Equation (35) can be solved for G_o and F and the value of λ. Since λ determines the minimum-cost it is surprising that this is only dependent upon the weighting: $P_c = P_{cd}^{-1} P_{cn}$ and the noise spectral factor : $Y_f = A^{-1} D_f$.

(b) Letting $F_{cn} = F_{nk} z^{-k}$ enables the second equation (36) to be solved explicitly for $H_o = F_{cd} B_k F - F_{nk} \lambda^{-1} F_s D_f$.

(c) The implied equation (38) which determines the characteristic polynomial for the system includes: $L_k = P_{cn} B_k F_{cd} - F_{nk} A P_{cd}$, D_f and F_s and these are by definition strictly-Hurwitz polynomials.

4.7 *Integral control and high frequency control weighting*

To introduce integral-action let $P_{cd} \triangleq 1 - z^{-1}$ and assume that $F_{cn} = F_{nk} z^{-k}$, so that (36) gives:

$$H_o = - F_{nk} \lambda^{-1} F_s D_f + F_{cd} B_k F \qquad (39)$$

Define $F_{cd} = F_d D_f$ then equation (39) gives:

$$H_o = D_f H_1 \quad \text{where} \quad H_1 \triangleq F_d B_k F - F_{nk} \lambda^{-1} F_s$$

Note that L_k in this case becomes:

$$L_k = (P_{cn}B_kF_dD_f - F_{nk}A(1-z^{-1}))$$ (40)

The remaining freedom in the weightings (P_{cn}, F_d and F_{nk}) should be used to ensure L_k is strictly Hurwitz. The controller now becomes:

$$C_o = (F_dB_kF - F_{nk}\lambda^{-1}F_s)^{-1}(1-z^{-1})^{-1}G_oF_d$$ (41)

If P_{cn} is preselected then G_o, F and F_s may be computed from (35).

Note that the above choice of F_{cd} does not restrict the freedom in selection of weights but does enable the expressions to be simplifed.

4.8 Maintenance of D_c strictly Hurwitz

If D_c is to remain strictly Hurwitz which is necessary for both stability and optimality the term $L_k \triangleq (P_{cn}F_{cd}B_o - F_{cn}P_{cd}A_o)z^k$ must remain strictly Hurwitz during self-tuning when the estimated B_o and A_o may be varying. This can always be achieved by appropriate choice of the weightings P_{cn}, P_{cd} and F_{cn}, F_{cd}. In fact a useful starting point for the adaptive algorithm is to choose L_k equal to the spectral-factor found in the equivalent LQG design problem. That is, let the initial calculated L_k satisfy:

$$L_k^*L_k = B_o^*F_{cd}^*P_{cn}^*P_{cn}F_{cd}B_o + A_o^*P_{cd}^*F_{cn}^*F_{cn}P_{cd}A_o$$

This minimum phase L_k polynomial enters the characteristic equation for the system giving the same initial closed-loop poles as in LQG design.

5. EXTENDED LEAST SQUARES IDENTIFICATION ALGORITHM

The extended least squares (ELS) identification algorithm is probably the most popular technique applied in self-tuning applications. It belongs to the class of recursive pseudolinear regression (RPLR) algorithms. It was developed by Young (1984 [13]) and Panuska (1980 [14].

The ELS algorithm for output prediction is determined by the following equations:

$$\hat{y}(t/t-1) = \varphi(t)\hat{\theta}(t-1)$$

$$\tilde{y}(t/t-1) = y(t) - \hat{y}(t/t-1) \qquad \text{(prediction error or innovations)}$$

$$\hat{\theta}(t) = \hat{\theta}(t-1) + \varphi(t)R(t)^{-1}\varphi(t)\tilde{y}(t/t-1)$$

$$R(t) = R(t-1) + \gamma(t)[\varphi(t)^T\varphi(t) - R(t-1)]$$

Note that the matrix $\varphi(t)$ denotes the regressor defined as:

$$\varphi(t) = [-y(t-1),...,-y(t-n_2) \; ; \; u(t-1) \, ,..., \, u(t-n_b) \; ; \; \bar{\varepsilon}(t-1) \, ,..., \, \bar{\varepsilon}(t-n_c)]$$

and this involves approximating past values of $\{\varepsilon(t)\}$ with the residuals $\bar{\varepsilon}(t) = y(t) - \varphi(t)\hat{\theta}(t)$. This scheme is sometimes called the Approximate Maximum-Likelihood (AML) estimation method. If the prediction error $\tilde{y}(t/t-1)$ is used in place of the residuals to approximate past $\{\varepsilon(t)\}$, the performance of the algorithm normally suffers (Norton, 1986 [15]) for little computational saving. This type of algorithm is sometimes called the Recursive Maximum-Likelihood 1 (RML1) method. The $\gamma(t)$ denotes a weighting scalar introduced in the quadratic prediction error criterion (Ljung, 1987 [16]).

The ELS algorithm has the alternative form:

$$\hat{y}(t/t-1) = \varphi(t)\hat{\theta}(t-1) \tag{42}$$

$$\tilde{y}(t/t-1) = y(t) - \hat{y}(t/t-1) \tag{43}$$

$$\hat{\theta}(t) = \hat{\theta}(t-1) + K(t)\tilde{y}(t/t-1) \tag{44}$$

$$K(t) = P(t-1)\varphi^T(t)/(\lambda(t) + \varphi(t)P(t-1)\varphi(t)^T) \tag{45}$$

$$P(t) = \frac{1}{\lambda(t)} \left[P(t-1) - \frac{P(t-1)\,\varphi(t)^T\varphi(t)\,P(t-1)}{(\lambda(t) + \varphi(t)P(t-1)\,\varphi(t)^T)} \right]. \tag{46}$$

where $\lambda(t) \triangleq (1 - \gamma(t))\gamma(t-1)/\varphi(t)$.

5.1 *Constant trace algorithms*

Most of the standard least squares and related algorithms turn themselves off as the samples tend to infinity. The algorithm gains reduce dramatically as the P matrix becomes small. To prevent this from happening a "constant trace" modification may be introduced.

Whenever the trace {P} falls below a certain level the P matrix is artificially increased. The approach is described in Goodwin, Hill and Palamiswani (1985 [19]).

5.2 *Improvement of the identification algorithm*

To improve the numerical conditioning of the equation UD factorization can be applied, by writing $P(t) = U(t)D(t)U^T(t)$ where U is a unit upper triangular factor and D denotes a diagonal matrix. The error covariance and gain update equations are the same as for the scalar case and the algorithm described by Chen and Norton (1987, [17] which uses UD factorization, and provides a mechanism for tracking slow and fast parameter variations, may therefore be used.

The following techniques are employed to improve robustness:

(i)   Near singularity of P(t) implies one of the elements of D(t) is near zero and this may easily be tested.

(ii)   Slow parameter changes are accommodated by choosing λ in the range $\lambda \varepsilon [0.95, 0.98]$.

(iii) When little new information is included in $\varphi(t)$ the P(t) updating is suspended until new parameter changes are detected.

(iv)   Rapid parameter changes are detected by a vector sequence based on the parameter-estimate corrections:

$$\Delta \hat{\theta}_i(t) = \hat{\theta}_i(t) - \hat{\theta}_i(t-1).$$

A low pass filtered sequence {q(t)} is formed, as in Hagglund's (1983 [13]) method.

$$q_i(t) = \alpha_{1i}q_i(t-1) + (1-\alpha_{1i}) \Delta \hat{\theta}_i(t), \ \alpha_1 \simeq 0.9 \sim 0.99.$$

When a parameter changes abruptly, the corresponding element in $q_i(t)$ will move away from zero and its trend can be detected using a statistic. The corresponding updating gain element for that parameter is increased once a threshold is reached.

6. SELF-TUNING H_∞ CONTROLLER

An H_∞ self-tuning controller can be constructed using either an explicit or implicit algorithm but the former approach will be taken below (Grimble 1984[11], 1986[12]).

Algorithm 6.1 H_∞ self-tuning controller

(i)   Initialise variables and set $\lambda = \lambda_0$. Choose initial cost weightings.

(ii)   Identify the system model (A, B, D_f) using ELS.

(iii)   Compute L_k using (28) and adjust weights so that L_k remains strictly Hurwitz.

(iv)   Solve the following linear equation for (F, G_o, F_s, λ):
$$P_c A F + z^{-k} G_o = P_{cn} F_s D_f \lambda^{-1}.$$

(v)   Compute: $H_o = F_{cd} B_k F - F_{nk} F_s D_f \lambda^{-1}.$

(vi)   Compute controller : $C_o = (H_o P_{cd})^{-1} (G_o F_{cd}).$

(viii)   Calculate the optimal control $u(t) = C_o e(t).$

(viii)   Go to (ii).

7. CONCLUSIONS

The proposed form of the GH_∞ control law is simple to compute and was based upon a minimax cost-function. The controller has the following inherent advantages:

(i)   Integral action may easily be introduced by choice of a dynamic weighting term in the cost function. Similarly blocking zeros may be introduced in the controller to attenuate disturbances by choice of weighting elements.

(ii)   A reference signal is included naturally in the system model.

(iii)   Control and error costing terms or sensitivity and control sensitivity terms are included in the cost-function.

(iv)   The degree of stability of the closed-loop system can be arranged via the choice of weighting terms.

REFERENCES

1.   Chiang, C-C and Chen, B-S., 1988, *Robust stabilization against nonlinear time-varying uncertainties,*, Int. J. Systems Sci., Vol. 19, No. 5, 747-760.

2.   Clarke, D.W. and Hastings-James, R., 1971, *Design of digital controllers for randomly disturbed systems*, Proc. IEE, Vol. 118, No. 10, 1503-1506.

3.   Clarke D.W. and Gawthrop, P.J., 1975, *Self-tuning controller*, Proc. IEE, Vol. 122, No. 9, 929-934.

4.   Kovio, H.N., 1980, *A multivariable self-tuning controller*, Automatica, Vol. 16, 351-366.

5.   Grimble, M.J., 1987, H_∞ *robust controller for self-tuning applications*, Int. J. Control, Part I, Vol. 46, No. 4, pp. 1429-1444, Part II, Vol. 46, No. 5, 1819-1840.

6.   Grimble, M.J., 1989, *Generalised H_∞ multivariable controllers*, IEE Proc. Vol. 136, Pt. D. No. 6, 285-297.

7.   Grimble, M.J., 1986, *Observations weighted controllers for linear stochastic systems*, Automatica, Vol. 22, No. 4, 425-431.

8.   Kucera, V., 1980, *Stochastic multivariable control - A polynomial equation approach*, IEEE Trans. on Auto. Contr., AC-25, 5, 913-919.

9.   Kwakernaak, H., 1985, *Minimax frequency domain performance and robustness optimization of linear feedback systems*, IEEE Trans. on Auto. Cont., AC-30, 10, pp. 994-1004.

10.   Grimble, M.J., 1986, *Optimal H_∞ robustness and the relationship to LQG design problems*, Int. J. Control, Vol. 43, No, 2, pp. 351-372.

11.   Grimble, M.J., 1984, *Implicit and explicit LQG self-tuning controllers*, Automatica, Vol. 20, 5, 661-669.

12.   Grimble, M.J., 1986, *Controllers for LQG self-tuning applications with coloured measurement noise and dynamic costing*, IEE Proc., Vol. 133, Pt.D., No. 1.

13.   Young, P., 1984, *Recursive estimation and time-series analysis - an introduction*, Springer-Verlag, New York,

14.   Panuska, V., 1980, *A new form of the extended Kalman filter for parameter estimation in linear systems with correlated noise,*, IEE Trans. on Auto. Contr., Vol. AC-25, pp. 229-235.

15. Norton, J.P., 1986, *An introduction to identification*, Academic Press, London.

16. Ljung, L., 1987, *System identification, theory for the user*, Prentice Hall, Information and System Sciences Series.

17. Chen M.J. and Norton, J.P., 1987, *Estimation technique for tracking rapid parameter changes*, Int. J. Control, 45, 4, 1387-1398.

18. Hagglund T., 1983, *New estimation techniques for adaptive control*, Research Report LUTFD2/(TFRT/1025)/1-120/(1983), Dept of Automatic Control, Lund Inst. of Tech., Lund, Sweden.

19. Goodwin, G.C., Hill, D.J. and M. Palamiswani, 1985, *Towards an adaptive robust controller*, Proc. IFAC Identification and Systems Param. Estimation Conference, York, pp. 997-1002.

20. Grimble, M.J. and Johnson, M.A., *Optimal Control and Stochastic Estimation : Theory and Applications*, Vols. I and II, John Wiley Chichester, 1988.

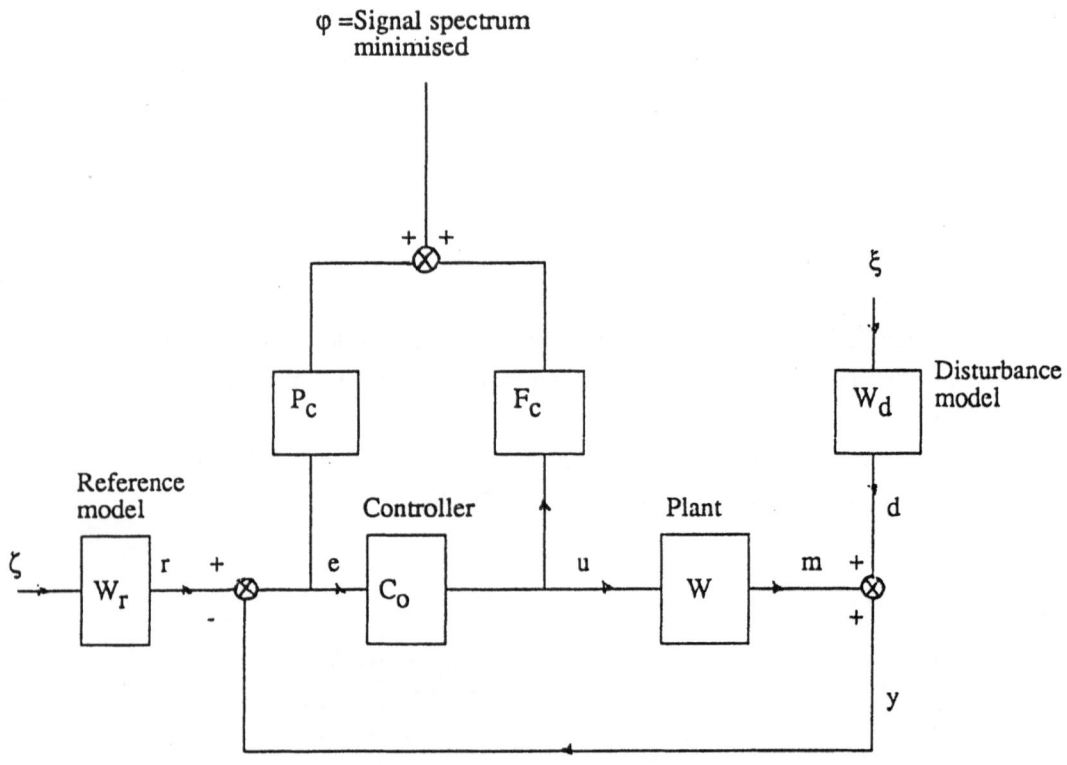

Fig. 1: *Closed-loop feedback system*

GNOSTICAL APPROACH TO ROBUST CONTROL

P.Kovanic

Institute of Information Theory and Automation,
Czechoslovak Academy of Sciences
182 08 Prague 8, Czechoslovakia

1 Summary

Data processing algorithms based on the gnostical theory of uncertain data possess high robustness with respect to both outlying data and changes of their statistical characteristics. There are several ways of substantial robustification of control systems by means of such algorithms. Effectiveness of this approach is demonstrated by examples.

2 Gnostical theory of uncertain data

Gnostical theory of uncertain data has been developed as an alternative of statistics for practical applications where nothing is known on a statistical model of the process, data are strongly disturbed, processes are non-stationary and there is lack of data to develop a statistical model. Algorithms based on the gnostical theory are inherently robust with respect to "outlying" data. They use no statistical assumptions related either to processes or to data. All necessary (gnostical) characteristics of the uncertainty are estimated directly from data. This is why a low sensitivity to changes of behaviour of disturbances is achieved by gnostical procedures. There are limits to the applicability of this theory connected with the validity of its axioms. However, these axioms express only basic algebraic requirements with respect to the nature of uncertainty.

3 Available gnostical algorithms

The following gnostical procedures have already been developed relevant for control applications:

- gnostical process monitors (treating time series, performing robust filtering of the process level, of its trend and of its acceleration, including robust diagnostics of the process);

- gnostical predictors for robust forecasting of disturbed processes;

- gnostical analysers of small data samples (for a detailed analysis of important data, such as estimating of probability of rare events, emergency limits for process control, random control of production quality, reliability studies, testing of homogeneity of objects and of their behaviour);

- gnostical identifiers of models (for robust identification of mathematical models of a process or object from disturbed observations).

These programs include some important auxiliary procedures such as estimation of scale parameters, data transformations, etc.

4   Robustification of control systems

There are several ways of robustifying control systems by means of gnostical procedures:

1) PID-control using gnostical filters to get robustly filtered proportional, integral and derivative signals;

2) robust filtering of the observed output level by the gnostical monitor pre-ceeding the input of an optimum linear control system;

3) control using a robust gnostical predictor;

4) robust adaptive control based on the on-line identification of the system model by means of a gnostical identifier;

5) gnostical formulation of the optimum control problem with corresponding synthesis.

Methods 1) – 4) combine the linear methods of synthesis of regulators with robust treat-ment of signals. Method 5) promises maximum effect but it opens new theoretical prob-lems because of nonlinearity: instead of the classical control error $e_c = z - z_0$ or $(z - z_0)/z_0$ (where z is the actual and z_0 the required output, $z, z_0 \in (0, \infty)$) a more complicated gnos-tical error function is used having the following form:

$$e_g = (q - 1/q)/(q + 1/q)$$

where $q = (z/z_0)^{2/s}$ and where s is a positive scale parameter. (Parameter s characterizes the intensity of random disturbances. It can be estimated from data).

Another important point is the quality of control. Within the framework of the gnos-tical theory there are functions of e_g available which evaluate the information loss and entropy gap caused by uncertainty of individual data. It could be interesting to optimize the control using these important criteria. It can be shown that it would result in high robustness.

5   Example 1

Examples demonstrate the effects of applying gnostics to control problems. Let us consider a continuous dynamic system of the type $1/(1 + p)^3$ controlled by an LQ-optimal discrete-time self-tuned controller. In addition to a slowly changing disturbance, an uncorrelated random component exists due to which the system output is observed as noisy. We protect the input of the controller against the observation errors by means of a filter. Fig.1 shows the case of a strong observation noise modelled by the absolute value of disturbing signal of the Cauchy type, the filter being a "specialized" one prepared by the Bayesian statistical approach under an a priori assumption of Cauchy distribution. (We shall denote this type of filter as "Bayes/Cauchy"). The same system and disturbances filtered by the simplest gnostical monitor is in Fig.2. The actions of the controller are more modest here and the quality of the control is not worse. This takes place in spite of the fact, that the statistical filter makes use of the additional a priori information about the type of the distribution function which is not assumed by the gnostical filter. What happens after a change in the noise distribution function? The case of an absolute value of a weak uniformly distributed signal is in Fig.3 with the same gnostical filter. It is obvious from the Fig.4 that the control quality is not affected even by a substantial increase of variance of the uniformly distributed noise in the case of the gnostical filter. However, as seen in Fig.5, the system using the Bayes/Cauchy filter is unstable with this noise. The same happens when the noise is the absolute value of a strong Gaussian disturbance. The controller protected by the gnostical filter still works well (Fig.6) while it fails with the Bayes/Cauchy filter (Fig. 7) as well as without filter (Fig.8).

6 Example 2

This example is related to the application of an identifier within an adaptive control system. What is shown here is as follows: a gnostical identification procedure due to its robustness approaches the true parameters of the identified system in a much shorter time interval and with much smaller maximum errors than an unrobust (e.g. least squares) identification method. The signal under consideration is a series of real data representing rather complicated vibrations of a steam generator of an atomic power station. The problem is of the diagnostical type: to discover and classify changes of vibration modes as symptoms of dangerous states of the object. One of such test includes analysis of the difference between the actual and predicted values of the observed quantity. Necessary one–step–ahead predicted values are obtained by an AR–model of the 24–th order. The coefficients of this predictor are identified (estimated) by two methods:

a) ordinary least-squares procedure;

b) gnostical identification procedures according to the algorithm described in detail in [1].

To evaluate the quality of the identification process we introduce the notion of "noise amplification" A of a predictor having the form

$$y_{t+1} = \sum_{i=1}^{M} c_i y_{t-i+1}$$

(where y_j is the value observed in the j-th time interval, M is the order of the AR–model, c is the estimate of the i-th coefficient) by the equation

$$A = \sum_{i=1}^{M} c_i^2$$

It is interesting to analyse the time dependence of coefficient A resulting from applying both of the methods mentioned:

Tab.1 Noise amplification for two identification methods

Time (sec)	Number of data	Noise amplification A of the method	
		least squares	gnostical
0.2	100	65000	4.63
0.4	200	7300	1.76
0.6	300	1160	1.51
0.8	400	300	1.48
1.0	500	114	1.42
1.2	600	49.9	1.46
1.4	700	9.42	1.42
1.6	800	5.75	1.44
1.8	900	2.69	1.42
2.0	1000	1.72	1.39

As this comparison shows, application of the robust gnostical identifier should result in better quality of the control.

References

[1] Kovanic P.: A new theoretical and algorithmical basis for estimation, identification and control, Automatica IFAC, 22, 6 (1986), 657-674

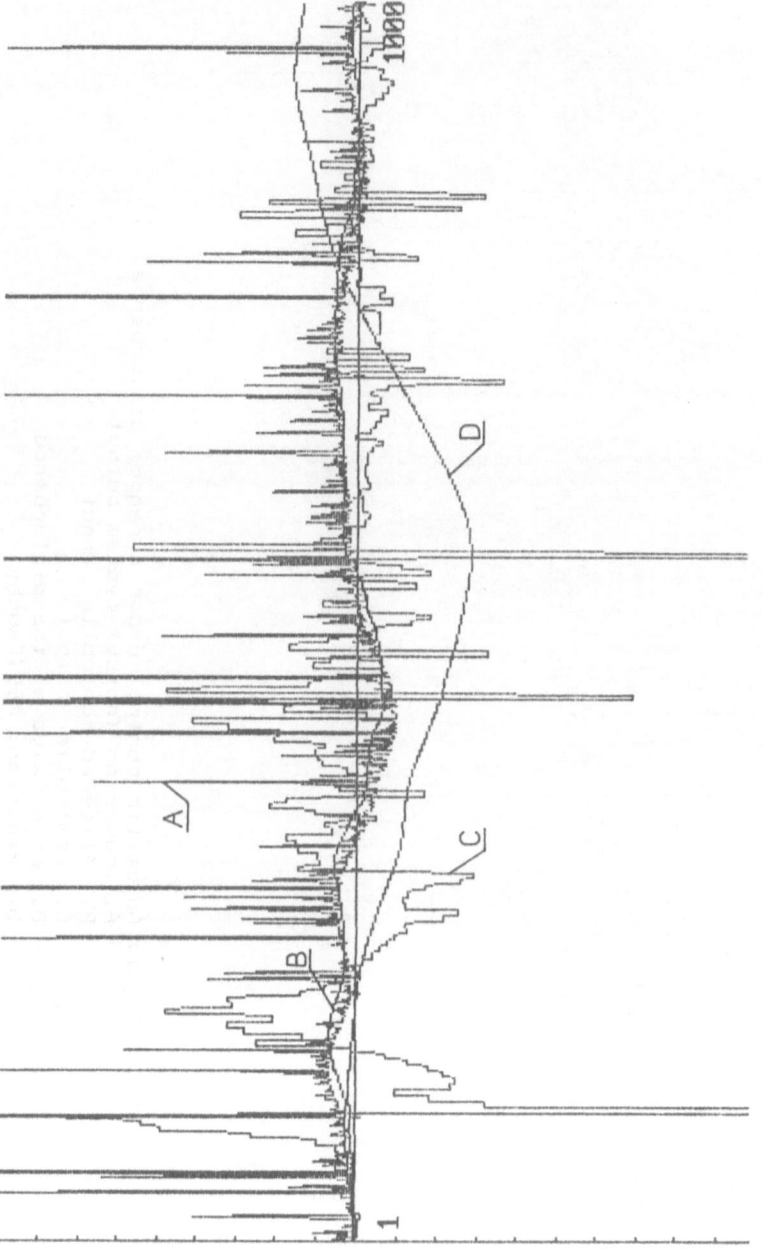

Fig.2: Automatic control under a random disturbance
A...observed (noisy) system output
B...filtered controller input
C...controller output
D...mean value of the disturbance
Distribution: ABS(Cauchy)
Filter: Gnostical

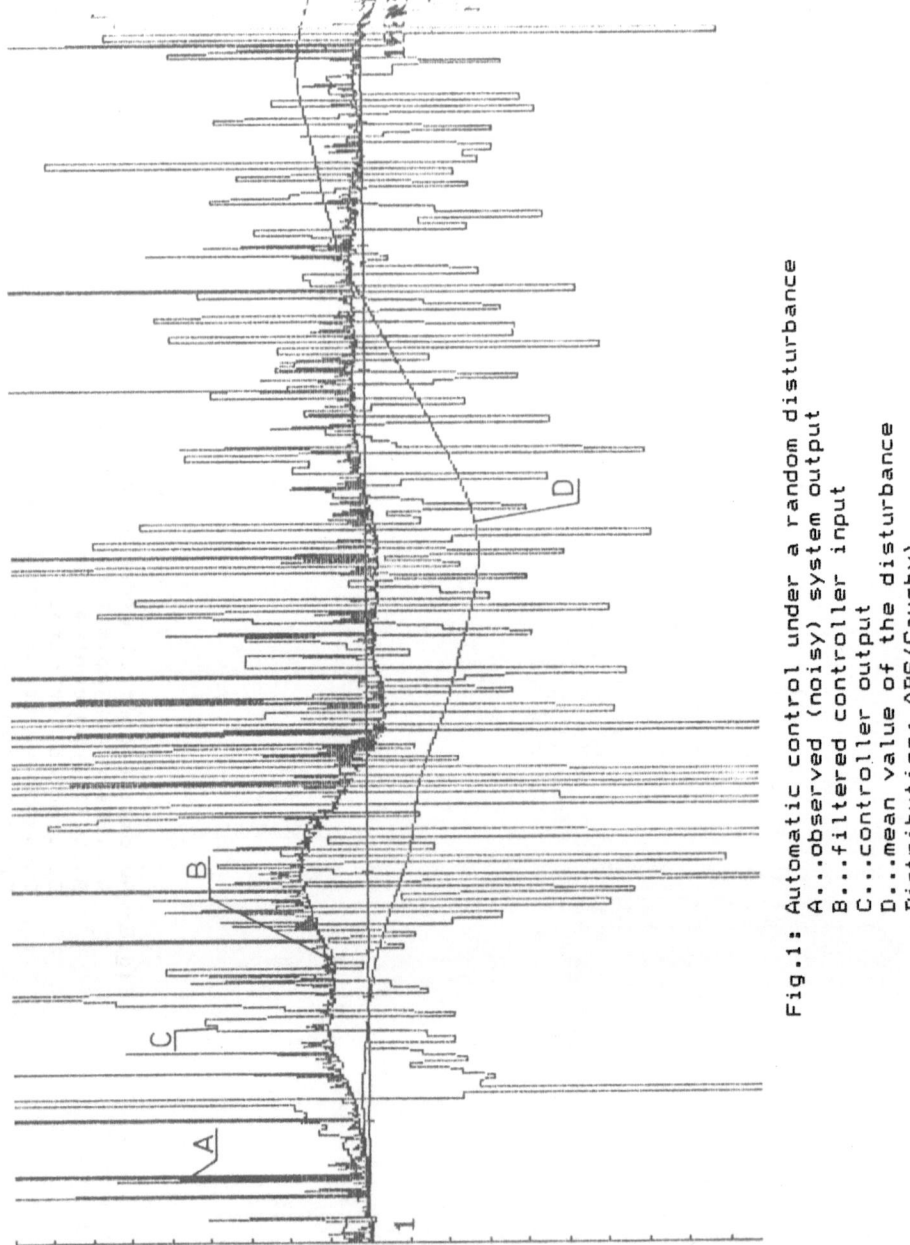

Fig.1: Automatic control under a random disturbance
A...observed (noisy) system output
B...filtered controller input
C...controller output
D...mean value of the disturbance
Distribution: ABS(Cauchy)
Filter: Bayes/Cauchy

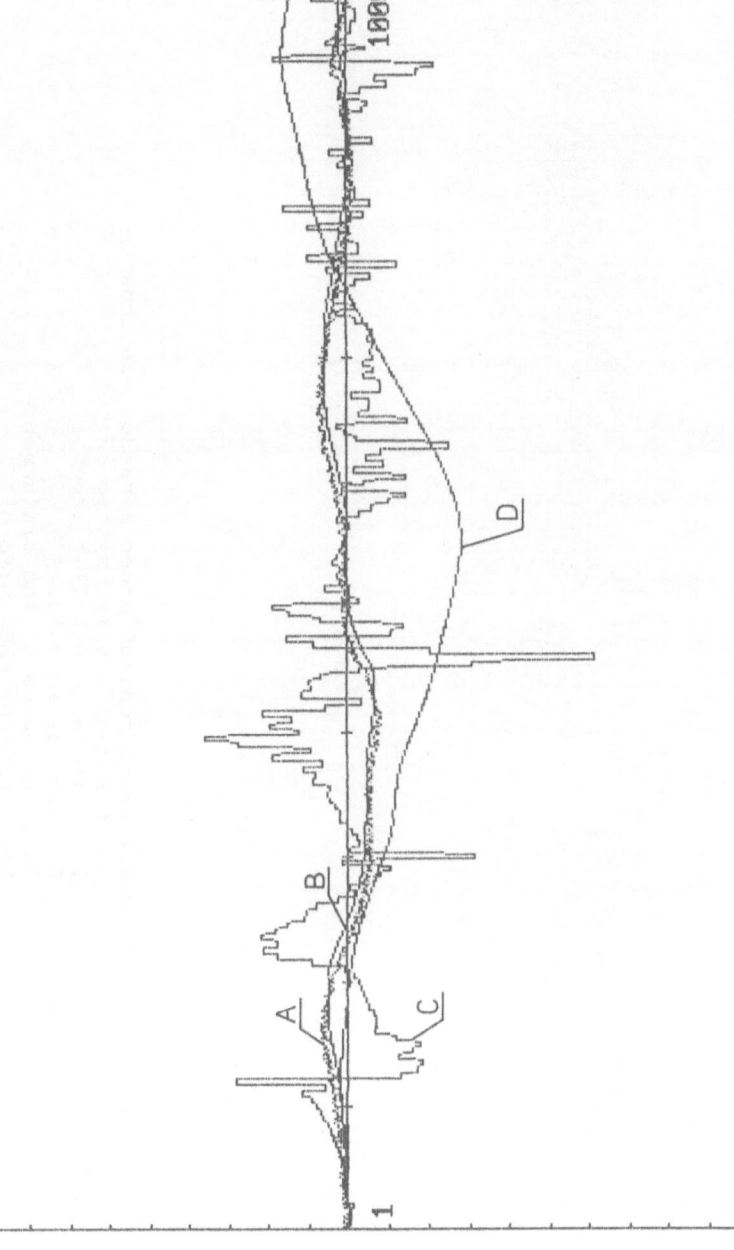

Fig.3: Automatic control under a random disturbance
A...observed (noisy) system output
B...filtered controller input
C...controller output
D...mean value of the disturbance
Distribution: ABS(uniform)
Filter: Gnostical

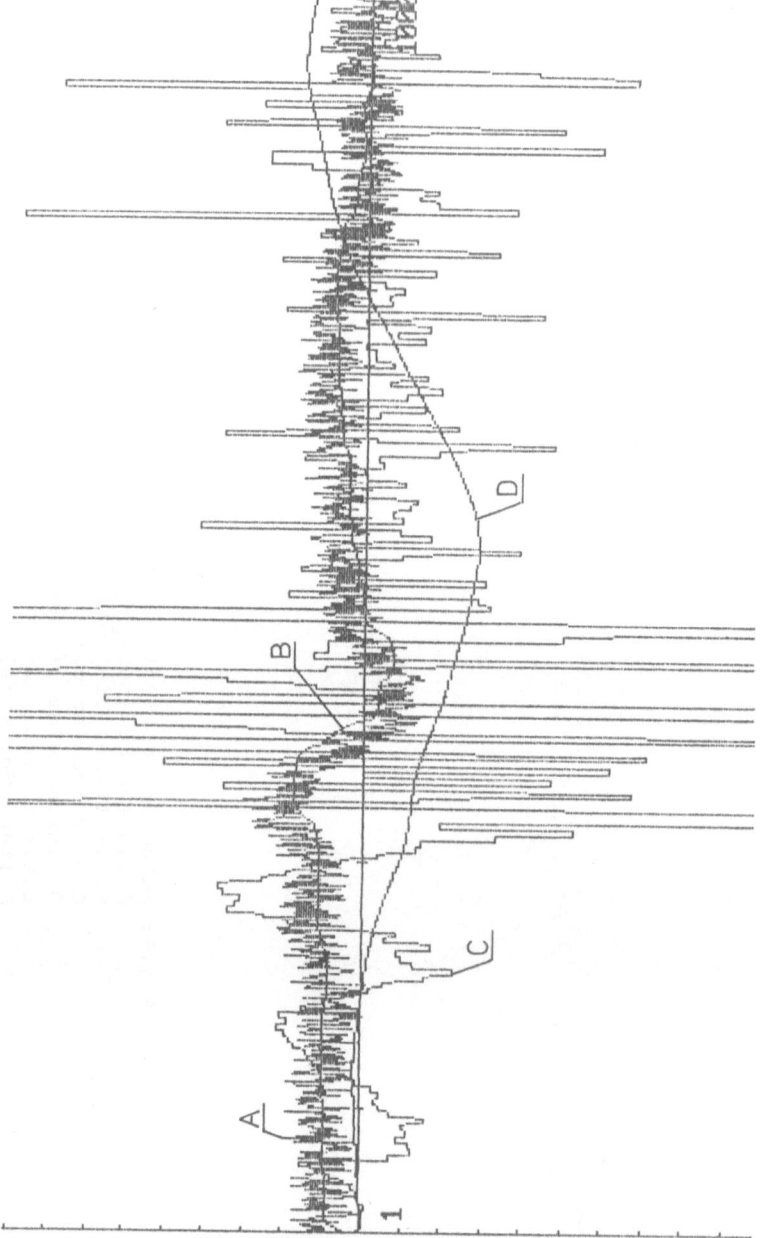

Fig.4: Automatic control under a random disturbance
A...observed (noisy) system output
B...filtered controller input
C...controller output
D...mean value of the disturbance
Distribution: ABS(uniform)
Filter: Gnostical

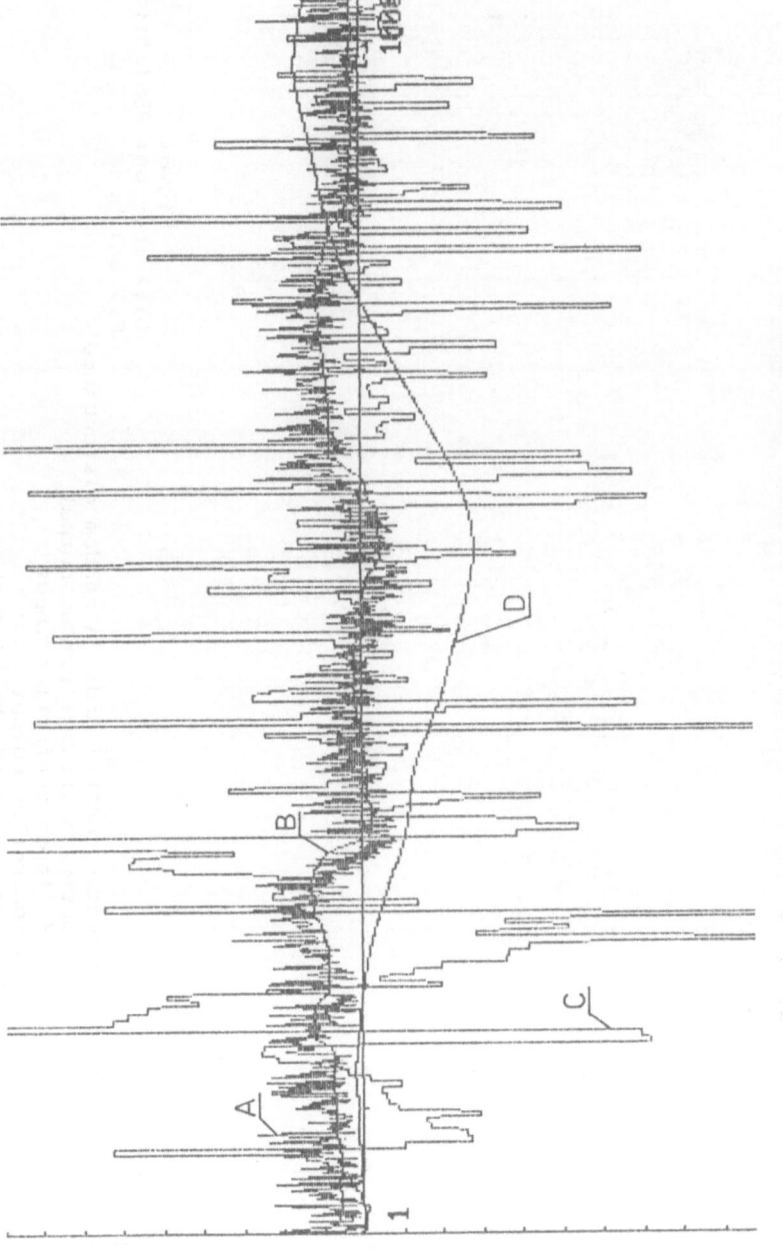

Fig.6: Automatic control under a random disturbance
A...observed (noisy) system output
B...filtered controller input
C...controller output
D...mean value of the disturbance
Distribution: ABS(Gauss)
Filter: Gnostical

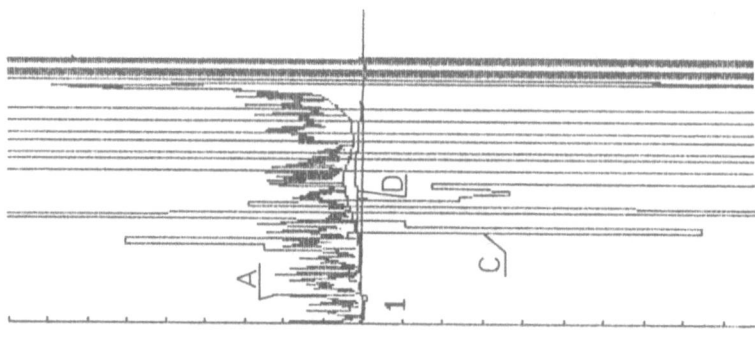

Fig.8:
Distribution: ABS(uniform)
Filter: None

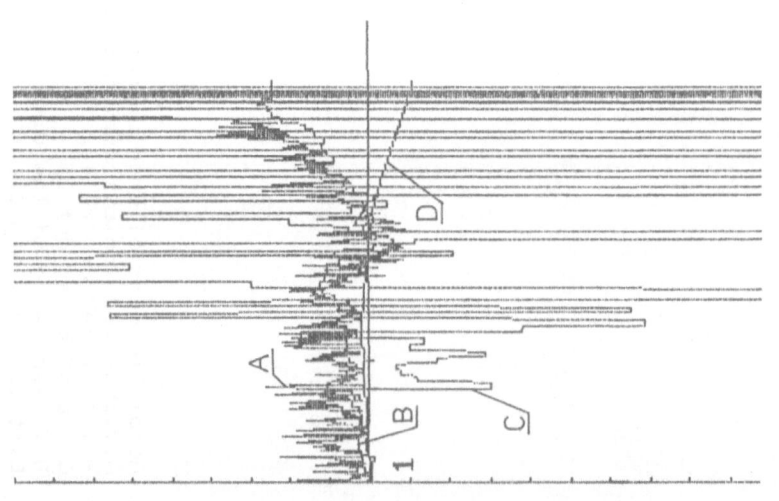

Fig.7: Automatic control under a random disturbance
A...observed (noisy) system output
B...filtered controller input
C...controller output
D...mean value of the disturbance
Distribution: ABS(Gauss)
Filter: Bayes/Cauchy

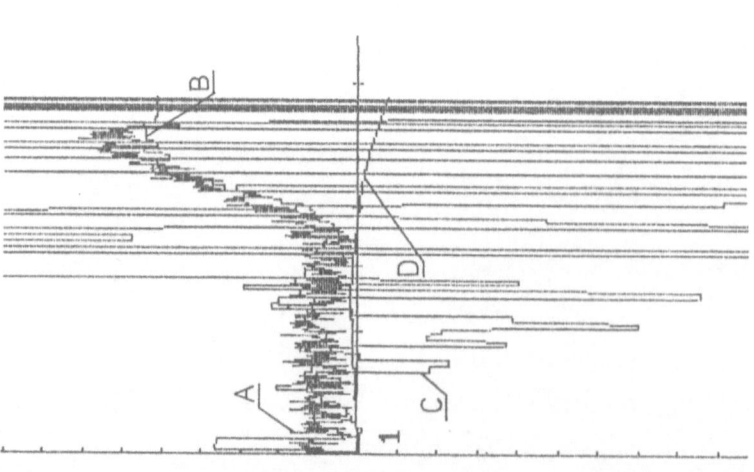

Fig.5:
Distribution: ABS(uniform)
Filter: Bayes/Cauchy

True Digital Control: A Unified Design Procedure
for Linear Sampled Data Control Systems

P.C. Young, A. Chotai and W. Tych

Centre for Research on Environmental Systems
Lancaster University
England

Summary

The paper describes the True Digital Control (TDC) design philosophy for linear, single input, single output (SISO) systems described by the backward shift (z^{-1}) and delta (δ) operator transfer function model; and outlines a Computer Aided Control System Design (CACSD) procedure based upon this design philosophy. The control system design analysis used in the CACSD procedure is based on the definition of suitable Non-Minimum State Space (NMSS) forms for the z^{-1} and δ models, which allow for state variable feedback (SVF) control involving only the measured input and output variables, together with their stored past values. The resulting "Proportional-Integral-Plus" (PIP) control systems then provide either SVF pole assignment control or optimal LQG control without resort to the complexity of state reconstructor (observer) design. The paper outlines the major stages in the TDC design: from model identification and parameter estimation; through PIP control system design and the evaluation of these designs in the presence of uncertainty, to their implementation in fixed gain, self-tuning of self-adaptive form.

1. INTRODUCTION: TRUE DIGITAL CONTROL (TDC)

Prior to 1960, most control systems were based on analog methodology. Where sampled data design was required, it was normally obtained by the digitisation of continuous-time analog designs using z transform theory and sample-hold circuitry. During the 1960's, the advent of "modern control" system design based on state-space systems theory held the promise of inherent digital designs, but most research was still dominated by continuous-time concepts. The rapid development of the digital computer during the 1970's certainly stimulated the move towards Direct Digital Control (DDC) implementation, but it failed to make too many inroads into the underlying faith of most traditional control systems designers in continuous system design methods. Indeed, it was only in the research on adaptive and self-tuning control systems where an all digital approach seemed to be finding favour, almost certainly because the recursive identification and estimation schemes required for adaptive implementation are rather simpler for discrete-time models than for their continuous-time equivalents. And even today, despite the tremendous developments in microcomputers over the 1980's and the now common use of microprocessors for the implementation of control systems, the majority of practical control systems in industry are still based on the mechanical digitisation of the continuous-time designs, such as the ubiquitous three term controller. In this chapter, we try to promote an alternative philosophy; one based on the idea of True Digital Control (TDC), as propounded in a number of our previous publications (Young et al, 1987a,b; Young et al, 1988; Young, 1989; Chotai et al, 1990a).

The TDC philosophy rejects the idea that a digital control system should be initially designed in continuous-time terms. Rather it demands that the designer considers the design from a digital, sampled data standpoint, even when fast sampling, near continuous-time operation is required. The design procedure consists of the following three major steps:-

(1) **Identification and estimation** of backward shift (z^{-1}) or delta operator (δ) operator digital control models based on the analysis of either planned experimental data; or via model reduction from data generated by (usually high order) continuous or discrete-time simulation models.

(2) **Off-line z^{-1} or δ operator TDC system design and initial evaluation** based on the models from step (1) using an *iterative* application of an appropriate discrete-time design methodology, coupled with sensitivity analysis based on Monte-Carlo simulation.

(3) **Implementation, testing and evaluation** on the real process, in the adaptive case using the on-line versions of the recursive algorithms used in step (1).

In subsequent Sections of this Chapter, we will describe this design strategy in more detail. But first, it is necessary to outline the mathematical background to the various stages in TDC system design.

2. TF MODELS IN THE z^{-1} AND δ OPERATORS

In all model-based control system design procedures, the form of the models and the associated theoretical background is of paramount importance. Most continuous-time procedures are based either on the Laplace transform transfer function model or state-space differential equations. It is not surprising, therefore, that TDC designs are based on the discrete-time equivalents of these models. It is possible to unify the analysis in this Chapter in terms of a general operator (see e.g. Goodwin, 1988). However, it is felt that the analysis is more transparent if we consider the two major forms of the transfer function model in the discrete-time domain: the well known backward shift $(z^{-1}$, or equivalently the forward shift, z transform) operator form; and the alternative, but less well known, delta $(\delta$ or discrete differential) operator form.

Backward Shift (z^{-1}) TF Model

The general, discrete-time, z^{-1} operator transfer function (TF) representation of an nth order single input, single output (SISO) discrete-time system, with a sampling interval of Δt time units, is normally written in the following form,

$$y(k) = \frac{B(z^{-1})}{A(z^{-1})} u(k) \tag{1}$$

where $A(z^{-1})$ and $B(z^{-1})$ are the following polynomials in z^{-1},

$$A(z^{-1}) = 1 + a_1 z^{-1} + \ldots + a_n z^{-n}$$
$$B(z^{-1}) = b_1 z^{-1} + b_2 z^{-2} + \ldots + b_m z^{-m}$$

In general, no prior assumptions are made about the nature of the transfer function $B(z^{-1})/A(z^{-1})$, which may be marginally stable, unstable, or possess non-minimum phase characteristics. However, if the input-output behaviour of the system is characterised by any pure time (transport) delay of τ sampling intervals, then this is accommodated by assuming that the first τ coefficients of the $B(z^{-1})$ polynomial, i.e. $b_1, b_2, \ldots, b_\tau$, are all zero.

The Discrete Differential (δ) Operator TF Model

A interesting alternative to the z^{-1} operator TF model is the following "discrete differential operator" model, which was revived recently under the title δ operator by Goodwin (1988) and Goodwin et al.(1988)[1],

$$y(k) = \frac{B(\delta)}{A(\delta)} u(k) \tag{2}$$

where $A(\delta)$ and $B(\delta)$ are polynomials of the following form,

$$A(\delta) = \delta^p + a_1 \delta^{p-1} + \ldots + a_p$$
$$B(\delta) = b_1 \delta^{p-1} + \ldots + b_p$$

with $p = \max(n,m)$ and the δ operator, for the sampling interval Δt, defined as follows in terms of the backward shift operator[2],

---

1. This nomenclature is by no means universal ; in the finite difference literature, for example Δ is often used as the simple backward difference operator and δ retained for central differences.

$$\delta = \frac{1 - z^{-1}}{z^{-1} \Delta t}$$

or, in terms of the forward shift operator z,

$$\delta = \frac{z - 1}{\Delta t} \quad ; \ i.e. \quad \delta x(k) = \frac{x(k+1) - x(k)}{\Delta t}$$

which is more convenient for the current analysis.

Remarks

(1) As $\Delta t \to 0$, the Delta operator reduces to the derivative operator $(s = d/dt)$ in continuous time (i.e $\delta \to s$).

(2) Given a polynomial of any order n in the z operator, this will be exactly equivalent to some polynomial in δ, also of order n.

As a consequence of (2), we can easily move between the z and δ operator domains. For example, a δ design is best implemented in practice by converting it back to the z^{-1} domain, so avoiding direct differencing. Also, the δ operator model coefficients are related to forward z operator coefficients by the vector matrix equations shown in Fig.1

One attraction of the δ operator model to those designers who prefer to think in continuous-time terms is that it can be considered as a direct approximation to a continuous-time system. For example, it is easy to see that the unit circle in the complex z plane maps to a circle with centre $-1/\Delta t$ and radius $1/\Delta t$ in the complex δ plane; so that, as $\Delta t \to 0$, this circular stability region is transformed to the left half of the complex s plane . For very rapidly sampled systems, therefore, the δ operator model can be considered in almost continuous-time terms, with the pole positions in the δ plane close to those of the 'equivalent' continuous-time system in the s plane; and with the TF parameters directly yielding information on factors such as the approximate natural frequency and damping ratio.

3. PIP CONTROL DESIGN FOR THE z^{-1} OPERATOR MODEL

The special non-minimal state-space (NMSS) representation which we associate with the z^{-1} transfer function model (1) is defined by the following state vector,

---

2. There are, of course other possible δ operator TF models based on the more sophisticated approximations to the differential operator.

$$
\begin{bmatrix} a_1 \\ a_2 \\ a_3 \\ a_4 \\ \cdot \\ \cdot \\ \cdot \\ a_P \end{bmatrix}_\delta
=
\begin{bmatrix} \frac{1}{\Delta t} & & & & & & \\ \frac{P\text{-}1C_1}{\Delta t^2} & \frac{1}{\Delta t^2} & & & \mathbf{0} & & \\ \frac{P\text{-}1C_2}{\Delta t^3} & \frac{P\text{-}2C_1}{\Delta t^3} & \frac{1}{\Delta t^3} & & & & \\ \cdot & \cdot & \frac{P\text{-}3C_1}{\Delta t^4} & \frac{1}{\Delta t^4} & & & \\ \cdot & \cdot & \cdot & \cdot & & & \\ \cdot & \cdot & \cdot & \cdot & & & \\ \cdot & \cdot & \cdot & \cdot & & & \\ \frac{1}{\Delta t^P} & \frac{1}{\Delta t^P} & \cdot & \cdots & \cdots & \cdot & \frac{1}{\Delta t^P} \end{bmatrix}
\begin{bmatrix} a_1 \\ a_2 \\ a_3 \\ a_4 \\ \cdot \\ \cdot \\ \cdot \\ a_P \end{bmatrix}_z
+
\begin{bmatrix} \frac{_PC_1}{\Delta t} \\ \frac{_PC_2}{\Delta t^2} \\ \frac{_PC_3}{\Delta t^3} \\ \cdot \\ \cdot \\ \cdot \\ \cdot \\ \frac{1}{\Delta t^P} \end{bmatrix}
$$

Fig. 1(a) Denominator Coefficients

$$
\begin{bmatrix} b_1 \\ b_2 \\ b_3 \\ b_4 \\ \cdot \\ \cdot \\ \cdot \\ b_P \end{bmatrix}_\delta
=
\begin{bmatrix} \frac{1}{\Delta t} & & & & & & \\ \frac{P\text{-}1C_1}{\Delta t^2} & \frac{1}{\Delta t^2} & & & \mathbf{0} & & \\ \frac{P\text{-}1C_2}{\Delta t^3} & \frac{P\text{-}2C_1}{\Delta t^3} & \frac{1}{\Delta t^3} & & & & \\ \cdot & \cdot & \frac{P\text{-}3C_1}{\Delta t^4} & \frac{1}{\Delta t^4} & & & \\ \cdot & \cdot & \cdot & \cdot & & & \\ \cdot & \cdot & \cdot & \cdot & & & \\ \cdot & \cdot & \cdot & \cdot & & & \\ \frac{1}{\Delta t^P} & \frac{1}{\Delta t^P} & \cdot & \cdots & \cdots & \cdot & \frac{1}{\Delta t^P} \end{bmatrix}
\begin{bmatrix} b_1 \\ b_2 \\ b_3 \\ b_4 \\ \cdot \\ \cdot \\ \cdot \\ b_P \end{bmatrix}_z
$$

Fig. 1(b) Numerator Coefficients

$$ _pC_r = \frac{p!}{(p\text{-}r)!\, r!} \qquad \text{(binomial coefficients)} $$

Fig. 1 Relationships between the parameters of the delta operator(δ) and forward shift (z) transfer functions : (a) denominator parameters; (b)numerator parameters

$$x^T = [y(k)\ y(k-1)\ ...\ y(k-n+1)\ u(k-1)\ ...\ u(k-m+1)\ z(k)]$$

where $z(k)$ is an "integral of error" term at sampling instant k,

$$z(k) = z(k-1)\ +\ \{y_d(k) - y(k)\}$$

in which $y_d(k)$ is the reference or command input to the servomechanism system. The NMSS representation is then obtained directly in the following form,

$$x(k) = F\ x(k-1)\ +\ g\ u(k-1)\ +\ d\ y_d(k) \tag{3}$$

where,

$$F = \begin{bmatrix} -a_1 & -a_2 & .... & -a_{n-1} & -a_n & b_2 & b_3 & ..... & b_{m-1} & b_m & 0 \\ 1 & 0 & .... & 0 & 0 & 0 & 0 & ..... & 0 & 0 & 0 \\ 0 & 1 & .... & 0 & 0 & 0 & 0 & ..... & 0 & 0 & 0 \\ . & . & & . & . & . & . & & . & . & . \\ 0 & 0 & .... & 1 & 0 & 0 & 0 & ..... & 0 & 0 & 0 \\ 0 & 0 & .... & 0 & 0 & 0 & 0 & ..... & 0 & 0 & 0 \\ 0 & 0 & .... & 0 & 0 & 1 & 0 & ..... & 0 & 0 & 0 \\ 0 & 0 & .... & 0 & 0 & 0 & 1 & ..... & 0 & 0 & 0 \\ . & . & & . & . & . & . & & . & . & . \\ . & . & & . & . & . & . & & . & . & . \\ 0 & 0 & .... & 0 & 0 & 0 & 0 & ..... & 1 & 0 & 0 \\ a_1 & a_2 & .... & a_{n-1} & a_n & -b_2 & -b_3 & ..... & -b_{m-1} & -b_m & 1 \end{bmatrix}$$

$$g = [\ b_1\ 0\ .....\ 0\ 1\ 0\ 0\ .....\ 0\ 0\ -b_1\ ]^T$$

$$d = [\ 0\ 0\ .....\ 0\ 0\ 0\ 0\ .....\ 0\ 0\ 1\ ]^T$$

The definition of this particular NMSS state vector is significant: it means that the State Variable Feedback (SVF) control law involves only the directly measurable input and output signals, together with their past values, all of which can be stored in the digital computer. As a result, any SVF control system design (e.g. pole assignment or optimal Linear Quadratic) does not need to resort to the complication of a state reconstructor (i.e. observer), since the *effective* output matrix in control terms is the I_{m+n} identity matrix. The conditions for controllability of the NMSS model are given by the following theorem.

Theorem (Wang and Young, 1987). Given a single input-single output system described by (1), the non-minimal state space representation (6), as described by the pair $\{F,g\}$, is completely controllable if, and only if, the following two conditions are satisfied:

(i)   the polynomials $A(z^{-1})$ and $B(z^{-1})$ are coprime

(ii) $b_1 + b_2 + \ldots\ldots + b_m \neq 0$

The coprimeness condition is equivalent to the normal requirement that the transfer function model (1) should have no pole-zero cancellations. The second condition avoids the presence of a zero at unity which would cancel with the unity pole associated with the integral action.

3.1 The Control Algorithm

In the context of the NMSS system representation, the automatic control objective is to design an SVF control law with gain vector $k = [f_0, f_1, \ldots, f_{n-1}, g_1, \ldots, g_{m-1}, k_1]^T$, i.e.,

$$u(k) = -k^T x(k) = -f_0 y(k) - f_1 y(k-1) - \ldots - f_{n-1} y(k-n+1) - g_1 u(k-1) - \ldots - g_{m-1} u(k-m+1) - k_1 z(k)$$

such that either the closed loop poles are at preassigned positions in the complex z-plane; or the system is optimised in an Linear-Quadratic (LQ) sense (or Linear-Quadratic-Gaussian (LQG) in the stochastic situation).

In the pole assignment case, the closed-loop system block diagram takes the Proportional-Integral-Plus (PIP) form shown in Fig.2. The closed-loop TF associated with this block diagram can be written,

$$y(k) = \frac{D(z^{-1})}{CL(z^{-1})} y_d(k)$$

where,

$$D(z^{-1}) = k_1 B(z^{-1})$$
$$CL(z^{-1}) = (1-z^{-1})[G(z^{-1})A(z^{-1}) + F(z^{-1})B(z^{-1}) + f_0 B(z^{-1})] + k_1 B(z^{-1})$$

The closed-loop characteristic polynomial $CL(z^{-1})$ can now be expanded and the coefficients for like powers of z^{-i} equated to those the desired closed loop characteristic polynomial $d(z^{-1})$,

$$d(z^{-1}) = 1 + d_1 z^{-1} + d_2 z^{-2} + \ldots\ldots + d_{m+n} z^{-(n+m)}$$

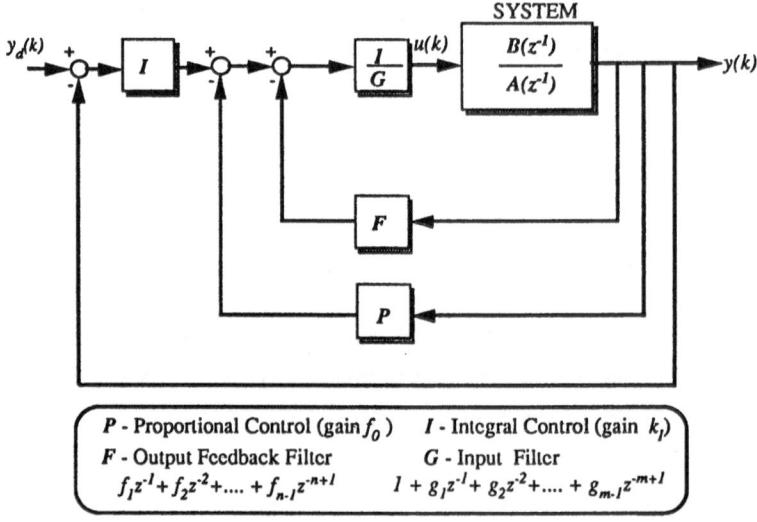

Fig.2 Block Diagram of the Backward Shift (z^{-1}) Operator PIP Servomechanism Control System.

which is chosen to ensure the assignment of the closed loop poles to designer-specified positions in the complex z plane. This results in the following set of linear, algebraic equations,

$$\sum . k = S_c \text{-} f \tag{4}$$

where k is the SVF control gain vector, i.e.,

$$k = [\,f_0\,,f_1\,,\,....,f_{n\text{-}1}\,,g_1\,,\,...\,,g_{m\text{-}1}\,,k_I\,]$$

while S_c and f are the following vectors,

$$S_c = [d_1\,,d_2\,,\,......\,,d_{n+m\text{-}1}\,,d_{n+m}\,]^T$$

$$f = [a_1\text{-}1\,,\,a_2\text{-}a_1,\,......\,,\,a_n\text{-}a_{n\text{-}1},\,\text{-}a_n\,,\,0\,,\,0\,,\,......\,,\,0]^T$$

and $\sum (n,m)$ is the following $(n+m)\text{x}(n+m)$ matrix (see Young et al.; 1987b),

$$\sum(n,m) = \begin{bmatrix}
b_1 & 0 & \cdots & 0 & 1 & 0 & \cdots & 0 & b_1 \\
b_2-b_1 & b_1 & \cdots & : & a_1-1 & 1 & \cdots & : & b_2 \\
: & b_2-b_1 & \cdots & : & : & a_1-1 & \cdots & 0 & : \\
: & : & 0 & : & : & & 1 & & : \\
b_m-b_{m-1} & : & \cdots & b_1 & : & : & \cdots & a_1-1 & b_m \\
-b_m & b_m-b_{m-1} & \cdots & b_2-b_1 & a_n-a_{n-1} & : & \cdots & : & 0 \\
0 & -b_m & \cdots & : & -a_n & a_n-a_{n-1} & \cdots & : & : \\
: & 0 & \cdots & : & 0 & -a_n & \cdots & : & : \\
: & : & & : & : & 0 & & : & : \\
: & : & & b_m-b_{m-1} & : & : & & a_n-a_{n-1} & : \\
0 & 0 & \cdots & -b_m & 0 & 0 & \cdots & -a_n & 0
\end{bmatrix}$$

Provided the controllability conditions of the above theorem are satisfied, this set of linear simultaneous equations can be solved to yield the unique set of SVF control gains which define the vector k.

In the LQ or LQG situation, the SVF gains are computed to minimise the following quadratic cost function,

$$J = \sum_{i=1}^{i=\infty} x(k)^T Q\, x(k) + q_u u(k)^2 \tag{5}$$

where Q is an $(n+m)\times(n+m)$ diagonal matrix, with elements defined as follows,

$$Q = diag.[q_1\, q_2 \,\cdots\, q_n\, q_{n+1}\, q_{n+2}\cdots\, q_{n+m-1}\, q_{n+m}]$$

with,

$$q_1 = q_2 = \cdots = q_n = q_y;$$
$$q_{n+1} = q_{n+2} = \cdots = q_{n+m-1} = q_u$$

and,

$$q_{n+m} = q_z$$

Here q_y, q_u and q_z are the *partial* weightings on the output, input and integral-of-error variables in the NMSS vector x. These partial weightings are defined as,

$$q_y = \frac{W_y}{n} \quad ; \quad q_u = \frac{W_u}{m} \quad \text{and} \quad q_z = W_z \tag{6}$$

so that the total weightings on the output $y(k)$, input $u(k)$ and integral-of-error $z(k)$ are W_y, W_u, and W_z. These three weighting variables are then chosen in the usual manner to achieve the desired closed loop performance (see the example later in Section 6.2).

The optimum SVF gains for the selected weighting values W_y, W_u, and W_z are computed by means of the well known iterative algorithm for computing the steady state solution of the associated, discrete-time, matrix Riccati equation, given the NMSS system description (F, g) and weighting matrices (Q, R); see e.g. Kuo, 1980, pp. 617 ff. Note that, in this case, $R = q_u$ because of the special, NMSS formulation.

3.2 Simple Example

Consider an undamped oscillatory system with the following z^{-1} transfer function,

$$y(k) = \frac{1 + 0.4\,z^{-1}}{1 - 1.6\,z^{-1} + z^{-2}} \; u(k-1) \tag{7}$$

The NMSS model in this case is given by,

$$x(k) = \begin{bmatrix} 1.6 & -1 & 0.4 & 0 \\ 1 & 0 & 0 & 0 \\ 0 & 0 & 0 & 0 \\ -1.6 & 1 & -0.4 & 1 \end{bmatrix} x(k-1) + \begin{bmatrix} 1 \\ 0 \\ 1 \\ -1 \end{bmatrix} u(k-1) + \begin{bmatrix} 0 \\ 0 \\ 0 \\ 1 \end{bmatrix} y_d(k)$$

where,

$$x(k) = [y(k)\,,\, y(k-1)\,,\, u(k-1)\,,\, z(k)]^T$$

Using the PIP control design method, with the SVF control law defined above, the closed loop system will be of fourth order. However, the desired closed loop polynomial can be chosen of lower dimension than this: e.g. if $d(z^{-1}) = 1 + 0.4z^{-1}$, it can can cancel out the effects of the numerator polynomial and result in dead beat response. Solving the above equations in this case, we obtain,

$$f_0 = 1.6 \ , \ f_1 = -1 \ , \ g_1 = 0.4 \ , \ k_1 = 1$$

With these SVF gains, the closed loop transfer function is given by,

$$y(k) = \frac{z^{-1} + 0.4 \, z^{-2}}{1 + 0.4 \, z^{-1}} \, y_d(k)$$

i.e.,

$$y(k) = z^{-1} y_d(k)$$

which provides the required deadbeat response.

Of course, this example is chosen for illustration: it may well not be a satisfactory control design in practical terms, since it is well known that such dead-beat designs can require excessive control effort and can be very sensitive to model parameter uncertainty. In practice, therefore, more reasonable closed loop pole locations would normally be specified, with the order of $d(z^{-1})$ set equal to the order of the closed loop characteristic polynomial and the parameters selected to yield a compromise between fast response, noise rejection and robust performance. Alternatively, an LQG design based on the NMSS model could be utilised, with the relative weighting on the input, output and integral of error states providing the required compromise.

4. PIP CONTROL DESIGN FOR THE δ OPERATOR MODEL

Following a similar design philosophy to that used for the z^{-1} operator model, the δ operator TF model can be represented by the following NMSS equations,

$$\delta x(k) = F \, x(k) + g \, v(k) + d \, y_d(k) \tag{8}$$

where,

$$
F = \begin{bmatrix}
-a_1 & -a_2 & \cdots & -a_{P-1} & -a_P & b_2 & b_3 & \cdots & b_{P-1} & b_P & 0 \\
1 & 0 & \cdots & 0 & 0 & 0 & 0 & \cdots & 0 & 0 & 0 \\
0 & 1 & \cdots & 0 & 0 & 0 & 0 & \cdots & 0 & 0 & 0 \\
\cdot & \cdot & & \cdot & \cdot & \cdot & \cdot & & \cdot & \cdot & \cdot \\
0 & 0 & \cdots & 1 & 0 & 0 & 0 & \cdots & 0 & 0 & 0 \\
0 & 0 & \cdots & 0 & 0 & 0 & 0 & \cdots & 0 & 0 & 0 \\
0 & 0 & \cdots & 0 & 0 & 1 & 0 & \cdots & 0 & 0 & 0 \\
0 & 0 & \cdots & 0 & 0 & 0 & 1 & \cdots & 0 & 0 & 0 \\
\cdot & \cdot & & \cdot & \cdot & \cdot & \cdot & & \cdot & \cdot & \cdot \\
0 & 0 & \cdots & 0 & 0 & 0 & 0 & \cdots & 1 & 0 & 0 \\
a_1 & a_2 & \cdots & a_{p-1} & a_p & -b_2 & -b_3 & \cdots & -b_{p-1} & -b_p & 0
\end{bmatrix}
$$

$$g = [\ b_1\ 0\ .....\ 0\ 1\ 0\ 0\ .....\ 0\ 0\ -b_1\ ]^T$$

$$d = [\ 0\ 0\ .....\ 0\ 0\ 0\ 0\ .....\ 0\ 0\ 1\ ]^T$$

In this formulation, the control variable is denoted by $v(k)$, which is defined as follows in terms of the control input $u(k)$,

$$v(k) = \delta^{P-1}\ u(k)$$

with the associated state vector $x(k)$ defined as,

$$x(k) = [\delta^{P-1}y(k),\ \delta^{P-2}y(k),\ ..\ ,\ \delta y(k),\ y(k),\ \delta^{P-2}u(k),\ ..\ ,\ \delta u(k),\ u(k),\ z(k)]^{\ T}$$

In these equations, $z(k)$ is, once again the "integral of error" state, which is now defined in terms of the the inverse delta operator, or digital integrator δ^{-1}, i.e.,

$$z(k) = \delta^{-1}\ \{y(k) - y_d(k)\}$$

and, as before, $y_d(k)$ is the reference or command input at the kth sampling instant.

4.1 The Control Algorithm

As in the z^{-1} operator case, the SVF control law is defined in terms of the state variables. In this δ operator situation, however, these are the output and input and their discrete differentials up to the appropriate order, as well as the integral of error state $z(k)$. This control law can be written in the form,

$$v(k) = -\ k^T x(k)$$

where now,

$$k^T = [\ f_{p-1}\ f_{p-1}\ \cdots f_0\ g_{p-2}\ g_{p-3}\ \cdots g_0\ k_I\ ]$$

is the SVF control gain vector for the NMSS δ operator model form.

A block diagram of δ operator PIP control system is shown in Fig. 3. Note that, in order to avoid noise amplification, the system shown here would normally be converted into the equivalent z^{-1} operator form and implemented accordingly. The control structure then resembles the z^{-1} operator version, with simple feedback of the sampled input and output signals.

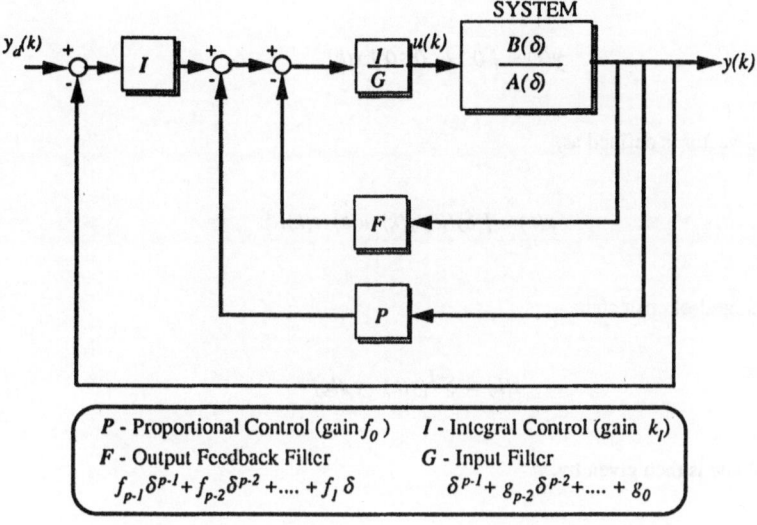

P - Proportional Control (gain f_0)   I - Integral Control (gain k_I)
F - Output Feedback Filter   G - Input Filter
$f_{p-1}\delta^{p-1}+f_{p-2}\delta^{p-2}+....+f_1\delta$ $\delta^{p-1}+g_{p-2}\delta^{p-2}+....+g_0$

Fig.3 Block Diagram of the Delta (δ) Operator PIP Servomechanism Control System.

4.2 A Simple Example

Consider a second order process described by the following δ operator TF model,

$$y(k) = \frac{\delta+b}{\delta^2}\, u(k) \qquad (9)$$

This represents the parallel connection of a digital integrator $1/\delta$ and a double integrator b/δ^2, and is characterised by a single parameter b. For a short sampling interval Δt, this process will behave in a similar manner to the equivalent continuous-time system defined by an algebraically identical TF, but with δ replaced by the Laplace operator s. Later, in Section 7.2, we consider the self-tuning and self-adaptive control of this process for a sampling interval of 0.01 units, using adaptive implementations of a simple PIP control system developed below.

The NMSS state space model in this case takes the form,

$$\delta x(k) = \begin{bmatrix} 0 & 0 & b & 0 \\ 1 & 0 & 0 & 0 \\ 0 & 0 & 0 & 0 \\ 0 & 0 & -b & 0 \end{bmatrix} x(k) + \begin{bmatrix} 1 \\ 0 \\ 1 \\ -1 \end{bmatrix} v(k) + \begin{bmatrix} 0 \\ 0 \\ 0 \\ 1 \end{bmatrix} y_d(k)$$

$$y(k) = [\ 0 \ \ 1 \ \ 0 \ \ 0 \ ] \ x(k)$$

where the NMSS vector is defined as,

$$x(k) = [\ \delta y(k) \ \ y(k) \ \ u(k) \ \ z(k)]^T$$

and $z(k)$ is the integral-of error state,

$$z(k) = \delta^{-1}\{y(k) - y_d(k)\}$$

The SVF control law is then given by,

$$v(k) = - k^T x(k)$$

where,

$$k^T = [f_1 \ f_0 \ g_0 \ k_1]$$

Note that in this δ operator case,

$$v(k) = \delta u(k)$$

and the closed loop system will be fourth order. As a result , let us consider a pole assignment design with all four poles assigned to -10 in the complex δ domain, i.e. the desired closed loop characteristic polynomial $d(\delta)$ will be defined as,

$$d(\delta) = \delta^4 + 40\delta^3 + 600\delta^2 + 4000\delta + 10000$$

This specification yields the following equations for the SVF control gains,

$$k_1 = 10000/b \ ; \ f_0 = (4000 - k_1)/b \ ; \ f_1 = (600 - f_0)/b \ ; \ g_0 = 40 - f_1$$

In a rapidly sampled situation, this model can be considered directly in continuous-time terms; so that the design specification is to have the system respond like a model system with unity steady state gain, composed of four first order systems, each with a time constant of 0.1 time units, i.e 'fast' critically damped response. For this specification, a short sampling interval of 0.01 time units, for example, will provide performance which can be compared directly to that of an equivalent, digitised continuous-time system.

An LQ design could also be considered in this δ operator example. It is clear, however, that the weightings in the equivalent cost function to (5) have a different meaning here and must be chosen accordingly, recognising that the elements of the state vector in the present situation are the numerical derivatives of the input and output variables.

5. RECURSIVE IDENTIFICATION AND ESTIMATION

The TDC concept assumes that recursive identification and estimation tools are available for modelling the system under study. Any such tools can be utilised but the Instrumental Variable (IV) approach, in its various forms, is probably the most flexible and easy to use in the TDC context. Recursive IV algorithms are fairly robust in the face of measurement noise and system disturbances. But, unlike other, competing methods such as maximum likelihood, prediction error minimisation or extended least squares, they do not demand simultaneous estimation of the noise characteristics, unless this is required for stochastic control system design purposes. IV methods of model order identification and parameter estimation are well known in the literature on time-series analysis and, for a general introduction to IV estimation, the reader should consult Soderstrom and Stoica (1983) and Young (1984).

For illustrative purposes, let us consider first the following backward shift version of the TF model (1),

$$y(k) = \frac{B(z^{-1})}{A(z^{-1})} \ u(k) \ + \ \xi(k) \tag{10}$$

where $\xi(k)$ is a general disturbance introduced to account for all stochastic and non-measurable deterministic inputs to the system. There are two major approaches to control system design for this kind of model: *deterministic design*, such as the NMSS methods discussed in previous Sections of this Chapter, where the disturbances are not specifically modelled for the purposes of control system design; and *stochastic design*, where the disturbances are characterised quantitatively in some appropriate manner and the resulting "noise" models are then utilised in the control system design calculations.

5.1 Recursive Parameter Estimation for Deterministic Control System Design

In the case of deterministic design, the basic IV algorithm provides the most obvious approach to recursive identification and parameter estimation. However, Young (1984, 1985) has demonstrated the practical value of the more sophisticated *Simplified Refined Instrumental Variable* (SRIV) procedure which includes the basic IV algorithm as a special case. For example: with appropriate changes in the definition of the data vectors (see later), it can be applied directly to δ operator models; it provides very good estimates of TF model parameters when the input-output data has been obtained from single impulse response testing; and it provides an excellent approach to model order reduction (Tych and Young, 1990). It is this SRIV algorithm which we propose as the normal basis for identification and estimation within the 'deterministic' TDC context.

The SRIV algorithm involves the adaptive prefiltering of the data used in the IV algorithm. It can be justified qualitatively by considering the following special case of equation (10),

$$y(k) = \frac{B(z^{-1})}{A(z^{-1})} \, u(k) \; + \; e(k)$$

where $e(k)$ is a zero mean, serially uncorrelated sequence of random variables with variance σ^2; and the TF is assumed to be stable, i.e. the roots of the characteristic equation $A(z)=0$ all lie within the unit circle of the complex z plane. This equation can be written in the following alternative vector form, which is *linear-in-the-parameters* $\{a_i; b_j\}$ of the TF model,

$$y(k) = z(k)^T a \; + \; \eta(k)$$

where,

$$z(k)^T = [\; -y(k-1), \; \ldots \ldots , \; -y(k-n) \; u(k-1), \; \ldots \ldots , \; u(k-m)]$$

$$a = [\; a_1 \; a_2, \; \ldots \ldots , a_n \; b_1, \; \ldots \ldots , b_m]^T$$

and $\eta(k)$ is a noise variable defined as follows in relation to the original white noise $e(k)$,

$$\eta(k) = e(k) + a_1 e(k-1) + \; \ldots \ldots \; + a_n e(k-n)$$

Most estimation problems are posed in a manner such that the variable to be minimised has white noise properties. Thus, a sensible error function is the *response* or *prediction error*, $\hat{e}(k)$

$$\hat{e}(k) = y(k) - \frac{\hat{B}(z^{-1})}{\hat{A}(z^{-1})} u(k)$$

where $\hat{B}(z^{-1})$ and $\hat{A}(z^{-1})$ are estimates of the TF polynomials $A(z^{-1})$ *and* $B(z^{-1})$. Unfortunately, this is nonlinear in the unknown parameters and so cannot be posed directly in simple linear estimation terms. However, the problem becomes *linear-in-the-parameters* if we assume prior knowledge of $A(z^{-1})$ in the form of an estimate $\hat{A}(z^{-1})$; then the error equation can be written in the form,

$$\hat{e}(k) = \frac{1}{\hat{A}(z^{-1})} \left[\hat{A}(z^{-1}) y(k) - \hat{B}(z^{-1}) u(k) \right]$$

which can be rewritten as,

$$\hat{e}(k) = \hat{A}(z^{-1}) y^*(k) - \hat{B}(z^{-1}) u^*(k)$$

where,

$$y^*(k) = \frac{1}{\hat{A}(z^{-1})} y(k) \quad ; \quad u^*(k) = \frac{1}{\hat{A}(z^{-1})} u(k) \tag{11}$$

are "prefiltered" variables, obtained by passing $y(k)$ and $u(k)$ through the prefilter $\dfrac{1}{\hat{A}(z^{-1})}$.

With this reasoning in mind, the ordinary recursive IV algorithm can be applied iteratively to estimate the model parameter vector a, with the variables $y(k)$, $u(k)$ and the instrumental variable $\hat{x}(k)$ replaced, at each iteration, by their adaptively prefiltered equivalents $y^*(k)$, $u^*(k)$ and $\hat{x}^*(k)$, respectively , and with the prefilter parameters based on the parameter estimates obtained at the previous iteration (see Young, 1984, 1985). The main recursive part of the SRIV algorithm takes the form,

$$\hat{a}(k) = \hat{a}(k-1) + g(k) \{ y^*(k) - z^*(k)^T \hat{a}(k-1) \} \tag{i}$$

where,

$$g(k) = P(k-1)\hat{x}^*(k) [1 + z^*(k)^T P(k-1)\hat{x}^*(k)]^{-1} \tag{ii}$$

$$\tag{12}$$

and,

$$P(k) = P(k-1) + g(k) z^*(k)^T P(k-1) \tag{iii}$$

where $P(k)$ is related to the covariance matrix $P^*(k)$ of the estimated parameter vector $a(k)$ by the equation,

$$P^*(k) = \sigma^2 P(k) \tag{13}$$

and an estimate $\hat{\sigma}^2$ of the variance σ^2 can be obtained from an additional recursive equation based on the squared values of a suitably normalised recursive innovation sequence (see Young, 1984; p.100).

This complete recursive algorithm can be considered simply as a modification of the well known Recursive Least Squares (RLS) algorithm, with the data vector $z^*(k)$ replaced alternately by $\hat{x}^*(k)$. At the jth iteration, the prefiltered instrumental variable $\hat{x}^*(k)$ required in the definition of $\hat{x}^*(k)$ is generated by adaptively prefiltering the output of an "auxiliary model" of the following form,

$$\hat{x}^*(k) = \frac{1}{\hat{A}_{j-1}(z^{-1})}\, \hat{x}(k) \qquad ; \qquad \hat{x}(k) = \frac{\hat{B}_{j-1}(z^{-1})}{\hat{A}_{j-1}(z^{-1})}\, u(k) \tag{14}$$

$$\textit{adaptive prefilter} \qquad\qquad \textit{adaptive auxiliary model}$$

where $\hat{A}_{j-1}(z^{-1})$ and $\hat{B}_{j-1}(z^{-1})$ are estimates of the polynomials $A(z^{-1})$ and $B(z^{-1})$ obtained by reference to the estimates obtained by the algorithm at the end of the *previous* (j-1)th iteration. The prefiltered input and output variables are obtained in a similar manner. Such a *recursive-iterative* or *relaxation* approach normally requires only three to four iterations to convergence on a stationary set of model parameters.

The δ operator version of the SRIV algorithm is applied to the vector version of the δ operator TF model, i.e.[3],

$$\delta^p y(k) = z(k)^T a + \eta(k) \tag{15}$$

where now,

$$z(k) = [\,\delta^{p-1}y(k)\,,\, \delta^{p-2}y(k)\,,\, \ldots\ldots\,,\, y(k)\,,\, \delta^{p-1}u(k)\,,\, \ldots\ldots\,,\, u(k)\,]^{\,T}$$

with a and $\eta(k)$ defined accordingly. The resulting algorithm is algebraically identical to the z^{-1} version: it is simply necessary to define the data vectors appropriately, i.e.,

---

3. note that we choose here to define the "dependent variable" as the pth numerical derivative of $y(k)$: this seems a somewhat arbitrary choice since the output $y(k)$ (or indeed any of the other variables in the NMSS state vector) could have equally well been utilised. However, we have discerned some practical advantages in this particular choice, although further research on the matter is still required before any firm conclusions can be reached.

$$z^*(k) = [\delta^{p-1}y^*(k), \delta^{p-2}y^*(k), \ldots, y^*(k), \delta^{p-1}u^*(k), \ldots, u^*(k)]^T$$

(16)

$$\hat{x}^*(k) = [\delta^{p-1}\hat{x}^*(k), \delta^{p-2}\hat{x}^*(k), \ldots, \hat{x}^*(k), \delta^{p-1}u^*(k), \ldots, u^*(k)]^T$$

with $y^*(k)$ in the innovation term $\{y^*(k) - z^*(k)^T\hat{a}(k-1)\}$ replaced by $\delta^{p}y^*(k)$. Here, the star superscript again indicates that the variables have been adaptively prefiltered, this time by the δ operator prefilter shown in Fig.4.

In this δ operator case, of course, the prefiltering operation is essential for other reasons. In particular, the elements of $z^*(k)$ and $\hat{x}^*(k)$ can be obtained from the adaptive prefilter, thus avoiding direct multiple differencing of the input and output signals, with its attendant problems of noise amplification. Used in this manner, the prefilters are seen as direct descendants of the "state variable filters" proposed by the first author in the early nineteen fifties for the estimation of continuous-time system models (see below).

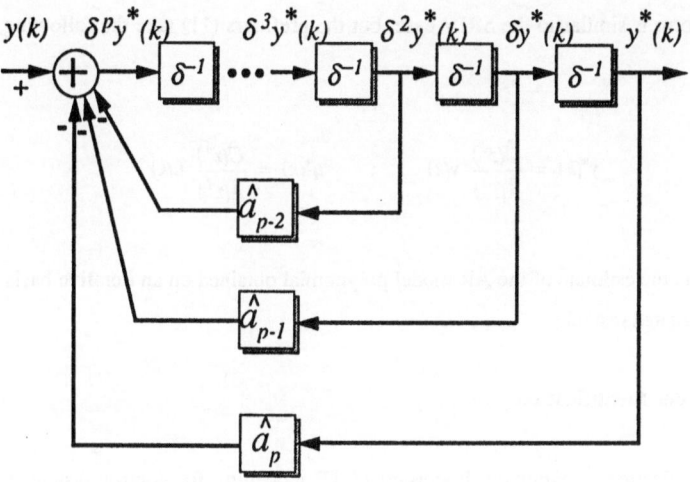

Fig.4 Block Diagram of the Prefilter in the SRIV recursive Algorithm
(shown operating on the output $y(k)$ signal)

5.2 Recursive Parameter Estimation for Stochastic Control System Design

As we point out later in Section 8, the PIP design procedure can be extended to the stochastic situation. In this context, the SRIV estimates will be asymptotically optimum in a minimum variance sense only if $e(k)$ has the properties of white noise; for coloured residuals, the estimates are sub-optimum in these terms but will normally have reasonable statistical efficiency. However, in the coloured noise case, two other possibilities exist, both of which are relevant for stochastic control system design:-

(i)   If the disturbances can be assumed to have rational spectral density, then the parameters of a TF model for the residuals (i.e. the noise process) can be estimated separately, under the assumption that they can be represented as either an AutoRegressive (AR) or an AutoRegressive-Moving Average (ARMA) process, by using one of the numerous recursive algorithms that are available for this purpose (e.g. Ljung and Soderstrom, 1983; Norton, 1986; Young, 1984). Normally, however, the AR model will suffice and, in this simple case, the RLS algorithm can be used to estimate the AR parameters.

(ii)   If minimum variance estimates are required in the coloured noise case, then the *Refined* or *Pseudo-Optimal* (RIV) algorithm can be utilised (Young, 1976; Young, 1984; Young, 1979; Young and Jakeman, 1979; Young and Jakeman, 1980; Soderstrom and Stoica (1983)[4]) . The simplest and most practical approach in this case is to assume that the noise can be modelled as an AR process. In this situation, the algorithmic form is similar to the SRIV case, but the prefilters (11) take the following, more complicated form,

$$y^*(k) = \frac{\hat{C}(z^{-1})}{\hat{A}(z^{-1})} \, y(k) \quad ; \quad u^*(k) = \frac{\hat{C}(z^{-1})}{\hat{A}(z^{-1})} \, u(k) \tag{17}$$

where $\hat{C}(z^{-1})$ is the estimate of the AR model polynomial obtained on an iterative basis by RLS estimation of the the estimated residuals.

5.3 IV Model Order Identification

Model order identification is extremely important in TF modelling for control system design. A successful identification procedure based on IV estimation (see Young, 1989) is to choose the model which minimises the following identification statistic,

$$YIC = log_e \left\{ \frac{\sigma^2}{\sigma_y^2} \right\} + log_e \{ NEVN \} \tag{18}$$

where,

σ^2 is the sample variance of the model residuals e(k)

σ_y^2 is the sample variance of the measured system output $y(k)$ about its mean value.

---

4. the latter authors refer to the RIV algorithm as IV4

while NEVN is the "Normalised Error Variance Norm" (Young et al, 1980) defined as,

$$NEVN = \frac{1}{np} \sum_{i=1}^{i=np} \frac{\sigma^2 p_{ii}}{\hat{a}_i^2}$$

Here, in relation to the TF models (1) and (2), np is the total number of parameters estimated, i.e. $n+m$ for model (1) and $2p$ for model (2); $\hat{a}_i^2$ is the estimate of the ith parameter in the parameter vector a; while p_{ii} is the ith diagonal element of the $P(N)$ matrix (so that $\sigma^2 p_{ii}$ is an estimate of the error variance associated with the ith parameter estimate after N samples).

The first term in (18) provides a normalised measure of how well the model explains the data: the smaller the variance of the model residuals in relation to the variance of the measured output, the more negative the first term becomes. Similarly, the second term is a normalised measure of how well the parameter estimates are defined for the npth order model: clearly the smaller the relative error variance, the better defined are the parameter estimates in statistical terms, and this is once more reflected in a more negative value for the term. Thus the model which minimises the YIC provides a good compromise between model fit and parametric efficiency: as the model order is increased, so the first term tends always to decrease; while the second term tends to decrease at first and then to increase quite markedly when the model becomes over-parameterised and the standard error on its parameter estimates becomes large in relation to the estimated values (in this connection, note that the square root of $\sigma^2 p_{ii} / \hat{a}_i^2$ is simply the relative standard error on the ith parameter estimate).

5.4 Remarks on SRIV and RIV Identification and Estimation

(i)   It is clear that the basic IV algorithm is a special case of the SRIV or RIV algorithms which applies where *equation error* minimisation is considered appropriate; whereas the SRIV algorithm provides for the alternative, and normally superior, *response error* minimisation. As a result, the SRIV algorithm should be used in all situations, except where the passband of the stochastic disturbances is similar to that of the system itself (i.e., in the case of AR noise, where $C(z^{-1}) \equiv A(z^{-1})$ or $C(\delta) \equiv A(\delta)$). In these latter, classic equation error circumstances, we see from equation (11) that no prefiltering is necessary, so that the basic IV algorithm is most appropriate.

(ii)   A more detailed discussion on the prefilters in the z^{-1} case is given in Young et al. (1988), including their use in overcoming low frequency noise effects such as drift, bias or load disturbances (also see later, Section 7.2) . Similar arguments apply in the δ operator case.

(iii) As we point out above, the prefiltering approach used in SRIV and RIV estimation is very similar to the "state variable filtering" method proposed previously (Young, 1979; Young and Jakeman, 1980) for continuous-time systems; which was itself an optimal generalisation of a previous, heuristic design where the prefilters were designed simply as band-pass filters chosen to span the passband of the system under study (Young, 1965a,b; Young, 1969a; see also Gawthrop, 1988; Goodwin, 1988). Moreover, simple discretisation of these earlier continuous-time designs (as required for their implementation in a digital computer) yields very similar implementations to the present δ operator algorithms. *However, we consider that the δ operator formulation is more satisfying from a TDC standpoint.*

(iv) Within the δ operator model context, the SRIV or RIV prefilters can also be considered optimal equivalents of the more arbitrary and heuristically defined prefilters proposed by Goodwin (1988). However, the performance of the SRIV algorithm is not particularly sensitive to the choice of prefilters and the SRIV estimates remain relatively efficient even if the filters are not optimal in a theoretical sense. And, as we shall see later in Section 7.2, sub-optimal prefilters designed on a more heuristic basis may prove essential in those practical situations where the theoretical assumptions cannot be satisfied; e.g. when the open loop system to be controlled is unstable.

(v) In both the z^{-1} and δ operator model cases, it will be noted that all the variables required for each recursive SRIV or RIV parameter estimation update are available from the prefilters at any *single* sampling instant. This means that the recursion does not need to be carried out at the same sampling interval as that used by the control algorithm; nor, indeed, does it even need to be at uniformly spaced intervals of time. This attractive feature, which was pointed out first by Young (1965b; 1969a), has important practical implications, particularly in self adaptive and self-tuning system design: for example, it can help overcome any problems associated with the computational time required for implementing the adaptive algorithms, and can also allow for easier allocation of time in multi-tasking operations. Moreover, by employing multiple sets of heuristically designed prefilters, each with different bandpass characteristics, it is possible to extract the information required for multiple SRIV parameter estimation updates at a single time instant.

6. CONTROL SYSTEM DESIGN EVALUATION

There are may ways of evaluating a control system design prior to implementation. Most designers would agree, for example, that such evaluation needs to include deterministic measures, such as the transient (step and impulse response etc.) and frequency (Bode, Nyquist, gain and phase margins etc.) response characteristics, which are normally considered as *sine qua non* for any reasonable assessment of a control system. Equally important, however, is the stochastic behaviour of the closed loop system and, in particular, its sensitivity to uncertainty in all its forms. Analytical methods are available for such an evaluation in the open loop (e.g. Matlab™ implementation in the Identification Toolbox; see Moler et al., 1987; and Ljung, 1988) but these have not been extended to the closed loop situation and, even in the open loop case, are often limited, either by the kind of assumptions that are required to obtain the analytical results, or by the need for limiting approximations. A more flexible approach, given the ready availability of high powered microcomputers, is the application of Monte-Carlo simulation analysis. It is this approach which we favour for general TDC system design.

6.1 Monte-Carlo Simulation Analysis

The design of the control system is based on the estimate of the parameter vector **a** which characterises the model of the controlled system. More precisely, since identification and estimation are statistical procedures, it is based on the probability distribution associated with the estimate, usually (under Gaussian assumptions) the mean value and associated covariance matrix. However, this uncertainty, which results from a combination of noisy measurements, unmodelled disturbances and model inadequacy, is usually difficult to handle in a purely analytical manner unless simplifying assumptions are made (e.g. linearisation allowing for the definition of approximate uncertainty bounds on the closed loop time or frequency responses). And since most of design procedures involve highly nonlinear transformations of the estimated model parameters, such simplifications may often provide either too optimistic or too pessimistic results.

The method of uncertainty evaluation which we propose here uses the covariance matrix estimate P^* associated with the estimate $\hat{a}$, under the assumption that the parameters are random variables with multivariable Gaussian distribution defined by the pair $\{\hat{a}, P^*\}$. This information, which is provided by the recursive estimation procedure, is used in the stochastic sensitivity analysis described below:

(i) calculate the TDC gain vector k, using the parameters $\hat{a}$ and the pole assignment or LQ-optimal algorithm (or any other suitable design method);

(ii) create a set of N quasi-randomly perturbed parameters $\tilde{a}_i$, $i=1, \ldots, N$, such that $\tilde{a}_i$ is generated from a Gaussian distribution defined by $N\{\hat{a}, P^*\}$;

(iii) for each $\tilde{a}_i$ calculate the required response or characteristic $\tilde{R}$ of the closed loop system defined by $\{\tilde{a}_i, k\}$: e.g transient and frequency responses, pole locations etc.

In practical terms, this is simply exploiting Monte-Carlo simulation to evaluate the sensitivity of the closed loop system when the parameters in the estimated parameter vector $\hat{a}$ are systematically perturbed according to the information contained in the estimate of the covariance matrix P^* obtained from the recursive estimation algorithm.

One possible modification of step (ii) can take into account the fact (Kendall and Stuart, 1963) that, since P^* is an estimate of the covariance matrix provided by the recursive algorithm, a measure of the distance of the realisations of the random variable a from its expected value (as defined by the quadratic form $\hat{a} P^* \hat{a}$) has an F distribution with $(np, N\text{-}np)$ degrees of freedom, where np is the size of $\hat{a}$ vector, i.e. the number of estimated parameters. In this case, one may generate the quasi-uniformly distributed variables from a confidence ellipsoid such that, say, 95% realisations are contained in it. Such a modification gives a better approximation to the envelope of the generated responses but, on the other hand, does not allow for further statistical results and is numerically much more demanding.

Of course, the distribution of the chosen response characteristic $\tilde{R}$ will be very much different from a normal distribution. We suggest the following possibilities:

(i) Closed Loop Pole Locations

In the case where $\tilde{R}$ represents the location of the closed loop poles, their realisations form a characteristic curved patterns on the Argand diagram which, for low levels of uncertainty, concentrate in the vicinity of unperturbed closed loop poles. These patterns are, in some senses, the stochastic analogues of the well known root loci and we shall refer to them as *stochastic root loci*. By simply counting the percentage of unstable realisations (with N sufficiently large, taking into account the number of parameters), one obtains a reasonable statistic which describes the robustness of design.

(ii) Closed Loop Step and Impulse Responses

In this case, multiple, overlaid plots of the response ensemble form "clouds" of both output and input realisations. It is then easy to assess both the "thickness" (a measure of dispersion) and other important characteristics; e.g. average or maximum overshoot, the percentage of cases which would cause saturation of the control inputs, etc.

(iii) Closed Loop Frequency Response

When $\tilde{R}$ is the frequency response of the closed loop, the method once again uses the overlaid plots of the chosen response measure, so allowing for comparisons with conventional frequency domain design techniques and providing an assessment of the uncertainty associated with the frequency response of the closed loop system. As the strict interval estimates of the closed loop frequency response can be obtained, it should be possible to establish links between this evaluation method and the H^∞ design method.

(iv) Other indices

When the Linear Quadratic algorithm is used to determine the controller gains, the results of the closed loop stochastic simulation may be used to compute the values of the objective function obtained in the stochastic realisations. The histogram plot of these values then provides information about the relative reliability of the deterministic optimal solution. Similarly, various additional parameters that are useful in assessing the closed loop behaviour may be computed: e.g. the "control effort" measured as the sum of squares of control input, or its accumulated value until some threshold phenomenon occurs (e.g. the output remains within some limits of the desired value etc.); the extremal values of the control input; the maximum overshoot etc.

A similar approach to the above is discussed elsewhere in this volume by Karny and Halouskova (1990) in the context of the preliminary design of self-tuning regulators (in effect, a subclass of the general class of design problems considered here). One difference between the two approaches lies in the fact that, in each of the realisations, Karny and Halouskova use a new controller gain vector, the assessment then being based on the pair $\{\tilde{a}_i, k_i\}$, where k_i is computed from $\tilde{a}_i$ (a natural choice in the context of an adaptive controller), while the approach presented here nominally uses the pair $\{\tilde{a}_i, k_i\}$. Clearly, extension of the approach to encompass this additional uncertainty is straightforward if it is required in the adaptive context. Another difference is that Karny and Halouskova compute only the ensemble characteristics of the objective function and base their design upon the distributions obtained in this manner. Here, we emphasise the many other measures that can prove useful for sensitivity evaluation. A combination of the two approaches, however, is clearly straightforward and offers both a flexible and powerful design procedure.

Finally, it should be added that a similar stochastic simulation method may be used to validate model reduction procedures based on the SRIV method applied to high order simulation models (Tych and Young, 1990): here, the model is obtained using the YIC model order identification procedure (Young, 1989), coupled with SRIV parameter estimation: the estimated reduced order covariance matrix is then used in a similar manner to that described above, where it replaces the covariance matrix obtained from normal identification and estimation studies. In such an application, it may well be necessary to introduce additional uncertainty to account for other stochastic influences which may not be present in the high order model simulation.

6.2. A Simple Example

As a simple illustrative example, let us consider the system described in Example 3.2., in which uncertainty is introduced by adding zero mean, white observational noise with variance 0.16 on to the output signal $y(k)$. The input and output signals, as obtained by simulating the system with a repeated unity step input, then form a basis for SRIV identification and estimation of this open loop system. The parameter vector $\hat{a}$ and its estimated covariance matrix P^* obtained in this manner are used to design two PIP controllers: one using the pole assignment (PA), and the other optimal (LQ) design. Here, the desired pole locations are chosen so that the step response of the resulting closed loop system is similar to that of the optimal controller, with the weights W_y, W_u, and W_z all set to unity. The estimated parameter covariance matrix P^* is now used to generate a set of perturbed parameters $\bar{a}_i$, $i=1, 2, ... , 100$, and the ensemble of closed loop characteristics are calculated for three response measures $\bar{R}$: step responses, stochastic root loci (closed loop system pole locations) and frequency responses.

Figs. 5 (a), (b) and (c) show the results obtained when this procedure is carried out for both types of controller using the same ensemble of $\bar{a}_i$ values in each case: the ensemble of step responses are shown in (a); the stochastic root loci in (b); and the ensemble of frequency response gain characteristics in (c). Note that, although the design specifications for the two designs are chosen so that the closed loop step responses are similar, the unperturbed closed loop pole positions are somewhat different for each design (all four poles close to the desired 0.14 for PA and at [0 0.202+/-0.44j 0.45] for LQ). This can influence the robustness of the designs: in this case the effect is not large, with the PA responses only a little more oscillatory (i.e. less robust) than those of the LQ design. In general, however, we have found that the LQ design is likely to be more robust unless the desired pole locations are chosen very carefully.

7. ADAPTIVE IMPLEMENTATION

On-line simplifications and modifications to the SRIV algorithm are straightforward. For example, the least squares version of this algorithm might well be appropriate in many circumstances. Also, in practice, the adaptive prefilters would normally be fixed gain, with parameters defined by prior experiment. In fact, as we shall see in the example below, the performance of the algorithm is not too sensitive to the prefilter parameters, provided that the pass-band encompasses the pass-band of the system being modelled (as first pointed out by Young, 1965b; and emphasised recently by Goodwin, 1988).

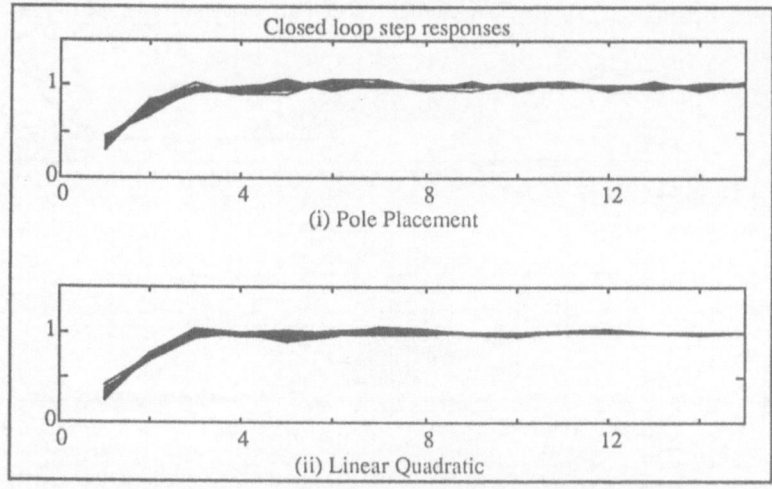

Fig. 5(a) ensemble of step responses for pole assignment
and optimal LQ designs

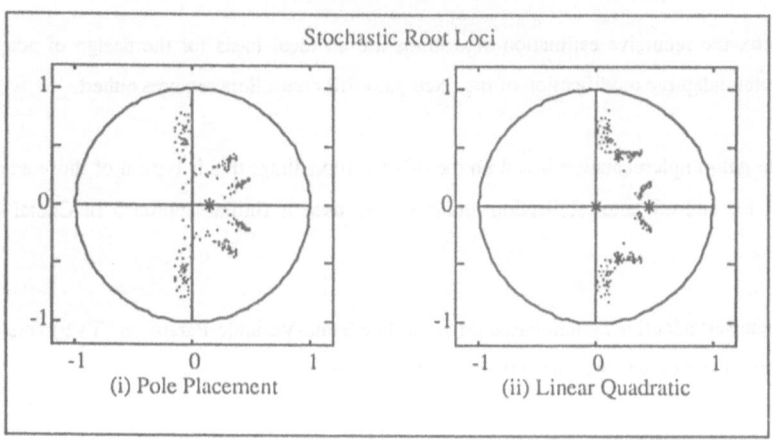

Fig. 5(b) stochastic root loci plots for
pole assignment and optimal LQ designs.

Fig .5 PIP Control system design evaluation using Monte-Carlo stochastic simulation based on 100 realisations for (i) pole assignment and (ii) optimal LQ designs: (a) ensemble of step responses; (b) stochastic root loci plots (Fig.5(c) appears over page)

98

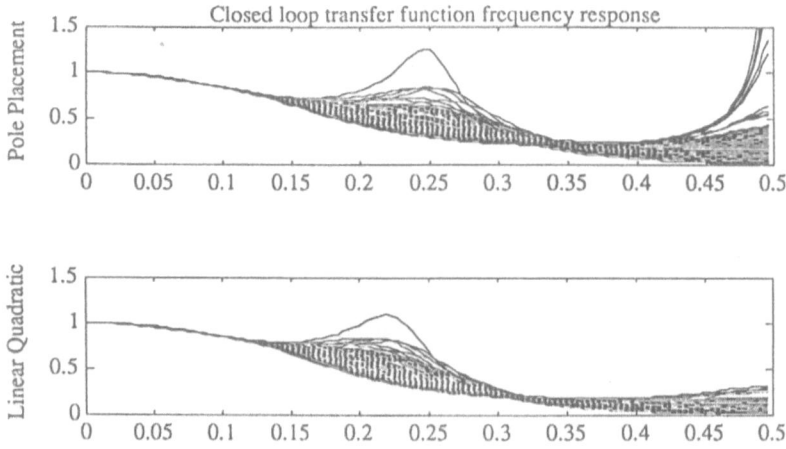

Fig.5(c) ensemble of frequency response gain plots for
pole assignment and optimal LQ designs.

In this on-line form, the recursive estimation algorithms are an ideal basis for the design of adaptive TDC systems. In particular, adaptive modification of the fixed gain PIP controllers requires either:

(a)   schedule gain implementation based on the off-line modelling: this is typical of those adaptive systems used for aircraft autostabilisation and we have used a similar approach in Chotai et al. (1990b); or

(b)   a complete self adaptive system based on an on-line Time-Variable-Parameter (TVP) version of the recursive estimation algorithms used in the off-line analysis.

For the type (b) approach, we propose the TVP version of the recursive least squares or recursive instrumental variable algorithms used in our previous research on adaptive systems (Young, 1969b, 1971, 1981; Young et al, 1988; Young, 1989; Chotai et al, 1990b). This allows for the exploitation of prior knowledge on parameter variation, which can range from a simple adaptive adjustment or on-line "tuning" of a scheduled gain system, to full adaptive control capable of tracking rapid parameter changes.

7.1 Modelling the Parameter Variations

In the above references, the TVP model parameter vector $a(k)$ is assumed to be decomposed as follows:

$$a(k) = T(k) \, a^*(k) \qquad (19)$$

where $T(k)$ is a time variable matrix defined on the basis of prior information; and $a^*(k)$ is a slowly variable parameter vector modelled as a random walk[5], i.e.,

$$a^*(k) = a^*(k-1) + \eta(k) \; ; \quad cov\{\eta(k)\} = Q\sigma^2 \qquad (20)$$

where Q is the "Noise Variance Ratio" (NVR) between the variance of the white noise input to the RW model and the observational white noise variance σ^2, under the usual assumption that these two white noise sources are statistically independent. This NVR value is an important design parameter in the implementation of the adaptive system (see example below), controlling, as it does, the tracking ability of the recursive estimator: for instance a combination of variable and constant parameters can be accommodated by setting the NVR's for the assumed constant parameters to zero and the other NVR values to suitable values based on the assumed temporal rates of variation (Young, 1984).

Combining the above equations, we see that $a(k)$ evolves as the Gauss Markov process,

$$a(k) = \Phi(k, k-1) \, a(k-1) + \Gamma(k) \, \eta(k) \qquad (21)$$

where,

$$\Phi(k, k-1) = T(k) \, [T(k-1)]^{-1} \; ; \quad \Gamma(k) = T(k) \qquad (22)$$

are time variable transition and input matrices which reflect the designer's prior knowledge of the potential parameter variation. This knowledge can be extensive, as in the adaptive control of an airborne vehicle (Young, 1981), where $T(k)$ was made a function of "air data" variables, such as dynamic pressure and Mach number; or fairly trivial, as in the adaptive temperature control of a heated bar system (Chotai et al, 1990b), where it related simply to fact that the system dynamics change instantaneously on reversal of the control signal.

---

5. although other models, such as the Integrated Random Walk (IRW - see Young, 1984), may sometimes be more appropriate.

7.2 Examples of Adaptive PIP Control System Design and Implementation

Practical examples of self-tuning and self-adaptive control based on the backward shift PIP controller have been reported in previous publications (see e.g. Young et al, 1988; Chotai et al, 1990b). Fig. 6, for example, shows typical results obtained in research on the self adaptive control of a Nutrient Film Technique (NFT) system used in glasshouse horticulture. Here the concentration of a solute in an NFT flow pilot plant is being controlled by an adaptive PIP system: it is clear that tight control is maintained throughout the test despite the sinusoidal leak (which is introduced to model the effects of nutrient uptake by plants which would be growing in the full scale system).

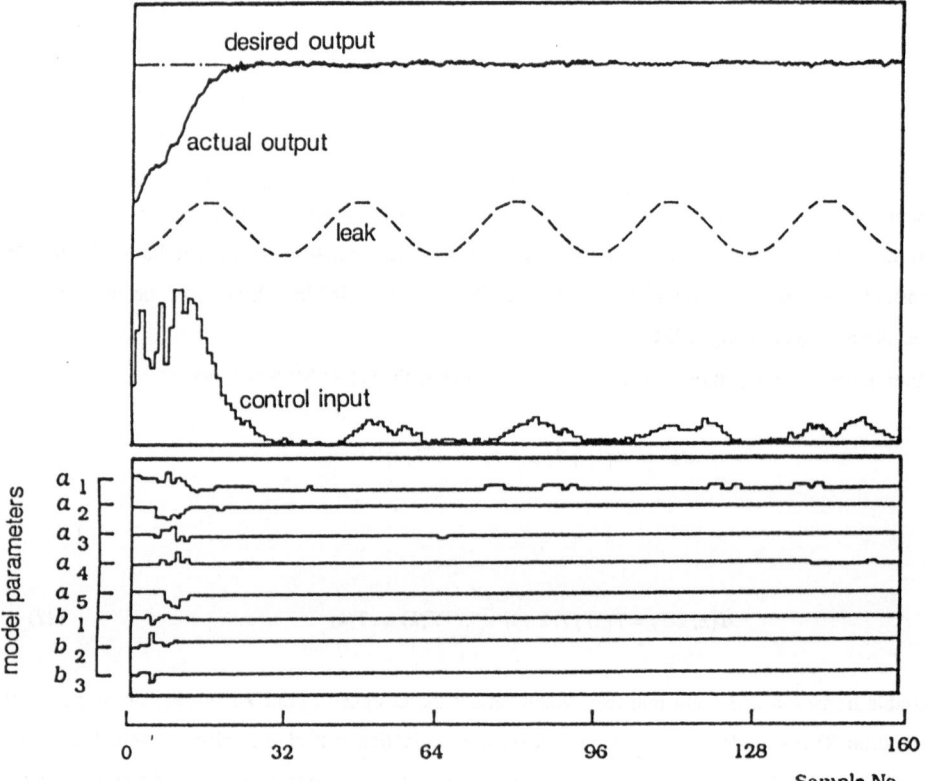

Fig. 6   Self-tuning control of solute concentration in the NTF pilot
plant based on a fifth order z^{-1} transfer function model

As a simple example of a self-tuning control based on the δ operator model, let us consider the double integrator process given in equation (9) with a sampling interval of 0.01 secs, using an adaptive version of the associated PIP design considered in Section 4.2. The adaptive system in this case is quite simple. First, the model is characterised by only the single unknown parameter b, so that the recursive estimation algorithm is in a scalar form. For additional simplicity, we also use the RLS version of the SRIV algorithm (i.e. prefiltering but no IV modification). Second, the prefilters are made time-invariant for implementational convenience, and are related to the desired *closed loop* pole assignment specifications: in particular, the prefilters for the input and output signals are specified as,

$$\frac{100\delta}{\delta^2 + 20\delta + 100}$$

with unity gain, two poles at -10 in the complex δ plane and a δ term in the numerator. This is a pragmatic choice: it assumes little knowledge of the open loop system (which is marginally stable in this case and so, in any case, is not a suitable basis for prefilter design), but it has a wide passband and so encompasses the frequency response characteristics of the open loop system, as required. The numerator is chosen to introduce high pass filtering and so remove any drift, bias or load disturbance effects from the signals: it is equivalent to the assumption that such behaviour can be modelled as a random walk in the δ domain (see discussion in Section 5 and Young et al, 1988). Finally, a simple random walk model is assumed for the parameter variation (i.e. $T(k)$ is set to the identity matrix I for all k); i.e. it is assumed that there is no *a priori* information about parameter variation. Thus the *NVR* value and the initial $P(0)$ matrix are the only parameters in the recursive algorithm to be specified by designer: $P(0)$ is set to 10^6I for the self-tuning case and to a lower value for the self-adaptive case; while the *NVR* is selected to accommodate the expected parameter variations, as discussed below.

Figs. 7(a), (b) and (c) show the results of a typical self-tuning run with $b=10$; $y_d(k)$ chosen as a repeated step waveform, and system noise introduced by adding a zero mean, white noise disturbance, together with a large, constant, output load disturbance added over the middle period of the simulation. Here, the *a priori* estimate $\hat{b}(0)$ is set to 5 with the associated $P(0)=10^6I$ (see footnote [6]); and the *NVR* set to zero, reflecting the assumption that the model parameter is constant. However, to avoid initial transients, the controller uses the prior value $\hat{b}(0)$ for the first 10 recursions, before switching to the appropriate recursive estimates after this initial period. Fig. 7(a) demonstrates excellent servo characteristics, with the system output rapidly tracking the changes in the command input. Fig. 7(b) shows how the the associated control input is counteracting the system disturbances, particularly the load change: note that the effect of this load change on the output variable in (a) is only just discernible because of this tightness of the control action. Finally (c) is a plot of the SRIV recursive estimate $\hat{b}(k)$ compared with true parameter value $b=10$.

Figs. 8(a), (b) and (c) show the δ operator PIP controller performing in the fully self-adaptive mode with a higher frequency repeated step command input. The b parameter is changed in value from 20 to 5 half way through the simulation period: $\hat{b}(0)$ is set to 18 and $P(0)$ to 100, reflecting assumed better *a priori* knowledge of the parameter; and the *NVR* is selected at a value of 0.000001 (equivalent to a forgetting factor of 0.999; see

---

6. strictly, if we had confidence in the *a priori* value, $P(0)$ should be set to a much lower value. However, this "diffuse prior" was chosen to see how fast the recursive estimate would converge.

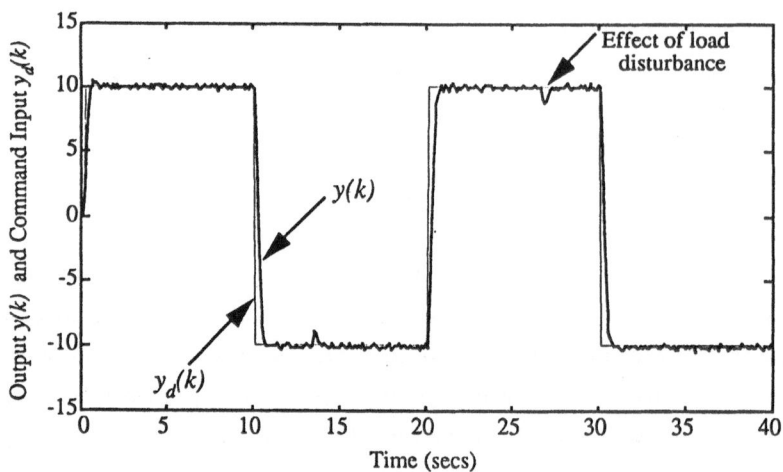

Fig. 7(a) response of system $y(k)$ to the repeated step
command input $y_d(k)$

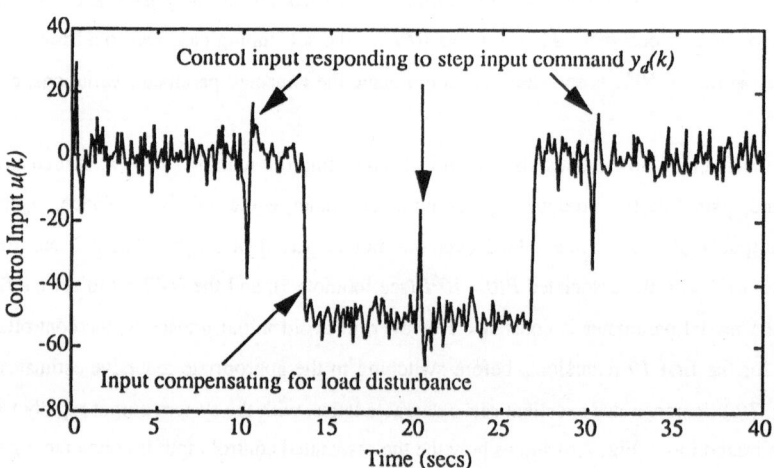

Fig. 7(b) control input $u(k)$, showing compensation for the large
load disturbance.

Fig. 7 Simulation results for a delta (δ) operator PIP self-tuning system applied to a
double integrator process with system noise and severe load disturbance: (a) response of
system to the repeated step command input; (b) control input signal, showing the rapid
compensation for the load disturbance; (c) SRIV recursive estimate compared with true
parameter value b=10 (Fig. 7(c) appears over page).

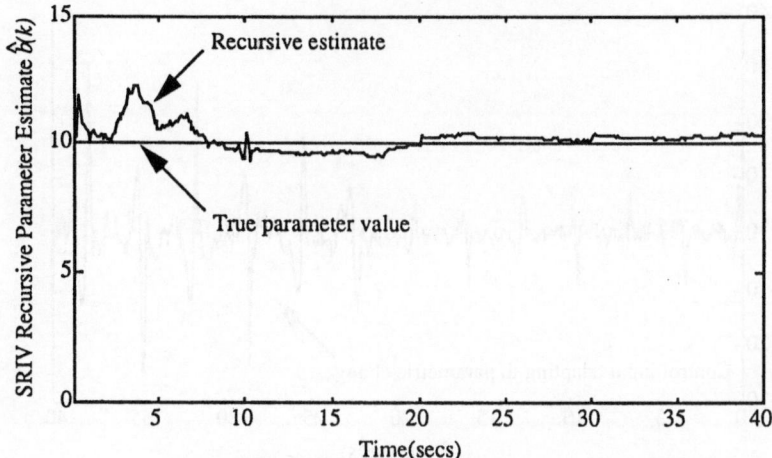

Fig. 7(c) SRIV recursive estimate compared with true
parameter value *b=10*.

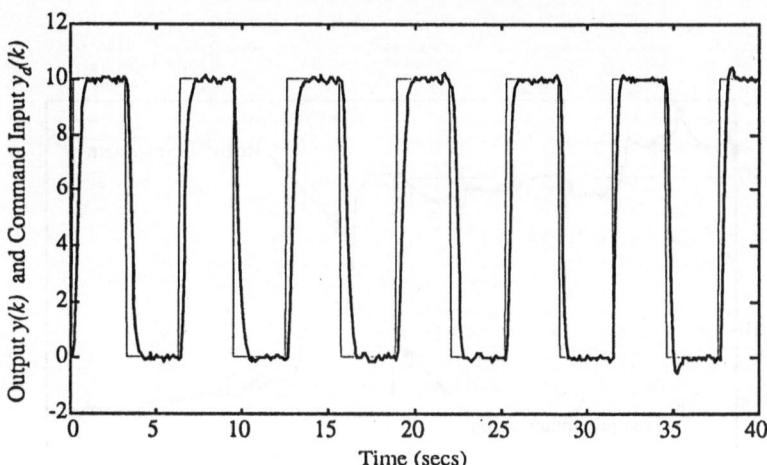

Fig. 8(a) response of system *y(k)* to the repeated step
command input $y_d(k)$

Fig. 8 Simulation results for a delta (δ) operator PIP self-adaptive system applied to a
double integrator process with system noise and a step change in the parameter from
b=20 to *b=5*: (a) response of system to the repeated step command input; (b) control
input signal, showing the adaption following the parametric change; and (c) adaptive
SRIV recursive estimation of the b parameter compared with true values (Figs. 8(b) and
(c) appear over page).

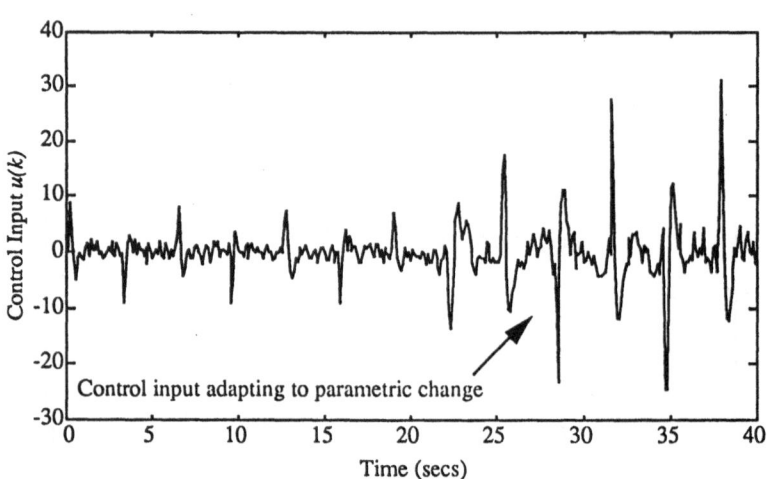

Fig. 8(b) control input *u(k)* adapting to the parametric change.

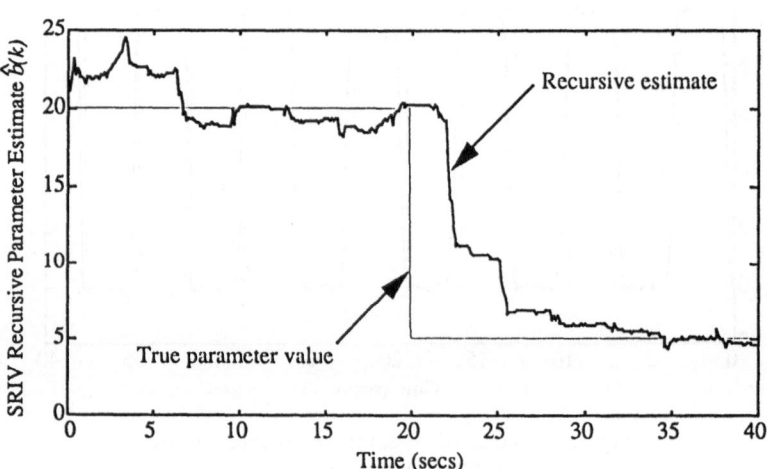

Fig. 8(c) adaptive SRIV recursive estimate compared with
true parameter value.

Young, 1984) to allow for parameter tracking. Once again, we see in (a) that excellent performance is maintained throughout, with rapid tracking of the more frequent step changes in the command input $y_d(k)$; while in (b), the notable changes in the magnitude of the control input indicate clearly the adaptive compensation for the changes in the system dynamics caused by the parametric change. The plot of the variations in the recursive estimate $\hat{b}(k)$ is given in 8(c), illustrating that although the parameter is not tracked exactly because of the noise effects, the estimation is quite adequate enough to maintain the tight servo control.

One advantage of the PIP controller is its relative robustness to one of the commonest forms of nonlinearity in practical systems; namely, input signal limiting. In particular, as shown in Fig.9 (a), (b) and (c), the PIP system avoids problems such as "integral windup". Here, the control input is limited to +/-2.0 units and, while the performance is clearly degraded, reasonable control is still maintained throughout the simulation experiment. Of course, this example is included here merely to demonstrate the robustness of the control system in this regard: in practice, the design specifications (e.g. closed loop pole assignments or LQ weighting parameters) would be selected either to avoid input signal saturation or yield acceptable performance despite saturation.

The reader may like to compare the results of this simulation example with those given in Goodwin (1988) and Astrom (1986), who consider the control of a deterministic, second order system with real roots. In the present case, we are controlling a more difficult double integrator system with system noise and load disturbance and still achieving excellent closed loop response characteristics.

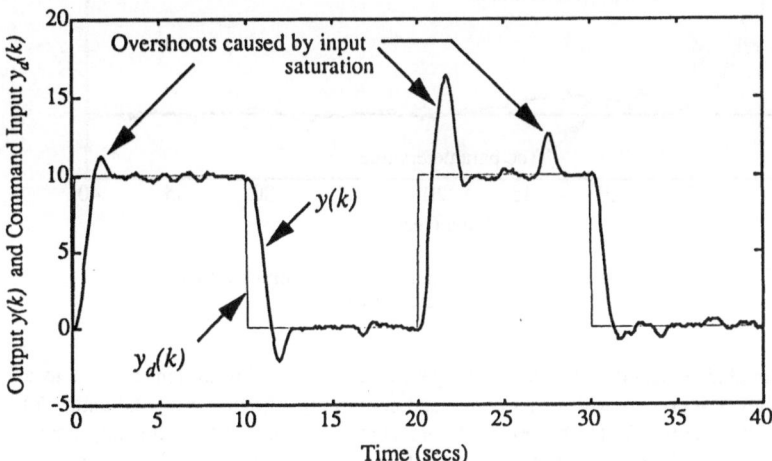

Fig. 9(a) response of system $y(k)$ to the repeated step command input $y_d(k)$, showing slower response and overshoot induced by input saturation

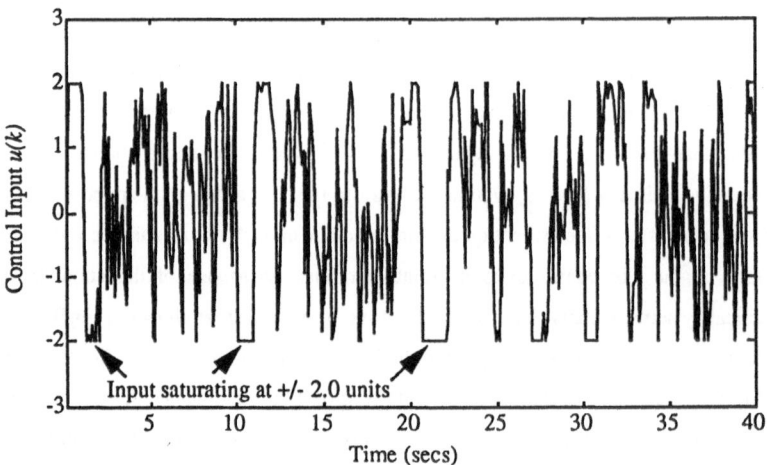

Fig. 9(b) control input $u(k)$, showing saturation during command
input changes.

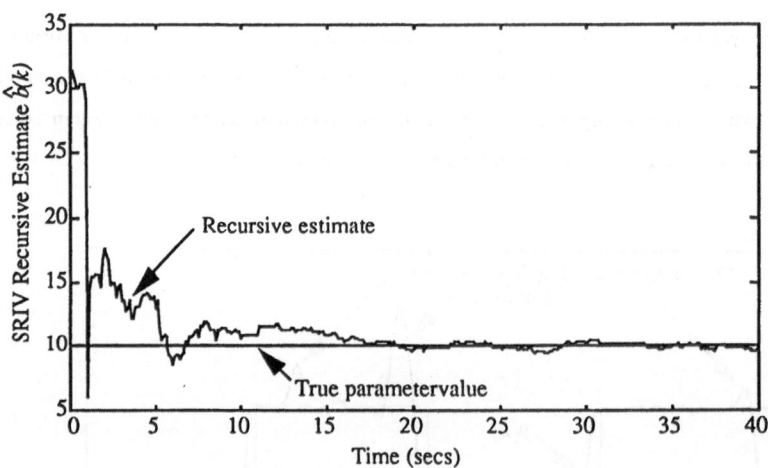

Fig. 9(c) adaptive SRIV recursive estimate compared with
true parameter value.

Fig. 9 Simulation results for a delta (δ) operator PIP self-tuning system applied to a double integrator process with system noise and control input limited to +/- 2.0 units: (a) response of system $y(k)$ to the repeated step command input $y_d(k)$, showing slower response and overshoot induced by input saturation (see previous page).; (b) control input $u(k)$, showing saturation during command input step changes; (c) SRIV recursive estimate compared with true parameter value.

8. DISCUSSION AND CONCLUSIONS

The TDC approach to control system design described in this paper provides a unified, *truly* digital implementation, since all stages in the design process are overtly digital, even when continuous-time (rapidly sampled) systems are being considered. Thus, on the basis of knowledge about the physical nature of the system under control, the designer can decide whether he wishes to choose either a relatively coarse sampling interval and think in completely digital terms; or, alternatively, a rapid sampling interval, where the δ operator model will provide a continuous-like controller in sampled data terms, without the need for digitisation from a continuous-time design.

The general approach to TDC system design outlined here within a deterministic, SISO framework, can be extended in a number of directions. Since the PIP (pole assignment or optimal LQ) control system design is posed in state-space terms, the stochastic formulation is obvious, involving simply application of the separation theorem and the introduction of optimal state estimation. Here, the RIV algorithms provide for full stochastic model identification and estimation; and an optimal (Kalman) filter is particularly simple to design since all the state variables are available for observation so that no state reconstruction is required (see also Young, 1979; Jakeman and Young, 1979).

The multivariable (MIMO) extension of TDC requires more thought. Again, model identification and estimation, although complicated, is possible using the multivariable RIV algorithm (Jakeman and Young, 1979; Young and Wang, 1986), suitably modified in the case of δ operator models. But the details of the control system design methodology are much less obvious. While pole assignment under a dyadic constraint (see Young and Willems, 1972) or LQ control system design are both straightforward, they do not address directly how the additional degrees of freedom should be used to solve problems such as decoupling. Initial research on this problem has yielded promising results (Wang, 1988) and the TDC design procedures are currently being extended in this direction.

Finally, it should be noted that the CACSD procedure outlined in this Chapter is being implemented in two versions: first, as an independent package written specifically for the PC/MS DOS machines, where it is fully interactive and provides the user with reasonable default values and menu selections; second, for PC/MSDOS machines and SUN™ Workstations, as a program which operates within the Matlab™ environment and takes advantage of this fact, both by using Matlab's existing facilities and toolboxes and by the the ability to access the variables (workspace) of the CACSD package (see Tych et al, 1990).

ACKNOWLEDGEMENTS

The research described in this paper was carried out under the U.K. Science and Engineering Research Council (SERC) Grant No. GR/E 8156.2, "Development, Evaluation and Implementation of True Digital Control (TDC)". The authors are grateful to SERC for this support.

REFERENCES

Astrom, K. J., L. Newman, and P. O. Gutman, (1986). A comparison between robust and adaptive control of uncertain systems. *IFAC Workshop on Adaptive Control, University of Lund, Sweden.*

Chotai, A., P.C. Young, and W. Tych. (1990a). A non-minimum state space approach to true digital control based on backward shift and and delta operator models; appears in *Proc. IASTED Conference. Lugano, Switzerland.*

Chotai, A., P.C. Young, and M.A. Behzadi. (1990b). The self-adaptive design of a nonlinear temperature control system, to appear in the Special Issue on *Self Tuning Control, Proc. I.E.E.* , Pt.D.

Gawthrop, P. J. (1988). Implementation of continuous-time controllers; appears in K. Warwick (ed.). *Implementation of Self Tuning Controllers.* Peter Perigrinus: London, 140-156 Goodwin, G .C. (1988). Some observations on robust stochastic estimation; appears in H. F. Chen (ed.). *Identification and System Parameter Estimation, 1988.* Pergamon:Oxford, 22-32

Goodwin, G. C., R. H. Middleton and M. Salgado. (1988). A unified approach to adaptive control; appears in K. Warwick (ed.). *Implementation of Self Tuning Controllers.* Peter Perigrinus: London, 126-139

Jakeman, A.J., and P.C. Young. (1979) Joint parameter-state estimation, *Electronics Letters,* 15, 582.

Karny, M. and A. Halouskova. (1990). Preliminary tuning of selftuners; this Volume

Kendall, M. and A. Stuart. (1963). *The Advanced Theory of Statistics.* Griffin and Co.: London.

Ljung, L. (1988). *System Identification Toolbox for Use with Matlab™.* The Mathworks Inc.: South Natick, M.A.

Ljung, L. and T. Soderstrom. (1983). *Theory and Practice of Recursive Estimation.* MIT Press: Cambridge, Mass.

Moler, C., J. Little and S. Bangert. (1987). *Matlab (Pro-Matlab)™ Users Guide.* The Mathworks Inc: South Natick, M.A.

Norton, J. P. (1986). *An Introduction to Identification.* Academic Press: London.

Soderstrom, T. and P.G. Stoica. (1983). *Instrumental Variable Methods for System Identification.* Springer-Verlag: Berlin.

Tych, W. and P.C. Young. (1990). A refined instrumental variable approach to model reduction for control systems design. Report No. TR81/(1990), Centre for Research on Environmental Systems, University of Lancaster.

Tych, W., Chotai, A., and Young, P.C. (1990). MicroCAPTAIN - True Digital Control (TDC) system design, *Proc. Workshop on Personal Computers in Industrial Control, Warren Spring Laboratory, Stevenage, Herts.,* England.

Wang, C. L. (1988). *New Methods for the Direct Digital Control of Discrete-Time Systems;* Ph.D. Thesis, Centre for Research on Environmental Systems, University of Lancaster.

Young, P. C. (1965a). The determination of the parameters of a dynamic process. *Radio and Electronic Engineer.* 29, 345-362.

Young, P. C. (1965b). Process parameter estimation and self adaptive control; appears in P.H.Hammond (ed.). *Theory of Self Adaptive Control Systems.* Plenum Press: New York.

Young, P. C. (1969a). *The Differential Equation Error Method of Process Parameter Estimation.*Ph.D Thesis, Department of Engineering, University of Cambridge.

Young, P. C. (1969b). *The use of a priori parameter variation information to enhance the performance of a recursive least squares estimator.* Tech. Note 404-90, Naval Weapons Center, California. *(53 pages).*

Young, P. C. (1971). *A second generation adaptive pitch autostabilisation system for a missile or aircraft.*Tech Note 404-109, Naval Weapons Center, California. *(100 pages).* Young, P. C. (1976). Some observations on instrumental variable methods of time-series analysis. *Int. Jnl. of Control.* 23, 593-612.

Young, P. C. (1979a). Parameter estimation for continuous-time models: a survey; appears in R. Isermann (ed.). *Identification and System Parameter Estimation.* Pergamon Press: Oxford, 1073-1086 (see also *Automatica,* 17, 459-469, 1981)

Young, P.C. (1979b).Self-adaptive Kalman filter, *Electronics Letters,* 15, 358.

Young, P. C. (1981). A second generation adaptive autostabilisation system for airborne vehicles. *Automatica.* 17: 459-469.

Young, P. C. (1984). *Recursive Estimation and Time Series Analysis.* Communication and Control Engineering. Springer-Verlag: Berlin.

Young, P. C. (1985). The instrumental variable method: a practical approach to identification and system parameter estimation; appears in H.A.Barker and P.C.Young (ed.). *Identification and System Parameter Estimation 1985, Vols 1 and 2.* Pergamon: Oxford, 1-16

Young, P. C. (1989). Recursive estimation, forecasting and adaptive control; appears in C.T.Leondes (ed.). *Control and Dynamic Systems.* Academic Press: San Diego, 119-166.

Young, P.C., and Willems, J.C. (1972) An apporoach to the linear multivariable servomechanism problem, *Int. Jnl. of Control*, **15**, 961-979.

Young, P. C. and A. J. Jakeman. (1979). Refined instrumental variable methods of recursive time-series analysis, Part 1: single input, single output systems. *Int. Jnl. Control.* **29**: 1-30.

Young, P. C. and A. J. Jakeman. (1980). Refined instrumental variable methods of recursive time-series analysis, Part 3: extensions. *Int. Jnl. Control.* **31**: 741-764.

Young, P. C., M. A. Behzadi, A. Chotai and P. Davis. (1987a). The modelling and control of nutrient film systems; appears in J. A. Clark, K. Gregson and R. A. Scafell (ed.). *Computer Applications in Agricultural Environments*. Butterworth: London, pp.21-43.

Young, P. C., M. A. Behzadi, C. L. Wang and A. Chotai. (1987b). Direct digital control by input-output, state variable feedback pole assignment. *Int. Jnl. Control.* **46**, 1867-1881.

Young, P. C., M.A. Behzadi, and A. Chotai. (1988) Self tuning and self adaptive PIP control systems; appears in K. Warwick (ed.). *Implementation of Self-Tuning Controllers*. Peter Perigrinus: London, 220-259

Polynomial LQ Synthesis for Self-tuning Control

K J Hunt

Control Group, Department of Mechanical Engineering
University of Glasgow, Glasgow G12 8QQ
Scotland

Abstract

This work is concerned with the polynomial equation approach to the synthesis of LQ controllers and, in particular, its application in the self-tuning control context. The theory presented provides not only feedback, but also optimal reference tracking and measurable disturbance feedforward compensation.

We address a number of practical issues arising from the basic theory. It is shown that the approach is applicable to both positional and incremental plant models without any changes to the basic theory. While the positional model is the most widely studied in the self-tuning control literature, the incremental type model has recently been shown to provide greater fidelity in some process control applications.

In addition, we show that the theory applies unchanged to delta-operator plant models as well as to the more widely studied delay-operator model. This provides a common platform for continuous and discrete time control since the delta operator model converges to the continuous plant model as the sample period reduces to zero. Thus, the delta operator model overcomes the numerical difficulties experienced with delay operator models when the sampling period is short.

The basic modelling framework used admits both stochastic and shape deterministic reference and disturbance models. These include drifting disturbances, and steps, ramps and sinusoids. Some practical measures to ensure the theory covers these situations are discussed.

Finally, various ways in which the underlying feedback system can be made more robust are described.

1 Introduction

Early attempts at using polynomial equations for the design of minimum variance regulators for inverse-stable (minimum phase) plants were made by Åström [1] and Peterka [21]. A solution for non-minimum phase plants was given by Peterka [22].

The *polynomial equation approach* to the synthesis of LQ regulators for general multivariable plants was developed by Kučera [19]. In contrast to the more traditional state-space approach, this is a purely algebraic method. Control synthesis reduces to the solution of linear polynomial equations whose coefficients are obtained by spectral factorisation. Kučera's output regulation solution was extended to the tracking of multivariable command signals by Šebek [30].

A more recent development is the inclusion of a feedforward compensator in the optimisation procedure. Feedforward can be used to improve the overall control performance when some disturbances are available for measurement. For the case of single input, single output plants the solution to such a modified optimal control problem was first given by Grimble [9] for stable disturbances. The solution for possibly unstable disturbances was subsequently given by Šebek et al [31]. A solution for a very general plant model, and for a cost function having dynamic weighting elements, has been obtained by Hunt [12].

Independently, Sternad and Söderström [28] obtained the solution for the stable disturbance case using an alternative proof technique. Sternad [27] then extended this approach to unstable disturbances.

The direct optimal measurable disturbance feedforward solution of Šebek et al was extended to multivariable plants by Hunt and Šebek [17]. Grimble [10] had previously obtained an indirect solution to the multivariable feedback/feedforward problem. This solution was obtained by reformulating the plant model as an equivalent, augmented, output regulation problem.

The use of the polynomial equation approach to LQ synthesis was first applied in the self-tuning control context by Zhao-Ying and Åström [32], [2]. The approach was further developed by Grimble [8] and Hunt et al [14]. LQ self-tuners incorporating measurable disturbance feedforward were developed by Hunt et al [15] for single input, single output plants, and by Hunt and Šebek [16] for multivariable plants. A full treatment for both scalar and multivariable systems is given by Hunt [13]. A detailed treatment of adaptive LQ feedforward controllers using the alternative polynomial derivation has been given by Sternad [25], [26], [27].

The purpose of the present paper is to discuss a range of issues concerned with the practical application of the polynomial LQ synthesis of self-tuning controllers.

Most of the work referred to above, and indeed the majority of the literature on self-tuning control, has been based around linear plant models in the delay operator q^{-1}. These models, consisting of an autoregressive part, a moving average part, and a control input, have the form

$$A(q^{-1})y(t) = B(q^{-1})u(t) + C(q^{-1})\phi_d(t)$$

where y is the plant output, u is the control input, and ϕ_d is a disturbance. This description is commonly referred to as an ARMAX or CARMA model.

However, much practical and experimental work (most notably by Clarke and co-workers [5] and Peterka [23]) has pointed to the utility of models which include a description of a drifting output disturbance:

$$y(t) = \frac{B(q^{-1})}{A(q^{-1})}u(t) + \frac{C(q^{-1})}{A(q^{-1})(1 - q^{-1})}\phi_d(t)$$

This type of model, the incremental model (or CARIMA model), more closely represents the kind of disturbances found on process plant.

In this work we show that the polynomial LQ synthesis approach can be applied to both positional and incremental models without alteration to the basic theory.

A further move in the plant modelling area is the introduction of the delta operator description as an alternative to the delay operator [7], [24]. The delta operator is defined as

$$\Delta = \frac{q-1}{T}$$

where T is the sample period. The delta operator overcomes some of the numerical difficulties experienced with delay operator models, particularly when the sampling rate is high. The delta operator also provides a mechanism for the unification of discrete and continuous time approaches since the delta operator model converges to the underlying continuous time differential operator model as the sampling interval approaches zero.

In this work we discuss the use of delta operator models for polynomial LQ synthesis and again show that this is covered without alteration to the basic theory.

Finally, a number of practical issues are discussed. It is shown that a useful range of reference and disturbance signals can be incorporated within the general modelling framework used. This includes drifting stochastic disturbances and shape deterministic signals such as steps, ramps and sinusoids. Ways in which the underlying optimal feedback system can be made more robust are also discussed.

Notation

All systems considered in this work are described by means of real polynomials in an indeterminate d. In the time domain d may be interpreted as the delay operator q^{-1}, the delta operator, or the inverse delta operator (the δ-operator), as required. In the frequency domain d is taken as the z-transform complex number. The reader is referred to the work by Kučera [19] for the background algebra.

The arguments of polynomials are usually omitted; a polynomial $X(d)$ is denoted by X. The adjoint of $X(d)$ is written as $X^*(d)$. In the delay operator case X^* is defined as $X^*(q^{-1}) \overset{\text{def}}{=} X(q)$, while in the δ-domain $X^*(\delta) \overset{\text{def}}{=} X(-T - \delta)$, with T the sample period. For any polynomial $X(d)$ we define $\langle X \rangle$ as the term independent of d. Stable polynomials are those with all zeros inside the stability region for the relevant operator basis. For the q^{-1} domain the stability region is inside the unit circle of the complex plane, and for the δ-domain the stability region is the finite region to the left of the line $Re(\delta) = -\frac{T}{2}$.

2 Canonical Model Structure

The discrete-time plant under consideration is assumed to be governed by

$$y(t) = W_p u(t) + W_x l(t) + W_d \phi_d(t) \tag{1}$$

where $y(t)$ is the output to be controlled, $u(t)$ is the plant control input, $\phi_d(t)$ is an unmeasurable disturbance, and $l(t)$ is a disturbance which is available for measurement. W_p, W_x and W_d are the transfer functions between the output and, repectively, the control signal, the measurable disturbance, and the unmeasurable disturbance. The open loop plant model is shown in Figure 1.

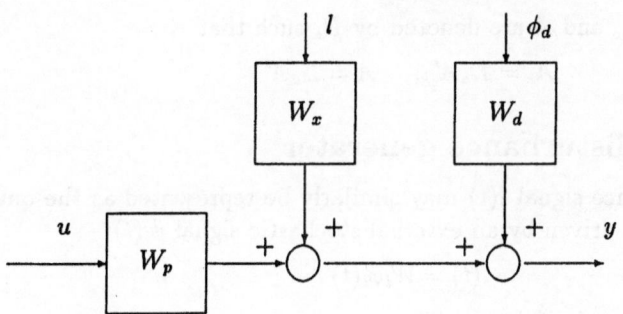

Figure 1: Open-loop system

In general, these sub-systems may have quite distinct dynamics. However, the plant model can *always* be expressed in a convenient canonical form by employing the least common denominator of the sub-systems. Denoting the least common denominator of W_p, W_x and W_d as A, these sub-systems may be expressed as

$$W_p = A^{-1}B \tag{2}$$

$$W_d = A^{-1}C \tag{3}$$

$$W_x = A^{-1}D \tag{4}$$

where A, B, C, and D are polynomials. Note that all systems (1) can be reduced to this common denominator form. In the case of plants having an integrating disturbance this will clearly lead to a model having an unstable common factor in A and B (provided the unstable factor does not also appear in the plant control-output transfer function W_p). This fact has implications for both estimation and control synthesis. The proper way to deal with this situation and to avoid these unstable common factors is covered in Sections 5 and 6.

2.1 Reference generator

For command tracking applications we introduce a reference signal $r(t)$. The signal $r(t)$ is represented as the output of a generating sub system W_r which is driven by an external stochastic signal $\phi_r(t)$:

$$r(t) = W_r \phi_r(t) \tag{5}$$

The sub system W_r is represented in polynomial form as

$$W_r = A_e^{-1} E_r \tag{6}$$

where A_e and E_r are polynomials.

The tracking error $e(t)$ is defined as

$$e(t) = r(t) - y(t) \tag{7}$$

Any common factors of A_e and A are denoted by D_e such that

$$A_e = D_e A'_{ec}, \quad A = D_e A' \tag{8}$$

2.2 Measurable disturbance generator

The measurable disturbance signal $l(t)$ may similarly be represented as the output of a generating sub system W_l driven by an external stochastic signal $\phi_l(t)$:

$$l(t) = W_l \phi_l(t) \tag{9}$$

W_l is represented in polynomial form as

$$W_l = A_l^{-1} E_l \tag{10}$$

where A_l and E_l are polynomials.

A wide range of practically useful reference and disturbance signals can be modelled using the above representation. This is further discussed in Section 7.4.

2.3 Assumptions

For the above plant we make the following assumptions

(A.1) Each sub-system is free of unstable hidden modes.

(A.2) The plant control-output transfer function W_p is strictly causal i.e. $\langle B \rangle = 0$, $\langle A \rangle = 1$.

(A.3) The polynomials C, E_r, and E_l are, without loss of generality, assumed to be stable.

3 Controller Synthesis Theory

In this section we formally state the theoretical solution to the optimal control problem. For a rigorous derivation and proof, including the optimisation problem solvability conditions, see Hunt [13]. A brief discussion of the solvability conditions is given below.

3.1 Control structure

The control law considered has three degrees of freedom in order that the measured output, the measured disturbance and the reference may be processed independently. The control law is given by

$$u(t) = -C_{fb} y(t) + C_r r(t) - C_{ff} l(t) \tag{11}$$

The feedback controller C_{fb}, the reference controller C_r and the feedforward controller C_{ff} may be expressed as ratios of polynomials as follows:

$$C_{fb} = C_{fbd}^{-1} C_{fbn} \tag{12}$$

$$C_r = C_{rd}^{-1} C_{rn} \tag{13}$$

$$C_{ff} = C_{ffd}^{-1} C_{ffn} \tag{14}$$

The closed-loop feedback system is shown in Figure 2.

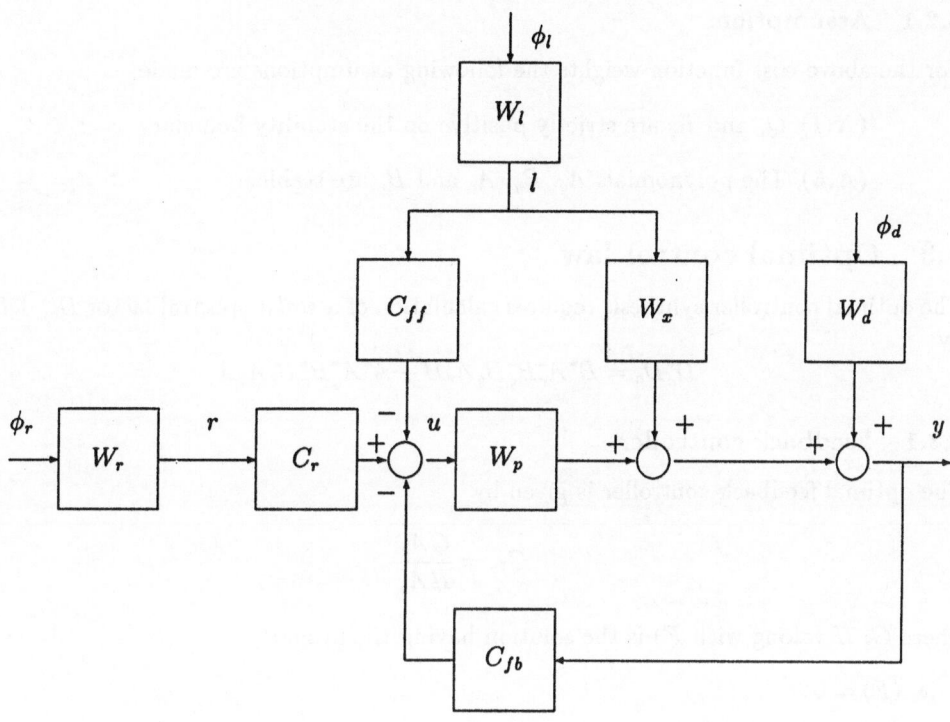

Figure 2: Closed-loop system

3.2 Cost function

The cost function which is used as the basis of the controller optimisation may be expressed in a variety of forms. In the time domain, the cost is defined as

$$J = \mathcal{E}\{(H_q e)^2(t) + (H_r u)^2(t)\} \tag{15}$$

where $\mathcal{E}$ denotes the expectation operator. Here, H_q and H_r are dynamic weighting elements which may be realised by rational transfer functions.

The cost function may also be expressed as

$$J = \langle Q_c \phi_e + R_c \phi_u \rangle \tag{16}$$

where ϕ_e and ϕ_u are the correlation functions of the tracking error and control input, respectively. In addition,

$$Q_c = H_q H_q^*, \quad R_c = H_r H_r^* \tag{17}$$

The weighting elements Q_c and R_c may be expressed as ratios of polynomials as

$$Q_c = \frac{B_q^* B_q}{A_q^* A_q}, \quad R_c = \frac{B_r^* B_r}{A_r^* A_r} \tag{18}$$

3.2.1 Assumptions

For the above cost function weights the following assumptions are made

(A.4) Q_c and R_c are strictly positive on the stability boundary.

(A.5) The polynomials A_q, B_q, A_r and B_r are stable.

3.3 Optimal control law

The optimal controller synthesis requires calculation of a *stable* spectral factor D_c, defined by

$$D_c^* D_c = B^* A_r^* B_q^* B_q A_r B + A^* A_q^* B_r^* B_r A_q A \qquad (19)$$

3.3.1 Feedback controller

The optimal feedback controller is given by

$$C_{fb} = \frac{G A_r}{H A_q} \qquad (20)$$

where G, H (along with F) is the solution having the property

- $\langle F \rangle = 0$

of the polynomial equations

$$D_c^* G + F^* A A_q = B^* A_r^* B_q^* B_q C \qquad (21)$$

$$D_c^* H - F^* B A_r = A^* A_q^* B_r^* B_r C \qquad (22)$$

By elimination of the coupling term F in the above equations it may be easily verified that G and H also satisfy the polynomial equation

$$A A_q H + B A_r G = D_c C \qquad (23)$$

Note, however, that the conditions required for (23) to yield a unique solution corresponding to the optimal controller are stricter than for the couple (21)–(22). This issue is discussed further in Section 3.4.

3.3.2 Reference controller

The optimal reference controller is given by

$$C_r = \frac{M A_r C}{E_r H A_q} \qquad (24)$$

where M (along with N and Q) is the solution having the property

- $\langle N \rangle = 0$

of the polynomial equations

$$D_c^* M + N^* A_q A_e = B^* A_r^* B_q^* B_q E_r \qquad (25)$$

$$D_c^* Q - N^* B A_r A_{ec}' = A^* A_q^* B_r^* B_r A' E_r \qquad (26)$$

By elimination of the coupling term N the polynomials M and Q can be shown to satisfy

$$D_e A_q Q + B A_r M = D_c E_r \qquad (27)$$

Again, the (more restrictive) conditions under which this single equation yields the unique optimal solution are discussed in Section 3.4.

3.3.3 Feedforward controller

The optimal feedforward controller is given by

$$C_{ff} = \frac{A_r(XC - GE_l D)}{H A_q E_l A} \qquad (28)$$

where X (along with Z and Y) is the solution having the property

* $\langle Z \rangle = 0$

of the polynomial equations

$$D_c^* X + Z^* A A_q A_l = B^* A_r^* B_q^* B_q D E_l \qquad (29)$$

$$D_c^* Y - Z^* B A_r A_l = A^* A_q^* B_r^* B_r D E_l \qquad (30)$$

By elimination of the coupling term Z the polynomials X and Y can be shown to satisfy

$$A A_q Y + B A_r X = D_c D E_l \qquad (31)$$

3.4 Solvability conditions

The precise conditions which must be satisfied for a solution to the optimal control problem to exist are derived and justified in Hunt [13]. A detailed discussion of the polynomial equation solvability conditions is also included. Here, we review only the conditions most relevant to the practical application of the theory.

3.4.1 Optimisation

For an acceptable solution to the optimisation problem to exist the following conditions must hold

(C.1) The polynomials A and B must have no unstable common factors. This condition is clear since any unstable disturbance modes must be controllable from the plant input.

(C.2) Any unstable factors of A_e must also be factors of A. Any unstable reference modes must appear in the forward path in order that they may be tracked.

(C.3) Any unstable factors of A_l must also be factors of both A and D. Again, any unstable measurable disturbance modes must be controllable from the plant input.

In fact, each of these conditions is just the polynomial manifestation of the internal model principle [6]; any unstable reference or disturbance modes must appear in the forward path.

3.4.2 Equation solution

We note that the condition (C.1) above

- The greatest common divisor of A and B must be stable

is also a necessary and sufficient condition for the existence of a unique solution to the three couples of polynomial equations (21–22), (25–26) and (29–30).

Stricter conditions apply in order to ensure unique solutions, corresponding to the optimal controller, of the single equations (23), (27) and (31). The most important condition required for these equations to yield the unique optimal controller is

- The polynomials A and B must be coprime.

This condition corresponds to the requirement that *all* poles (stable and unstable) of the disturbance transfer functions must also appear in the forward path.

4 Delta Operator Model

It is well known that plant models based upon the delay operator model in q^{-1} display undesirable characteristics when the sampling period is reduced. Discrete models with fast sampling will always have unstable zeros (non minimum phase behaviour) for plants with pole-zero excess greater than two [3]. This situation occurs even when the underlying continuous time plant has no right half plane zeros. Further, the numerator and denominator of the discrete time model converge to either zero or a power of $1 - q^{-1}$ when the sampling period approaches zero, and information on the plant dynamics is lost.

These factors lead to serious numerical difficulties for controllers based on delay operator models used with fast sampling. As an alternative, the difference, or *delta*, operator model has been introduced for control synthesis and parameter estimation purposes by Goodwin [7] and Peterka [24]. The delta operator Δ is defined by

$$\Delta = \frac{q - 1}{T} \tag{32}$$

where T is the sampling period. For a signal $x(t)$ the delta operator gives

$$\Delta x(t) = \frac{x(t+1) - x(t)}{T} \tag{33}$$

It may easily be verified that the following relation holds for the adjoint operator

$$\Delta + \Delta^* = -T\Delta^*\Delta \tag{34}$$

The standard plant model described by Equations (1–10) may be rearranged in terms of the delta operator. The key advantage of the delta operator formulation is that for very fast sampling the model converges to the underlying continuous time model. Thus, the numerical difficulties associated with fast sampling for the delay operator model are avoided. Using the delta operator model the continuous and discrete time approaches are unified in a single model, where the continuous time model is obtained in the limit as $T \to 0$.

Use of the delta operator model for polynomial LQ synthesis has previously been considered by Nagy and Ježek [20] and Katebi and Byrne [18]. A difficulty in this approach is that the adjoint operator defined by Equation (34) leads to a fractional result, which has implications for the synthesis procedure. As a solution to this problem Nagy and Ježek introduced the inverse delta operator, the δ-operator, defined by

$$\delta = \frac{1}{\Delta} \tag{35}$$

For this operator the adjoint relation becomes

$$\delta + \delta^* = -T \tag{36}$$

which leads to a polynomial result.

Thus, polynomial LQ synthesis using an alternative operator is based upon this δ-operator. As shown by Nagy and Ježek, the equations and solvability conditions for LQ synthesis using the δ-operator formulation are identical to those for the delay operator. The theory outlined in Section 3 therefore applies unchanged for δ-operator models.

The only difference is that, due to the differing adjoint relation (36), a modified numerical algorithm for the spectral factorisation of polynomials in δ must be used. Such an algorithm is given in the work by Nagy and Ježek.

5   Incremental Models

The practical utility of the incremental model has been demonstrated in many situations. This type of model is particularly well suited to self-tuning control applications in process control type problems. Here, it is very common to have constant offsets or drifting disturbances acting on the plant output. A control requirement is to reduce the steady state offset to zero in the face of such disturbances. Using integral action in conjunction with the incremental model has the desired effect in many self-tuning control algorithms.

Here, we show how the incremental model can be straightforwardly handled by the general theory presented in the preceding section.

The incremental model has an integrating disturbance acting on the plant output. We represent the integrator by the transfer function $\frac{1}{\nabla}$, where ∇ is a possibly rational function. For example, in the delay operator case ∇ is defined as

$$\nabla = 1 - q^{-1}$$

while in the δ-operator case

$$\frac{1}{\nabla} = 1 + \frac{1}{T}\delta$$

The incremental model may be written as

$$y(t) = \frac{B}{A'}u'(t) + \frac{D'}{A'}l(t) + \frac{C}{A'\nabla}\phi_d(t) \qquad (37)$$

The signals in this equation are as before, with u' a control input to the plant. A', B, C and D' are polynomials. Note that a straightforward transformation of this model to the form (A, B, C, D) would lead to an unstable common factor in A and B corresponding to the unstable poles in $\frac{1}{\nabla}$ (if these poles are not already in A'). The technique described in the following overcomes this problem.

Since the plant has a drifting output disturbance it is clear that integral action is required in the controller to offset this. In common with other approaches, we therefore force integral action into the controller by assuming a controller of the form

$$\frac{C_n}{C_d} = \frac{C'_n}{C'_d}\cdot\frac{1}{\nabla} \qquad (38)$$

At this point we make the conceptual distinction that what we know as the plant is that part of the system which is known *a priori* before controller synthesis takes place. Since we know that the controller must include integral action we therefore conceive of the plant as the original system (37) together with the controller integrator. This situation is illustrated in Figure 3.

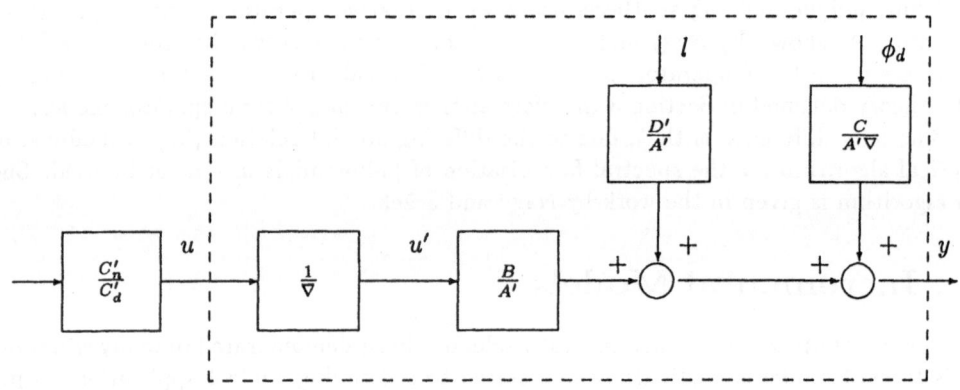

Figure 3: Incremental model - notional plant (dashed box)

With this understanding, the plant model (37) becomes

$$y(t) = \frac{B}{A}u(t) + \frac{D}{A}l(t) + \frac{C}{A}\phi_d(t) \qquad (39)$$

where, to make a consistent translation between (37) and (39),

$$\frac{B}{A'\nabla} \rightarrow \frac{B}{A} \qquad (40)$$

$$\frac{D'\nabla}{A'\nabla} \to \frac{D}{A} \tag{41}$$

$$\frac{C}{A'\nabla} \to \frac{C}{A} \tag{42}$$

and, in particular, A is chosen as the least common denominator of these transfer functions. Further, a pseudo control signal $u(t)$ is defined by

$$u(t) = u'(t)\nabla \tag{43}$$

Notice that with these definitions the incremental model immediately transforms to our standard plant model (1,2–4) used as the basis of the LQ controller synthesis theory. We may therefore apply the standard LQ synthesis theory to the transformed model. The design steps are as follows

1. Estimate A', B, C and D' using standard techniques (or fix C).

2. Form A, B, C, D using (40)–(42).

3. Solve LQ synthesis for (A, B, C, D), generate $u(t)$

4. Implement $u'(t) = \frac{1}{\nabla}u(t)$

Finally, we note that this transformation process creates no problems as far as the problem solvability conditions are concerned since

- The greatest common divisor of A and B remains stable (c.f. Condition C.1),

6   General Unstable Reference and Disturbance Models

The preceding section discussed the very specific case of an integrating disturbance acting on the plant output. We now generalise the theory to cover unstable disturbances (measurable and unmeasurable) and reference signals of a general nature. The aim here is to set up the design problem for a given plant in such a way that a meaningful optimisation problem is posed. This again amounts to ensuring that the optimal control problem solvability conditions (C.1)–(C.3) are satisfied.

Recall that the conditions (C.1)–(C.3) are a polynomial statement of the internal model principle; any unstable reference or disturbance modes must appear in the forward path of the feedback loop.

In the case of the incremental model our approach was to cascade an integrator (corresponding to the disturbance model) with the plant. In the general case we now cascade a general transfer function W_u with the plant, where

$$W_u = \frac{B_u}{A_u} \tag{44}$$

with B_u and A_u polynomials.

Our particular interest here is in inserting appropriate *unstable* factors into the feedback loop via this transfer function element to ensure that the problem solvability conditions are satisfied.

We use the terms A_u and B_u to design for any unstable reference and disturbance modes present in the given plant W_p', W_x, W_d, W_r, W_l. Appealing to the notion used above that the plant is that part of the system known *a priori* in advance of controller synthesis, the transfer function W_u is now considered part of the new notional plant. This situation is illustrated in Figure 4 where the pseudo control signal u is the notional input to the modified plant. To match our new plant with the canonical model (1) we have

$$W_p = W_u W_p' \tag{45}$$

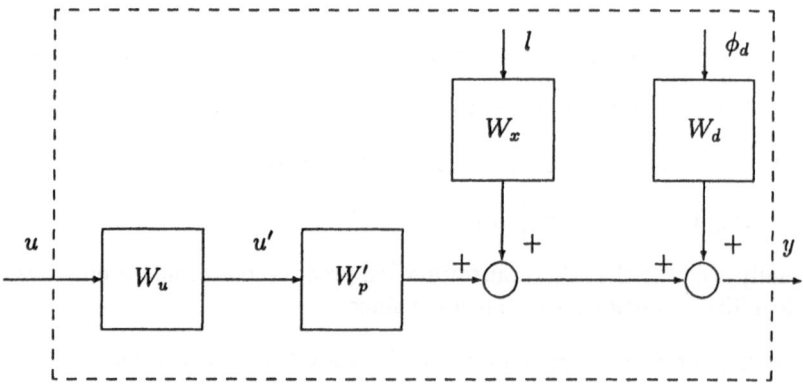

Figure 4: Notional plant (dashed box)

The controller is then designed for this new plant where the solvability conditions (C.1)–(C.3) have been satisfied by matching A_u and B_u with the relevant original unstable plant modes.

Finally, the true plant control signal u' is implemented as

$$u'(t) = W_u u(t) \tag{46}$$

The specific case of the incremental model discussed in Section 5 is obtained by selecting

$$W_u = \frac{B_u}{A_u} = \frac{1}{\nabla}$$

The approach outlined in this section for unstable reference and disturbance modes not appearing in the original plant control-output transfer function may be summarised as follows

1. Match A_u and B_u to unstable reference and disturbance modes not appearing in the original plant control-output transfer function to ensure (C.1)–(C.3) are satisfied.

2. Incorporate W_u as part of the plant.

3. Perform the synthesis outlined in Section 3 and generate $u(t)$.

4. Implement $u'(t) = W_u u(t)$

7 Robustness and Design Choices

7.1 Fixing the C polynomial

From Equations (1–2) and (12) it is clear that the closed-loop characteristic polynomial ρ is given by

$$\rho = AC_{fbd} + BC_{fbn} \tag{47}$$

Substituting for the optimal feedback controller transfer function (20) the characteristic polynomial becomes

$$\rho = AA_q H + BA_r G \tag{48}$$

Notice that this equation is equivalent to the polynomial equation (23) arising from the optimisation. Thus, the characteristic polynomial of the optimal feedback system is given by

$$\rho = AA_q H + BA_r G = D_c C \tag{49}$$

The optimal controller can therefore be viewed as a pole placement in $D_c C$. The polynomial D_c is obtained from the spectral factorisation (19). However, in a self-tuning control implementation the C polynomial is replaced by its estimate $\hat{C}$. It is well known that in practice it is very difficult to obtain accurate estimates of the C coefficients. Since C has a direct effect on the stability of the closed-loop system poor control performance can be obtained if one attempts to estimate this polynomial.

As an alternative, the C polynomial can be viewed as a controller design choice and fixed during implementation. In this way the difficulties associated with estimating C can be avoided. Some examples of the robustness improvement which can be achieved using this technique for polynomial LQ synthesis have been given by Sternad [27]. The approach has also been used in the GMV and GPC algorithms. Design guidelines have been given by Clarke and co-workers [4], [29], [5].

7.2 Feedforward and controller gain

Altering the dynamic weighting elements in the cost function varies the feedback gain of the controller. Increasing the gain will reduce stability margins and robustness against unmodelled high-frequency dynamics. Decreasing the gain will, however, reduce the disturbance rejection achieved by the controller. A trade-off between disturbance rejection and robustness therefore exists.

The use of feedforward compensation itself will alleviate this situation and enhance the robustness of the controller. Where measurable disturbances are available a higher disturbance rejection will be achieved for a given controller gain than that which could be achieved using feedback only. Sternad [27] has given examples which illustrate this improvement.

7.3   Dynamic weights

As discussed in Section 6, the dynamic weighting elements directly affect the frequency domain properties of the feedback loop. These weights can therefore be used in the design process to shape the frequency response of the closed-loop system to achieve enhanced robustness (see [13]). Grimble and Biss [11] also give design guidelines for the selection of dynamic weights to achieve this aim.

7.4   Reference and disturbance models

A range of reference and disturbance signals can be modelled using the framework of Section 2. The signals which can be modelled range from stochastic signals through to shape deterministic signals. This range of signals is covered by appropriate definition of the external generating signals ϕ_d, ϕ_r, ϕ_l, and the associated filters W_d, W_r and W_l. The following types of reference and disturbance signals are covered:

- Stationary stochastic signals are generated when the driving source (ϕ_d, ϕ_r or ϕ_l) is a zero-mean white noise sequence and the filter (W_d, W_r or W_d) is asymptotically stable.

- Random walk, or drifting, sequences are generated when the driving source is a zero-mean white noise sequence and the filter has a single pole on the stability boundary (factor $(1 - q^{-1})$ in the delay operator case, an infinite pole in the δ-operator case).

- Shape deterministic sequences are generated when the driving source is a sequence of random pulses and the filter has poles on the stability boundary; for steps a single pole, and for a sinusoid a pair of complex conjugate poles on the boundary.

7.5   Common factors in $\hat{A}$ and $\hat{B}$

In pole assignment based controllers there is a requirement to solve a linear polynomial equation having a similar form to Equation (23). When the closed-loop poles are chosen arbitrarily the equation to be solved is of the general form

$$AC_{fbd} + BC_{fbn} = T \tag{50}$$

where T is the chosen design polynomial. The solvability condition for this equation is that

- Any common factor of A and B must also be a factor of T.

In the self-tuning context this presents numerical difficulties for such pole assignment algorithms when the estimated A and B polynomials have roots which are close. It is well known that control performance can seriously deteriorate in this case.

Although the LQ controller can be interpreted as an optimal pole placement algorithm it should be stressed that this problem with common factors does not appear in the synthesis algorithm of Section 3. As noted in Section 3.4.2 the condition for solvability of the LQ problem, and Equations (21–22), is weaker than the above condition since we only require that

- The greatest common divisor of A and B must be stable.

Thus, provided they are stable, common factors are directly handled by the LQ synthesis procedure. Dealing with possible unstable common factors in $\hat{A}$ and $\hat{B}$ is a generic problem in self-tuning control.

Acknowledgements

Kenneth Hunt currently has the support of a Personal Research Fellowship from the Royal Society of Edinburgh.

References

[1] K. J. Åström. *Introduction to Stochastic Control Theory*. Academic Press, New York, 1970.

[2] K. J. Åström. LQG self-tuners. In *Proc. IFAC Workshop on Adaptive Control, San Francisco, USA*, 1983.

[3] K. J. Åström, P. Hagander, and J. Sternby. Zeros of sampled systems. *Automatica*, 20:31–38, 1984.

[4] D. W. Clarke. The application of self-tuning control. *Trans. Inst. M. C.*, 5:59–69, 1982.

[5] D. W. Clarke, C. Mohtadi, and P. S. Tuffs. Generalised predictive control - parts 1-2. *Automatica*, 23:137–160, 1987.

[6] B. A. Francis and W. M. Wonham. The internal model principle of control theory. *Automatica*, 12:457–465, 1976.

[7] G. C. Goodwin. Some observations on robust estimation and control. In *Proc. IFAC Symposium on Identification and System Parameter Estimation, York, England*, 1985.

[8] M. J. Grimble. Implicit and explicit LQG self-tuning controllers. *Automatica*, 20:661–669, 1984.

[9] M. J. Grimble. Feedback and feedforward LQG controller design. In *Proc. American Control Conference, Seattle, USA*, 1986.

[10] M. J. Grimble. Two-degrees of freedom feedback and feedforward control of multivariable stochastic systems. *Automatica*, 24:809–817, 1988.

[11] M. J. Grimble and D. Biss. Selection of optimal control weighting functions to achieve good H_∞ robust designs. In *Proc. IEE Conf. Control '88, Oxford, England*, 1988.

[12] K. J. Hunt. General polynomial solution to the optimal feedback/ feedforward stochastic tracking problem. *Int. J. Control*, 48:1057–1073, 1988.

[13] K. J. Hunt. *Stochastic Optimal Control Theory with Application in Self-tuning Control.* Springer - Verlag, Berlin, 1989.

[14] K. J. Hunt, M. J. Grimble, M. J. Chen, and R. W. Jones. Industrial LQG self-tuning controller design. In *Proc. IEEE Conf. on Decision and Control, Athens, Greece,* 1986.

[15] K. J. Hunt, M. J. Grimble, and R. W. Jones. LQG feedback and feedforward self-tuning control. In *Proc. IFAC World Congress, Munich, FRG,* 1987.

[16] K. J. Hunt and M. Šebek. Multivariable LQG self-tuning control with disturbance measurement feedforward. In *Proc. IFAC Symposium on Adaptive Systems in Control and Signal Processing, Glasgow, Scotland,* 1989.

[17] K. J. Hunt and M. Šebek. Optimal multivariable regulation with disturbance measurement feedforward. *Int. J. Control,* 49:373–378, 1989.

[18] M. R. Katebi and J. C. Byrne. Polynomial LQG control design in δ domain. In *Proc. IEE Conf. Control '88, Oxford, England,* 1988.

[19] V. Kučera. *Discrete Linear Control.* Wiley, Chichester, 1979.

[20] I. Nagy and J. Ježek. Polynomial LQ control synthesis for delta-operator models. In *Proc. IFAC Workshop on Adaptive Systems in Control and Signal Processing, Lund, Sweden,* 1986.

[21] V. Peterka. Adaptive digital regulation of noisy systems. In *Proc. IFAC Symposium on Identification and System Parameter Estimation, Prague, Czechoslovakia,* 1970.

[22] V. Peterka. On steady-state minimum variance control strategy. *Kybernetika,* 8:219–232, 1972.

[23] V. Peterka. Predictor based self tuning control. *Automatica,* 20:39–50, 1984.

[24] V. Peterka. Control of uncertain processes: applied theory and algorithms. *Kybernetika,* 22:1–102, 1986. Journal supplement.

[25] M. Sternad. Disturbance decoupling adaptive control. In *Proc. IFAC Workshop on Adaptive Systems in Control and Signal Processing, Lund, Sweden,* 1986.

[26] M. Sternad. *Optimal and Adaptive Feedforward Regulators.* PhD thesis, Department of Technology, Uppsala University, 1987.

[27] M. Sternad. The use of disturbance measurement feedforward in LQG self-tuners. In *Proc. IFAC Symposium on Adaptive Systems in Control and Signal Processing, Glasgow, Scotland,* 1989.

[28] M. Sternad and T. Söderström. LQG-optimal feedforward regulators. *Automatica,* 24:557–561, 1988.

[29] P. S. Tuffs. *Self-tuning control: algorithms and applications.* PhD thesis, Department of Engineering Science, Oxford University, 1984.

[30] M. Šebek. Direct polynomial approach to discrete-time stochastic tracking. *Problems of Control and Information Theory*, 12:293–302, 1983.

[31] M. Šebek, K. J. Hunt, and M. J. Grimble. LQG regulation with disturbance measurement feedforward. *Int. J. Control*, 47:1497–1505, 1988.

[32] Z. Zhao-Ying and K. J. Åström. A microprocessor implementation of an LQG self-tuner. Technical report, Lund Institute of Technology, 1981.

NUMERICAL ANALYSIS OF DIOPHANTINE EQUATIONS

Vladimír Kučera, Jan Ježek and Miloš Krupička

Institute of Information Theory and Automation,
Czechoslovak Academy of Sciences
182 08 Prague 8, Czechoslovakia

The use of algebraic tools in the analysis and synthesis of
linear control systems has revived an interest in linear Diophantine equations for polynomials. This paper makes a
survey of the major solution methods and reports on their
numerical properties.

1   Introduction

Throughout the paper we shall consider a linear Diophantine equation

$$AX + BY = C \tag{1}$$

for polynomials in one indeterminate z over the field of real numbers. Any pair of
polynomials X, Y which satisfy (1) is called a solution of (1). The degree of a polynomial
$A \neq 0$ will be denoted by $\deg A$; the zero polynomial has no degree.

Let D be a greatest common divisor of the polynomials A, B and write

$$\bar{A} = \frac{A}{D}, \quad \bar{B} = \frac{B}{D}.$$

Then equation (1) has a solution if and only if D is a divisor of C, that is, if and only if

$$\bar{C} = \frac{C}{D}$$

is a polynomial (Kučera 1979).

The equation (1) is linear, hence its general solution is the sum of a particular
solution of (1) and the general solution of the homogeneous equation

$$AX + BY = 0.$$

Specifically, if $\bar{X}, \bar{Y}$ is a particular solution of (1) then the general solution of (1) reads

$$X = \bar{X} + \bar{B}T, \quad Y = \bar{Y} - \bar{A}T$$

where T ranges over polynomials (Kučera 1979).

Therefore, if equation (1) is solvable, its solution class is infinite. If $A \neq 0$ the class contains exactly one solution pair X, Y for which either $Y = 0$ or $\deg Y < \deg \bar{A}$; it is the minimum-degree solution with respect to Y. Similarly, if $B \neq 0$ the class contains exactly one solution pair X, Y for which either $X = 0$ or $\deg X < \deg \bar{B}$; it is the minimum-degree solution with respect to X. Finally, if

$$\deg \bar{C} < \deg \bar{A} + \deg \bar{B}$$

the two minimum-degree solutions are identical.

2 Methods of Solution

Equation (1) can be solved in several ways. We shall distinguish matrix, polynomial and mixed methods of solution.

The matrix methods solve equation (1) by converting it into a system of linear equations over the field of coefficients; a typical example is the method of indeterminate coefficients (Volgin 1962) or that of polynomial interpolation (Antsaklis 1983). The polynomial methods reduce equation (1) to a polynomial equation that is much easier to solve; this group includes the method of the Euclidean algorithm (Kučera 1979) and the method of polynomial reductions (Ježek 1982). The mixed methods combine both matrix and polynomial operations; a typical representative is the method of state-space realisation (Emre 1980).

The best method of each group will be presented in detail and thoroughly analysed. To avoid trivia, we shall assume from now on that A, B and C are non-zero polynomials of degree p, q and r and write

$$A = a_0 + a_1 z + ... + a_p z^p$$

$$B = b_0 + b_1 z + ... + b_q z^q$$

$$C = c_0 + c_1 z + ... + c_r z^r.$$

The Method of Indeterminate Coefficients (Volgin 1962) is the best-known representative of the matrix methods. It leads directly to a system of linear equations over the field of coefficients.

Given the polynomials A, B and C we shall determine a solution pair X, Y that satisfies $\deg Y < p$. To this end we write

$$Y = y_0 + y_1 z + ... + y_{p-1} z^{p-1}.$$

If $r < p + q$, we put

$$X = x_0 + x_1 z + \ldots + x_{q-1} z^{q-1}$$

and if $r \geq p + q$ then

$$X = x_0 + x_1 z + \ldots + x_{r-p} z^{r-p}.$$

Substituting into (1) and equating the coefficients at like powers of z we obtain a system of linear equations, which we write in the matrix form

$$wM = N. \tag{2}$$

When $r < p + q$ it is a system of $p + q$ equations for $p + q$ unknowns

$$w = \begin{bmatrix} x_0 & x_1 & .. & x_{q-1} & y_0 & y_1 & .. & y_{p-1} \end{bmatrix}$$

having the coefficient matrices

$$M = \begin{bmatrix} a_0 & a_1 & . & . & . & a_p & & & & \\ & a_0 & a_1 & . & . & . & a_p & & & \\ & & & . & . & . & & & & \\ & & & & a_0 & a_1 & . & . & . & a_p \\ b_0 & b_1 & . & . & . & b_q & & & & \\ & b_0 & b_1 & . & . & . & b_q & & & \\ & & & . & . & . & & & & \\ & & b_0 & b_1 & . & . & . & b_q \end{bmatrix}$$

$$N = \begin{bmatrix} c_0 & c_1 & .. & c_r & 0 & ... & 0 \end{bmatrix}$$

When $r \geq p + q$ it is a system of $r + 1$ equations for $r + 1$ unknowns

$$w = \begin{bmatrix} x_0 & x_1 . & .. & x_{r-p} & y_0 & y_1 & .. & y_{p-1} \end{bmatrix}$$

having the coefficient matrices

$$M = \begin{bmatrix} a_0 & a_1 & . & . & . & a_p & & & \\ & a_0 & a_1 & . & . & . & a_p & & \\ & & & . & . & & & & \\ & & & a_0 & a_1 & . & . & . & a_p \\ b_0 & b_1 & & . & . & b_q & & & \\ & b_0 & b_1 & & . & . & b_q & & \\ & & & . & . & & & & \\ & & b_0 & b_1 & . & . & b_q & 0 \end{bmatrix}$$

$$N = \begin{bmatrix} c_0 & c_1 & .... & c_r \end{bmatrix}.$$

Equation (2) is equivalent to equation (1). The matrix M has full rank if and only if the polynomials A and B are relatively prime. In this case the method yields the minimum-degree solution of (1) with respect to Y.

When a solution X, Y such that $\deg X < q$ is of interest, we simply interchange the roles of X and Y in (1).

The Method of Polynomial Reductions (Ježek 1982) is the most efficient polynomial method. It consists of a sequence of operations, each reducing the degree of a certain polynomial.

Given the polynomials A, B and C we shall apply one of the following four operations. If $r \geq p$ we replace equation (1) by the equation

$$AX' + BY = C'$$

where

$$C' = C - A\frac{c_r}{a_p}z^{r-p} \tag{3}$$

is a polynomial of lower degree than C; the new equation is solvable if and only if (1) is and

$$X = X' + \frac{c_r}{a_p}z^{r-p}. \tag{4}$$

If $r \geq q$ we replace equation (1) by the equation

$$AX + BY' = C'$$

where

$$C' = C - B\frac{c_r}{b_q}z^{r-q} \tag{5}$$

is a polynomial of lower degree than C; the new equation is solvable if and only if (1) is and

$$Y = Y' + \frac{c_r}{b_q}z^{r-q}. \tag{6}$$

If $q \geq p$ we replace equation (1) by the equation

$$AX' + B'Y = C$$

where

$$B' = B - A\frac{b_q}{a_p}z^{q-p} \tag{7}$$

is a polynomial of lower degree than B; the new equation is solvable if and only if (1) is and

$$X = X' - Y\frac{b_q}{a_p}z^{q-p}. \tag{8}$$

If $p \geq q$ we replace equation (1) by the equation

$$A'X + BY' = C$$

where

$$A' = A - B\frac{a_p}{b_q}z^{p-q} \tag{9}$$

is a polynomial of lower degree than A; the new equation is solvable if and only if (1) is and

$$Y = Y' - X\frac{a_p}{b_q}z^{p-q}. \tag{10}$$

The reductions (3), (5), (7) and (9) are repeated for the new polynomials A', B' and C' as long as applicable. Two cases can occur. Either all polynomials but one have been reduced to zero; then the resulting equation has a solution $X' = 0$, $Y' = 0$ and the solution pair X, Y of (1) is obtained through the backward substitutions (4), (6), (8) and (10). Or only one polynomial can be reduced to zero; then the resulting equation and hence equation (1) has no solution.

If one applies the reduction (3) as many times as possible before applying any other reduction, the method yields the minimum-degree solution with respect to Y. On the other hand, if one applies the reductions (5) first the method will yield the minimum-degree solution with respect to X.

The Method of State-Space Realisation (Emre 1980) combines matrix and polynomial operations. It is based on an interpretation of B/A or A/B as the transfer function of some linear system.

Given the polynomials A, B and C we shall look for a solution pair X, Y that satisfies deg $Y < p$. To this end we write equation (1) in the form

$$X + \frac{B}{A}Y = \frac{C}{A} \qquad (11)$$

and denote by

$$B' = b'_0 + b'_1 z + \ldots + b'_{p-1} z^{p-1}$$
$$C' = c'_0 + c'_1 z + \ldots + c'_{p-1} z^{p-1}$$

the polynomials B and C reduced modulo A. We determine an observable state-space realisation (F, G, H) of the strictly proper rational function

$$R = \frac{B'}{A}$$

such that F has size p. The coefficients of polynomial Y are then determined by solving the system of linear equations

$$
\begin{bmatrix} G & FG & \ldots & F^{p-1}G \end{bmatrix}
\begin{bmatrix} y_0 \\ y_1 \\ \cdot \\ \cdot \\ \cdot \\ y_{p-1} \end{bmatrix}
=
\begin{bmatrix} c'_0 \\ c'_1 \\ \cdot \\ \cdot \\ \cdot \\ c'_{p-1} \end{bmatrix}
\qquad (12)
$$

while the polynomial X is obtained from (11) as the difference between the polynomial parts of C/A and BY/A.

The equation (12) greatly simplifies when the state-space realisation is taken in the observability standard form

$$
F = \begin{bmatrix} 0 & & & -\frac{a_0}{a_p} \\ 1 & & & -\frac{a_1}{a_p} \\ & \cdot & & \cdot \\ & & \cdot & \cdot \\ & & \cdot & \cdot \\ & & 1 & -\frac{a_{p-1}}{a_p} \end{bmatrix}, \quad G = \begin{bmatrix} b'_0 \\ b'_1 \\ \cdot \\ \cdot \\ \cdot \\ b'_{p-1} \end{bmatrix}
$$

$$
H = \begin{bmatrix} 0 & \cdots & 0 & \frac{1}{a_p} \end{bmatrix}
$$

Equation (12) is equivalent to equation (1). The matrix $\begin{bmatrix} G & FG & \cdots & F^{p-1}G \end{bmatrix}$ has full rank if and only if the polynomials A and B are relatively prime; the state-space realisation (F, G, H) is then reachable. In this case the method furnishes the minimum-degree solution with respect to Y.

When a solution X, Y such that $\deg X < q$ is of interest , we write equation (1) as

$$
\frac{A}{B}X + Y = \frac{C}{B}
$$

and proceed accordingly.

3 Numerical Analysis

The analysis of numerical properties of the three methods reported here is based on an analysis by Krupička (1984). We shall investigate the computational complexity of each method and its precision in the presence of round-off errors.

Each method was tested on a set of 953 examples. The polynomials A, B and C were systematically generated by means of their roots and had degrees $0, 1, ..., 9$.

The Count of Arithmetic Operations depends on the degrees of A, B and C. We shall analyse the number of additions (ADD), multiplications (MUL) and divisions (DIV) of real numbers. The operations with indices, which serve to organise the calculations, will be neglected. Whenever a system of linear equations is to be solved, we use the Gauss elimination method.

The formulas for the operation counts are involved, so we shall suppose that

$$
\deg A = n, \quad \deg B = n, \quad \deg C = 2n - 1
$$

and give the formulas valid for relatively prime A and B.

The Method of Indeterminate Coefficients:

$$
ADD = \frac{8}{3}n^3 + 2n^2 - \frac{5}{3}n
$$

$$MUL = \frac{8}{3}n^3 + 4n^2 - \frac{8}{3}n$$

$$DIV = 4n - 1$$

The Method of Polynomial Reductions:

$$ADD = \frac{5}{2}n^2 + \frac{9}{2}n$$

$$MUL = \frac{5}{2}n^2 + \frac{5}{2}n$$

$$DIV = 4n + 1$$

The Method of State-Space Realisation:

$$ADD = \frac{1}{3}n^3 + \frac{7}{2}n^2 - \frac{11}{6}n + 1$$

$$MUL = \frac{1}{3}n^3 + 4n^2 - \frac{4}{3}n$$

$$DIV = 5n + 1.$$

The Precision of the Solution depends on the properties of A, B and C and on the round-off errors. The most relevant property of the polynomials A, B and C is the relative position of their roots. We shall investigate the numerical behaviour of each method in the vicinity of the unsolvable case, when A and B have a common root which is not the root of C. Denoting d the distance of the nearest roots and k their multiplicity, then the nearness of A and B can be characterised by d^k; this number took values of $10^{-1}, 10^{-2}, ..., 10^{-16}$ during the experiments.

The effect of round-off errors was studied by varying the floating-point format. A built-in software feature was used to restrict the mantissa to a prespecified number of bits, namely $4, 8, 16, ..., 64$, which roughly corresponds to $1, 2, 5, ..., 19$ decimal digits.

The relative precision of the solution is expressed by the average coefficient of the error $C - AX - BY$ related to that of C. The maximal attainable relative precision is given by the length of the mantissa, $2^{-(b+1)}$, when using b bits.

The results are arranged in the tables below. Each row corresponds to a measure of nearness of A and B, each column corresponds to a length of the mantissa. The entries show the precision achieved, expressed by the number of decimal digits. When no decimal digit is correct, the entry is void.

The Method of Indeterminate Coefficients:

	1	2	5	10	19
10^{-1}	-	-	2	7	15
10^{-2}	-	-	2	7	15
10^{-4}	-	-	2	6	14
10^{-8}	-	-	2	5	10
10^{-16}	-	-	2	4	6

The Method of Polynomial Reductions:

	1	2	5	10	19
10^{-1}	-	-	3	8	17
10^{-2}	-	-	3	8	17
10^{-4}	-	-	2	6	14
10^{-8}	-	-	1	2	7
10^{-16}	-	-	-	-	-

The Method of State-Space Realisation:

	1	2	5	10	19
10^{-1}	-	-	3	8	16
10^{-2}	-	-	3	8	16
10^{-4}	-	-	2	6	14
10^{-8}	-	-	-	3	8
10^{-16}	-	-	-	-	3

4 Comparison and Conclusion

The method of indeterminate coefficients is straightforward and leads directly to a system of linear equations for the coefficients of the unknown polynomials. The method of polynomial reductions solves the polynomial equation by polynomial means and is not suitable for pencil-and-paper calculations, for it requires a large number of logical operations. The method of state-space realisation combines the two above: one unknown polynomial is obtained by solving a system of linear equations while the other results from polynomial manipulations.

The comparison of the methods with respect to the arithmetic complexity is quite clear. The least demanding is the method of polynomial reductions, where the operations count increases with the square of the degrees of the given polynomials. For the other two methods the arithmetic complexity is proportional to the cube of these degrees. The most demanding, however, is the method of indeterminate coefficients because it leads to a larger system of linear equations than the method of state-space realisation.

The comparison of the methods from the precision point of view is not that simple, however. The effect of the properties of the given polynomials is more pronounced than the effect of the method used. Provided the polynomials A and B have no multiple roots close to each other, the precision of all three methods is comparable. The most precise is the method of polynomial reductions, the least precise is the method of indeterminate coefficients. The ill-conditioned data, however, make the method of polynomial reductions fail even with a mantissa of full length. On the other hand, the method of indeterminate coefficients solves these cases in a realiable way with the mantissa as short as 16 bits. The method of state-space realisation shows no clean-cut tendency, it stays between the two preceding methods.

In conclusion, the method of polynomial reductions is fast but sensitive, the method of indeterminate coefficients is robust but slow, and the method of state-space realisation is universal but second-best in each aspect.

5 References

1. Antsaklis, P. J., 1983, Polynomial matrix interpolation in control equations: The Diophantine equation, Proc. Conf. Information Sciences and Systems, The John Hopkins University, Baltimore, USA.

2. Emre, E., 1980, The polynomial equation $QQ_c + RP_c = \Phi$ with application to dynamic feedback, SIAM J. Contr. Optimiz., 18, 611-620

3. Ježek, J., 1982, New algorithm for minimal solution of linear polynomial equations, Kybernetika, 18, 505-516.

4. Krupička, M., 1984, Algorithms for the solution of a linear polynomial equation (in Czech), Thesis, Faculty of Electrical Engineering, Czech Technical University, Prague, Czechoslovakia.

5. Kučera, V., 1979, Discrete Linear Control: The Polynomial Equation Approach (Chichester: Wiley).

6. Volgin, L. N., 1962, The Elements of the Theory of Controllers (in Russian) (Moscow: Soviet Radio).

Simple Predictive LQ Controller Design

Vladimír Havlena, Department of Control,
Faculty of Electrical Eng., Czech Technical
University, Charles Sq. 13, 121 35 Prague 2

May 17, 1990

Abstract: LQ optimal controller design method based on lower-band matrix input-output description of the plant on a finite prediction horizon and dyadic reduction algorithm for quadratic form optimization is described. Manipulated input amplitude and/or slew rate limits can be taken into consideration.

1 Introduction, problem formulation

Let the plant dynamics at a given set point be described by a linear n-th order AR/ARMA model

$$y(t) = \sum_{i=1}^{n} a_i y(t-i) + \sum_{i=0}^{n} b_i u(t-i) + \sum_{i=1}^{n} c_i e(t-i) + \sum_{i=0}^{n} d_i v(t-i) + k + e(t) \quad (1)$$

(AR model can be obtained by omitting the c_i terms) where $y(t)$ is plant output, $u(t)$ is manipulated input, $v(t)$ is measurable disturbance input and $e(t)$ is Gaussian white noise $N(0, \sigma^2)$ independent of $y(t-i), i > 0$ and $u(t-i), v(t-i), i \geq 0$.

The time indexing is such that the value of $y(t)$ can be affected by the inputs $u(t), v(t)$. The control law may use $y(t-1)$ as the latest available sample of the output for the calculation of $u(t)$. Delays d_u, d_v can also be included in the model; they are not considered here to simplify the exposition. The values of manipulated input on a T step horizon

$$u = [u(t_0 + 1), ..., u(t_0 + T)]^T \quad (2)$$

are determined to minimize the criterion

$$J = \sum_{t=t_0+1}^{t_0+T} \{q(t) \, [\hat{y}(t|t_0) - w(t)]^2 + r(t) \, \Delta u(t)^2\} \tag{3}$$

where

$$\Delta u(t) = u(t) - u(t-1)$$

and

$$\hat{y}(t|t_0) = E\{y(t)|P(t_0), D(t_0)\}$$

is a predicted value of output $y(t)$ based on the parameter values estimates $P(t_0)$ at time t_0 (certainty equivalence principle) and data $D(t_0)$.

Weighting factors $q(t), r(t)$ are user's knobs for tuning the controller according to his subjective idea of optimal performance. Penalizing the increments of manipulated input ensures input damping during the transients as well as zero steady-state error in the case of non-zero reference following and/or disturbance compensation (integral action is introduced to the controller).

Note that the control strategy based on long range output prediction [Clarke, 1987] results in feedback control law in the first step $u(t_0 + 1)$ only. The following values $u(t_0 + 2), \ldots, u(t_0 + T)$ are calculated in open loop. However, only the first value of control is applied to the plant and then the whole control vector is recalculated (receding horizon strategy). The first value of the control vector is independent of the optimization strategy (open/closed loop) [Peterka, 1989]. Moreover, as all the values of control vector u are calculated explicitly, actuator amplitude and/or slew rate limits can be considered in the optimization procedure [Havlena, 1988]. The information about input limitation in future steps spreads backward the time and may affect the value of $u(t_0 + 1)$. These features are relevant to the "deterministic part" of the plant dynamics and inputs, however, responses to pre-programmed set point or load changes with subsequent in- put limitation may improve considerably.

2   Basic Algorithm

The algorithm for the minimization of (3) will be developed here. First we will modify the increments weighting criterion (3) to position weighting one. Let us denote.

$$\hat{y} \;=\; [\hat{y}(t_0 + 1|t_0), \ldots, y(t_0 + T|t_0)]^T$$

$$y = [y(t_0 + 1), \ldots, y(t_0 + T)]^T$$
$$w = [w(t_0 + 1), \ldots, w(t_0 + T)]^T$$
$$\Delta u = [\Delta u(t_0 + 1), \ldots, \Delta u(t_0 + T)]^T$$
$$Q = \text{diag}[q(t_0 + 1), \ldots, q(t_0 + T)]$$
$$R = \text{diag}[r(t_0 + 1), \ldots, r(t_0 + T)]$$

Criterion (3) can be written as

$$J = (\hat{y} - w)^T Q (\hat{y} - w) + \Delta u^T R \Delta u$$

Let S be a "difference" matrix

$$S = \begin{bmatrix} 1 & 0 & \ldots & 0 \\ -1 & 1 & \ldots & 0 \\ & & \ddots & \\ 0 & \ldots & -1 & 1 \end{bmatrix}$$

The input increments vector Δu can be obtained as

$$\Delta u = S u - u_0$$

where

$$u_0 = [u(t_0), 0, \ldots, 0]^T$$

and the criterion is modified to

$$J = (\hat{y} - w)^T Q (\hat{y} - w) + (S u - u_0)^T R (S u - u_0) \tag{4}$$

Let us arrange the data $D(t_0)$ necessary for the optimal control design into vectors

$$\tilde{y} = [y(t_0 - n + 1), \ldots, y(t_0)]^T$$
$$\tilde{u} = [u(t_0 - n + 1), \ldots, u(t_0)]^T$$
$$\tilde{v} = [v(t_0 - n + 1), \ldots, v(t_0)]^T$$
$$\tilde{e} = [e(t_0 - n + 1), \ldots, e(t_0)]^T \tag{5}$$

and let k be the vector of constants k. The plant model (1) on the horizon $t_0 + 1, \ldots, t_0 + T$ gives

$$\left[\begin{array}{c|c} A_t | A_p \end{array}\right] \left[\begin{array}{c} \tilde{y} \\ y \end{array}\right] = \left[\begin{array}{c|c} B_t | B_p \end{array}\right] \left[\begin{array}{c} \tilde{u} \\ u \end{array}\right] + \left[\begin{array}{c|c} C_t | C_p \end{array}\right] \left[\begin{array}{c} \tilde{e} \\ e \end{array}\right]$$

$$+ \left[\begin{array}{c|c} D_t | D_p \end{array}\right] \left[\begin{array}{c} \tilde{v} \\ v \end{array}\right] + k \qquad (6)$$

where

$$\left[\begin{array}{c|c} A_t | A_p \end{array}\right] = \left[\begin{array}{cccccc|cccccc}
-a_n & . & . & . & -a_1 & & 1 & 0 & . & . & . & 0 \\
0 & -a_n & . & . & . & & -a_1 & 1 & 0 & & . & 0 \\
. & . & . & . & . & & . & . & . & . & . & . \\
0 & & 0 & -a_n & & & . & . & -a_1 & 1 & 0 & . & 0 \\
0 & . & . & . & .0 & & -a_n & . & . & -a_1 & 1 & . & 0 \\
. & . & . & . & . & & . & . & . & . & . & . \\
0 & . & . & . & 0 & & . & . & -a_n & . & . & -a_1 & 1
\end{array}\right]$$

is a $(T, T+n)$ matrix consisting of a (T, n) upper triangular matrix A_t and (T, T) monic stationary lower band matrix A_p,

$$\left[\begin{array}{c|c} B_t | B_p \end{array}\right] = \left[\begin{array}{cccccc|cccccc}
b_n & . & . & . & b_1 & & b_0 & 0 & . & . & . & 0 \\
0 & b_n & . & . & . & & b_1 & b_0 & 0 & . & . & 0 \\
. & . & . & . & . & & . & . & . & . & . & . \\
0 & & 0 & b_n & & & . & . & b_1 & b_0 & 0 & . & 0 \\
0 & . & . & . & .0 & & b_n & . & . & b_1 & b_0 & . & 0 \\
. & . & . & . & . & & . & . & . & . & . & . \\
0 & . & . & . & 0 & & . & . & b_n & . & . & b_1 & b_0
\end{array}\right]$$

and the matrices $[C_t | C_p], [D_t | D_p]$ are of the same structure. From (7) the predicted output $\hat{y}$ is

$$A_p \hat{y} = B_p . u + D_p \hat{v} + s \qquad (7)$$

where $\hat{v}$ is the vector of predicted disturbance values (if no model of the disturbance dynamics is available, simple random walk model $v(t_0 + i | t_0) = v(t_0)$ can be used) and s represents the initial state response of the process at time t_0

$$s = -A_t \tilde{y} + B_t \tilde{u} + C_t \tilde{e} + D_t \tilde{v} + k \qquad (8)$$

The criterion (5) can be written in factorized form (from now on, the notation introduced in [Peterka, 1986b] will be used) as

$$J = \left[\begin{array}{c} u \\ x \end{array}\right]^T M^T D M \left[\begin{array}{c} u \\ x \end{array}\right] = \left| D; M \left[\begin{array}{c} u \\ x \end{array}\right] \right| \qquad (9)$$

where the matrices D, M are built as follows

$$J = \left\| \begin{bmatrix} O \\ O \\ R \\ Q \end{bmatrix} ; \begin{bmatrix} I & O \\ O & 1 \\ S & -u_0 \\ A_p^{-1}B_p & A_p^{-1}(D_p\hat{v} + s) - w \end{bmatrix} \begin{bmatrix} u \\ 1 \end{bmatrix} \right\| \tag{10}$$

This structure is a precondition for the application of the dyadic reduction algorithm [Peterka, 1986a, 1986b] to convert the matrix M to a monic upper triangular matrix M of the form

$$J = \left\| \begin{bmatrix} D_u \\ d \end{bmatrix} ; \begin{bmatrix} M_u & m_u \\ O & 1 \end{bmatrix} \begin{bmatrix} u \\ 1 \end{bmatrix} \right\| \tag{11}$$

The minimum value of the criterion

$$J^* = d \tag{12}$$

will be reached for optimal control vector u^* given by

$$M_u u^* + m_u = 0 \tag{13}$$

which can easily be solved as M_u is a monic upper triangular matrix. If the optimal criterion value is of no interest, only the first T columns of M in (11) should be transformed and the second line in D and M may be omitted.

3 Control Input Structuring

Providing the user with the weights $r(t), q(t)$ as a tuning knob is not always convenient, as too many degrees of freedom are given with few or no guidelines how to treat them. Control input structuring [Peterka, 1990] admits nonzero input increments only in a limited number of time points during the optimization horizon. It is also a suitable tool for non-minimum phase system input stabilization and/or manipulated input bandwidth limitation. Control input structuring is also reported to increase closed loop robustness to parameter uncertainty. Control input vector with only T_s different values of control on the control horizon of the length T can be described as

$$u = Eu_s \tag{14}$$

where u_s is a T_s-vector

$$u_s = [u_s(1), \ldots, u_s(T_s)]^T \tag{15}$$

and E is a (T, T_s) "expansion" matrix of the form

$$
E = \begin{bmatrix}
1 & 0 & . & . & . & 0 \\
. & . & . & & & . \\
1 & 0 & & & & 0 \\
0 & 1 & 0 & . & . & 0 \\
& . & & & & \\
0 & 1 & 0 & . & . & 0 \\
. & . & . & . & . & . \\
0 & & & & 0 & 1
\end{bmatrix}
$$

Criterion (11) then can be modified to

$$
J = \left\| \begin{bmatrix} O \\ O \\ R \\ Q \end{bmatrix} ; \begin{bmatrix} I & O \\ O & 1 \\ SE & -u_0 \\ A_p^{-1} B_p E & A^{-1}(D_p \hat{v} + s) - w \end{bmatrix} \begin{bmatrix} u_s \\ 1 \end{bmatrix} \right\| \tag{16}
$$

with the dimension T_s of u_s significantly lower then the dimension of u (as low value as $T_s \cong n$ may be reasonable). The minimization procedure does not change. To stabilize the control input in the case of nonminimum phase plant, expansion matrix of the form

$$
E = \begin{bmatrix}
1 & 0 & . & . & . & 0 \\
0 & 1 & 0 & . & . & 0 \\
. & . & . & . & . & . \\
0 & . & . & 0 & 1 & 0 \\
0 & . & . & . & 0 & 1 \\
. & . & . & . & . & . \\
0 & . & . & . & 0 & 1
\end{bmatrix}
$$

is suitable. Only T_s steps of the control input may vary, then the control input is constant on the remaining part of the optimization horizon.

Note that the post multiplication by the expansion matrix corresponds to summing up the appropriate columns of the matrix multiplied.

4   Manipulated input limitation

Constrained optimization task

$$
u^* = \arg \min_{u \in \mathcal{U}} J \tag{17}
$$

with "box" constraint describing the actuator amplitude or slew rate limits

$$U = \{u; u_{\min}(t) \le u(t) \le u_{\max}(t), t = t_0 + 1, \ldots, t_0 + T\} \qquad (18)$$

can also be solved. The solution using Choleski factorization of the criterion is given in [Havlena, 1988]. The solution based on the above described LD factorization will be presented here.

To simplify the formulas, notation $u_k = u(t_0 + k)$ will be used in this chapter. Let B be the set of indices of variables, for which lower or upper bound is active. To initialize the algorithm in a reasonable way, the function sat(.) defined as

$$sat(u) \quad \begin{aligned} &= u_{\min} &&\text{for} && u && < u_{\min} \\ &= u && \text{for} && u_{\min} && \le u \le u_{max} \\ &= u_{\max} && \text{for} && u_{\max} && < u \end{aligned} \quad \text{is used (with vector ar-}$$

guments it is applied component wise). The following algorithm, based on [Fletcher, Jackson, 1974] with guaranteed finite number of iteration provides a solution of (20):

(i) - initialization

Set $r = 0$ (iteration counter)
$u^{(0)} = \text{sat}(u^*)$
$B = \{1, 2, ..., T\}$ (fix all variables)

(ii) - major iteration loop

Set $g = -$ grad $q(u^{(r)})$
Select a fixed variable $u_l, l \in B$, for which

$g_l < 0$ and $u_l > u_{max}(t_0 + l)$ or

$g_l > 0$ and $u_l < u_{min}(t_0 + l)$
(i.e. variable u_l does not fulfill Kuhn-Tucker conditions)

If such a variable exists,

$set B = B \div \{l\},$ (make this variable free)

go to to (iii).   (start minor loop)

If such a variable does not exist,

finish.   (solution achieved)

(iii) - minor iteration loop
Find

$$u^{(r)*} = \arg \min_{u_l, l \notin B} q(u)$$

(see Appendix B)

Find maximum $\alpha_{\lim} \in\; <0, 1>$ such that

$$u^{(r)} + \alpha \left(u^{(r)*} - u^{(r)} \right)$$

does not cross the constraint. Let $k_{\lim}$ be the index, for which a new bound has become active.

If $\alpha_{\lim} = 1, u^{(r)*}$ is admissible. Set

$$u^{(r+1)} = u^{(r)*}$$

$$r = r + 1, \quad \text{and}$$

go to (ii)   (new major iteration).
If $\alpha_{\lim} < 1, u^{(r)*}$ is not admissible. Set

$$u^{(r)} = u^{(r)} + \alpha_{\lim} \left(u^{(r)*} - u^{(r)} \right)$$

$$B = B \bigcup \{ k_{\lim} \} \quad \text{(fix variable)}$$

go to (iii)   (cont. minor iteration).
The algorithm consists of two loops. In major iteration loop, the values of criterion for solutions $u^{(r)}$ establish a decreasing (non increasing in the case of positive semi definite criterion matrix) sequence. The minor loop finishes in less then n_f steps for each major iteration, as some free variable is fixed in each minor iteration. These properties can be used to prove that the

algorithm terminates in a finite number of steps [Fletcher, Jackson, 1974].
Moreover, if the algorithm is initialized with the value

$$u^{(0)} = sat(u^*)$$

i.e. the value of unconstrained optimal control cut at the constraints (which
is often used in practice), major iteration loop can be terminated at any iteration (e.g. due to computation time limits in real time application) and the
control obtained always results in a lower (or equal) value of criterion. To
obtain a numerically robust algorithm, the constrained optimization is performed using LD factorization of the criterion again. Some computational
details are given in Appendix.

5   Conclusion

Simple and effective LQ optimal controller design based on AR/ARMA
input-output description of the plant on a finite prediction horizon and
dyadic reduction algorithm for quadratic form optimization has been developed. Some modifications of the basic algorithm including control input
structuring and limitation have been described. The resulting algorithm is
extremely simple and effective, which makes it suitable for application in
adaptive control.

6   Appendix

To compute the minimum value of a quadratic form

$$q(u) = \left\| \begin{bmatrix} D_u \\ d \end{bmatrix} ; \begin{bmatrix} M_u & m_u \\ 0 & 1 \end{bmatrix} \begin{bmatrix} u \\ 1 \end{bmatrix} \right\| \tag{19}$$

with respect to free variables $u_k = u(t_0 + k)$, $k \notin B$, sort the vector n into
two parts consisting of n_f free variables u_f and n_b bound (fixed) variables
u_b. Suppose the matrices D M of the form

$$q(u_f, u_b) = \left\| \begin{bmatrix} D_f \\ D_b \\ d \end{bmatrix} ; \begin{bmatrix} M_{ff} & M_{fb} & m_f \\ 0 & M_{bb} & m_b \\ 0 & 0 & 1 \end{bmatrix} \begin{bmatrix} u_f \\ u_b \\ 1 \end{bmatrix} \right\| \tag{20}$$

Then the optimal value of the criterion

$$\min_{u_f} q(u_f, u_b) \tag{21}$$

for given u_b is achieved for u_f^* given by

$$M_{ff}u_f^* + M_{bf}u_b + m_f = 0 \qquad (22)$$

The gradients of the criterion required in the optimization algorithm can be computed as

$$
\begin{aligned}
\operatorname{grad}_{u_f} q(u_f, u_b) &= \frac{\partial q}{\partial u_f} = M_{ff}^T D_f \left(M_{ff}u_f + M_{fb}u_b + m_f\right) \\
\operatorname{grad}_{u_b} q(u_f, u_b) &= \frac{\partial q}{\partial u_b} = M_{bb}^T D_b \left(M_{bb}u_b + M_b\right)
\end{aligned}
$$

Now we shall describe how a given quadratic criterion LD decomposition of the form (20) can be updated effectively after an entry of the free variables vector u_f has been moved to bound variables vector u_b and vice versa. Let the matrix M for given n_f vector u_b and n_b vector u_b be monic upper triangular and $u_k, k \notin B$ a free variable. To obtain the factorization (20) for $B = B \bigcup k$, move u_k from u_f to the first position u_b and the corresponding column i_k of matrix M to the n_f-th position by which the triangularity of M has been lost. To reconstruct M as a monic upper triangular matrix, move the i_k-th rows of matrices D, M to the n_f-th position. The value of criterion has not been affected by these changes. The matrix M of the form

$$
M =
\begin{bmatrix}
1 & x & x & x & . & x & x & x & . & . & x \\
. & . & . & . & . & . & . & . & & & . \\
 & & 1 & x & . & x & 0 & x & & & x \\
 & & & 1 & . & x & 0 & x & & & x \\
 & & & & . & . & . & . & & & . \\
 & & & & & 1 & 0 & x & . & . & x \\
 x & x & . & x & 1 & x & . & . & x \\
 & & & & & & & . & . & . & . \\
 & & & & & & & & 1 & x \\
 & & & & & & & & & 1
\end{bmatrix}
\qquad (23)
$$

can be easily recalculated to a monic upper triangular matrix using several steps of dyadic reduction algorithm. The partitioning of vector u should be updated to

$$
\begin{aligned}
n_f &= n_f - 1 \\
n_b &= n_b + 1
\end{aligned}
$$

In a similar way a bound variable u_l, $l \in B$ can be set free. To obtain the LD decomposition for $B = B \div \{l\}$, move u_l from u_b to the last position in u_f and the corresponding column i_l of matrix M to the $(n_f + 1)$-th position. To reconstruct M as a monic upper triangular matrix in an effective way, move the i_l-th rows of matrices D, M to the $(n_f + 1)$-th position. The matrix M of the form

$$M = \begin{bmatrix} 1 & x & x & x & . & x & x & x & . & . & x \\ . & . & . & . & . & . & . & . & . & . & . \\ & & & 1 & 0 & . & 0 & 0 & x & & x \\ & & & x & 1 & . & x & x & x & & x \\ & & & & & . & . & . & & & . \\ & & x & & & & 1 & x & x & . & . & x \\ & & x & & & & & 1 & x & . & . & x \\ & & & & & & & & . & . & . & . \\ & & & & & & & & & 1 & x \\ & & & & & & & & & & 1 \end{bmatrix} \quad (24)$$

can be again easily recalculated to a monic upper triangular matrix using several steps of dyadic reduction algorithm and the partitioning of vector u should be updated to

$$n_f = n_f + 1$$
$$n_b = n_b - 1$$

Note that in the case a bound variable u_l is removed from u_b the dyadic reduction algorithm should be applied to non zero elements below the main diagonal in the $(n_f + 1)$-th column up wards to utilize the zero entries in the $(n_f + 1)$-th row to minimize the number of dyadic reduction steps.

References

[1] Clarke, D. W., C. Mohtadi and P. S. Tuffs: Generalized Predictive Control. Automatica 23, No. 2, 1987.

[2] Fletcher, R., Jackson, M. P.: Minimization of a Quadratic Function of Many Variables Subject only to Lower and Upper Bounds. J. Inst. Maths. Applics. 14, p. 159, 1974.

[3] Havlena, V.: Control with Set Point Optimization. PhD. thesis (in Czech), Dept. of Control, Fac. of El. Eng., Czech Technical University, Prague, 1988.

[4] Peterka, V.: Algorithms for LQG self-tuning control based on input-output Delta models. 2-nd IFAC Workshop on Adaptive Systems in Control and Signal Processing , Lund, Sweden, 1986a.

[5] Peterka, V.: Control of Uncertain Processes: Applied Theory and Algorithms. Supplement to Kybernetika, volume 22, No. 3 - 6, 1986b.

[6] Peterka, V.: Predictive and LQG Optimal Control: Equivalences, Differences and Improvements. IFAC Symposium on Adaptive. Control and Signal Processing , Glasgow, 1989.

[7] Peterka, V.: Adaptation of LQG control design to engineering needs. Joint U.K. - Czechoslovak Sem. Advanced Methods in Adaptive Control for Industrial Application, Prague, 1990.

TOWARD HIGHER SOFTWARE EFFECTIVENESS IN THE FIELD OF ADAPTIVE CONTROL

Petr Nedoma

Institute of Information Theory and Automation,
Czechoslovak Academy of Sciences
182 08 Prague 8, Czechoslovakia

1 Introduction

Works of recent years have proved that today's "matrix environments" (represented by MATLAB and derivatives) are state-of-art of programs for Computer Aided Design of Control Systems (CADCS). Many software packages has appeared in the field and some of them has reached a high commercial success (being the most known e.g. ACSL, MATRIX-X, PC-MATLAB, CTRL-C, CC, SIMNON, LSAP etc.- see Boom, Herget, 1988 for references). But, till now, no of them can be selected as optimum for CADCS.

The environment itself will be referred to as "shell" system - cnf. Schmidt 1988. It provides a set of commands that allows to the researcher to code his algorithms, to form interfaces to his functions (often written in different programming languages) and to run his algorithms. The result is usually a modular set of functions covering algorithms of a stage of CADCS cycle. Software design of such a type is often referred to as "toolbox" approach.

Special purpose CADCS tools are then designed using the toolboxes. Those are used by terminal users to solve practical problems of CADCS. Design of such a type will be referred as "package" approach (often used in more general meaning).

The numerical contents of the shell system need not be too rich and can be even reduced to basic numerical operations. But, an effective interface to the user's routines must be provided in the case. The numerical problems can be completely let to user's routines and the shell system can contain more services then it now offers.

The idea was employed during a decade of development of a package SIC+KOS (cnf. Kulhavy, Nedoma 1988) in the group of Adaptive Control of the Institute of Information Theory and Automation, Czechoslovak Academy of Sciences, Prague. The package covers software support for CAD of (adaptive) control systems of stochastic processes.

Two years ago has decided to build a shell system of a new generation (with the working name PLUS) that will be used as an experimental environment for the future development. There are many underlying reasons for the design. The main of them is connected with the observation that the packages commercially available are usually not oriented to a higher level research, do not support an effective simulation environment and cannot be flexible changed (the source code is not available).

The article tries to contribute to the discussion about more powerful shell systems of a next generation. The discussion is related with a research group in the field of CADSC.

2 Data Structures

The simplest data model which is in wide-spread use for for CADCS is the complex matrix. Usage of one data model gives the power and simplicity to the shell systems. It is surprising what work has been done with it.

But, many problems of CADCS should be based on more complicated data structures. A technical support should reach the level of C language. The shell system should handle the definitions of data structures , data input and output and access to substructures on the level of command language.

The primarily role of the shell system is to pass data structures (with descriptions) to the user routines. Formalization of data description and interface coding is inevitable. The data description should contain an indicator of semantic contents of data.

In the context, each data structure should be allowed to reside on an external storage. In such a case the designer will be allowed to handle the situation of large dimensions that is not only organizational problem. A tool is needed for access to data structures located in a database.

Let us express the meaning that only appropriate support of data structures will allow structuring of toolboxes to match the semantic of the applications. But, the additions should be done in such a way not to loose the power of MATLAB.

3 Extended Modularity - Simulation

Simulation is needed during any project phase - being the most frequent task in the field of research. But the current simulation support corresponds rarely to the requirements of the research work.

The toolbox approach results usually in several powerful modular functions, on- and off-line comments and demonstration batch files. A combination of toolboxes is, up to now, very similar to the use of a general-purpose programming language - with the same possibilities of undiscovered errors. The designer is forced to study many details of the methods employed, he needs to built a "main" batch file merging demo-files etc. Coding of a wise simulation program is complicated.

The MATLAB functions as tools are great in their simplicity and flexibility. But something similar is missing at the "main" program level. The following approach of "systems" (used in a form in KOS, fully in PLUS, mentioned in Maciejowski 1988) can solved the problem.

The system is a data/programming unit maintained by the shell system. It contains local data definitions done optionally in dialogue with the user. Some of the definitions are marked as "ports". Those can be connected with data structures outside the system

- forming a new unit in such a way which behaves again as a system. A high degree of modularity is achieved - any complex unit behaves from outside exactly as a single one.

The connections are maintained during any computation phase (unlike in the case of functions). Connections are based on data overlay. The ports can be divided into inputs, outputs and parameters according to signal flow.

Inside a system, a "simulation task" can be defined. It contains all commands and expression needed to carry out computations in a subsequent simulation run. Task are classified according to their role in simulation as initial, dynamic and terminal. Their definition contains sampling frequency and initial time offset, too.

Simulation means to run a simulation program in a cycle of simulation (or real) time. Before it starts, all simulation tasks are checked an (optionally) sorted according to signal flow. Then, the simulation program is compiled into a metalanguage and user's program modules are loaded into memory and repeatedly computed. The experimental simulation environment offers graphics and other services. The simulation run can be interrupted and the restarted at any moment. During the interruption, all capacity is available of the shell system (even another simulation run can be started in other system).

The main advantage of systems lies in information hidding and in the fact that everything is coded by the original package designer: data definitions, default values, use of the toolbox functions, consistent checking of the use of the methods during computations etc. The user must only to connect ports between systems.

4 User's Interfaces

While the culture of toolboxes has reached a high level, tools for preparing packages are not supported adequately by shell systems. But, packages are very important for industrial applications, for repeated solutions and for unexperienced users. The packages should be solved uniformly to be able cooperate in an more complex CADCS environment. More research results would be submitted to practice if appropriate tools are available during the design of toolboxes.

Using the system approach, the toolbox designer have a direct control of the process of data allocation and entry and should be (in an optimum design) responsible for checking of all data changes. He can easily code dialogues about data values. A simple informational system can be directly employed in the process description.

The design of systems is very similar to the design of packages. If the approach of systems will be accepted, a formalization of interfaces is inevitable. The rules can be organizational but a common support should be included from the part of the shell system.

As an example of such a support, we shall mention a command DIALOGUE implemented in KOS, PLUS. From the point of view of a designer, the command only marks points where data should be collected. It is composed from various options and names of variables and text files. A menu is formed on the screen based on the selection.

The shell system supports all details of the menu display and data transfer. The user changes data directly at the place where they appear on the screen. No knowledge of commands is required from the user.

The amount of information displayed depends upon the user's skill expressed as novice - advanced - expert. The range of information displayed is automatically modified according to how deeply the user wants to change defaults prepared by the designer (expressed in the same scale). The menu must be modified according to connection of systems - the questions solved before cannot appear again on the menu A simple device is present used to answer user's questions (based on the menu).

In such a way, quite simple but effective support can be given to the designer. Using a similar approach, the shell system system should support of building of coherent interfaces in more fields of research.

5 Conclusion

The shell system approach seems to be able to serve as a base for almost of general - purpose CADCS packages. However, additional features should be integrated at least for research purposes.

There are lot of problems not employed adequately in CADCS - advanced graphics, databases, artificial intelligence, new programming principles etc. The CADCS remains active research field. Let the next generation of shell systems be solved uniformly and adequately to the task.

6 References

Ad van den Boom (1988). CADCS Developments in Europe. In *Proceedings 4th IFAC Symposium on Computer Aided Design of Control Systems*, Beijing China, August 1988.

C.J. Herget (1988). Survey of Existing Computer Aided Design in Control System Packages.in the United States of America. In *Proceedings 4th IFAC Symposium on Computer Aided Design of Control Systems*, Beijing China, August 1988.

R. Kulhavý (1988). *PC-SIC, User's Guide, Version 2.0*. Institute of Information Theory and Automation, Czechoslovak Academy of Sciences, Prague.

R. Kulhavý and P. Nedoma (1988). *PC-KOS, User's Guide, Version 3.0*. Institute of Information Theory and Automation, Czechoslovak Academy of Sciences, Prague.

J.M. Maciejowski (1988). Data Structures and Software Tools for the Computer Aided Design of Control Systems. In *Proceedings 4th IFAC Symposium on Computer Aided Design of Control Systems*, Beijing China, August 1988.

C. Moler, J. Little, S. Bangert (1977). *PC-MATLAB for MS-DOS PC, Version 3.2-PC*. The MathWorks, Inc. 1987.

P. Nedoma (1990). *Projekt monitoru PLUS pro pro softwarovou infrastrukturu UTIA* (in Czech). Technical Report. To be published.

M. Rimwall (1988). Interactive Environments for CADCS software. In *Proceedings 4th IFAC Symposium on Computer Aided Design of Control Systems*, Beijing China, August 1988.

Chr. Schmidt (1989). Techniques and Tools for CADCS. In *Proceedings 4th IFAC Symposium on Computer Aided Design of Control Systems*, Beijing China, August 1988.

NOVEL CONTROL ARCHITECTURES

George W. Irwin

Department of Electrical and Electronic Engineering,

Queen's University, Belfast BT9 5AH.

1. INTRODUCTION

Advances in microelectronics have had a significant impact on the implementation of digital controllers and a variety of processor hardware has been employed by the control engineer to meet the discipline of real-time operation including microprocessors, application specific integrated circuits (ASICs), digital signal processors (DSPs) and digital signal controllers (DSCs).

The availability of this technology has allowed developments in adaptive control theory to be incorporated in new industrial controllers like the Novatune from Asea [1], a self-tuning regulator using a minimum-variance control law and a recursive least-squares parameter estimator. It is interesting to contrast that early product with controllers from Eurotherm and Control Techniques Process Instruments, which were reported recently [2]. The Expert 440, for example, costs about £750 and claims auto-tuning of any plant which is open-loop stable. Three microprocessors carry a one-shot tuner, an adaptive process estimator, a controller designer, the controller itself and an 'expert system' overseer. In addition to conventional PID control, the Expert 460 can provide two alternative adaptive control laws viz. implicit control with a variable prediction horizon or a generalised explicit pole-placement controller.

Despite the undoubted potential of embedded digital controllers for industrial applications, an increasing number of concurrent processors have appeared in the market, some aimed at front-end signal processing applications, some designed for more general purpose parallel computing systems. [3], [4]

Increased computational speed is of course the primary benefit of parallel processing. This allows faster systems to be controlled and gives the engineer the choice of added complexity in the control algorithm. Easy expansion within a uniform hardware and software base is another feature, since more processors can be added as required with clear implications for reducing development and maintenance costs. Parallel processing also offers a closer relationship between a block diagram description of a control system and the corresponding hardware implementation which may in the future ease the path from design to implementation. Lastly, fault tolerance can be realised in a parallel processing system by organising the computation to be distributed, so that an operational failure results in performance degradation rather than a complete failure. Reference 5 provides a good overview of recent developments in the field of parallel processing for real-time control.

The present chapter will describe recent work on parallel Kalman filtering, in particular fine-grained systolic algorithms and their hardware realisation using transputers. The next section provides some background on systolic arrays. Algorithms for regular and square root covariance Kalman filtering are defined in section 3, while section 4 describes the systolic array architectures. Section 5 discusses mapping of the systolic arrays onto transputers. Results are included to illustrate the effectiveness of the mapping method and the potential speedups for a simple application. The chapter ends with a brief comment on the broader implications of these results.

2.   SYSTOLIC ARRAYS BACKGROUND

The concept of a systolic array was first proposed by Kung and Leiserson [6] in order to exploit the high speed switching potential of current VLSI technology. These are arrays of largely individual processing cells, each of which has some local memory and is connected only to its nearest neighbours in the form of a regular lattice. On each cycle of a system clock, every cell in the array receives data from its neighbouring cells and performs a specific processing operation on it. The resulting data is stored within the cell and the result passed to a neighbouring cell on the next clock cycle. Consequently, each item of data is passed from cell to cell across the array in a particular direction and the term 'systolic' refers to this rhythmical movement of data which is analogous to the regular pumping action of the heart. All data is input to, or output from, the array through the boundary cells.

Originally arrays of inner step processors were devised for a number of matrix computations including matrix x vector multiplication, matrix x matrix multiplication and LU decomposition. This stimulated research into VLSI architectures for a whole range of real time matrix algebraic operations [7], largely driven by the needs of signal and information processing applications. The consensus emerging from this work is that the structures should be regarded as 'systolic algorithms' rather than specific hardware designs. These can be implemented in practice by mapping these algorithms onto a more general purpose system. This is the philosophy of the work to be described in the sections which follow next.

3. KALMAN FILTER ALGORITHMS

3.1 Regular Covariance Filter

The linear, discrete-time system is described by

$$\underline{x}(k + 1) = A(k)\underline{x}(k) + B(k)\underline{u}(k) + \underline{w}(k)$$

$$\underline{z}(k) = C(k)\underline{x}(k) + \underline{v}(k)$$

where $\underline{x}(k)$ is the (n x 1) state vector, $\underline{z}(k)$ is the (m x 1) measurement vector (m $\leq$ n) and $\underline{u}(k)$ is the (p x 1) control or deterministic forcing vector. A(k), B(k) and C(k) are known matrices of appropriate dimensions. Also $\underline{w}(k)$ and $\underline{v}(k)$ are zero-mean independent Gaussian white-noise sequences with known covariances W(k) and V(k), respectively. Further, W(k) is assumed to be a symmetric non-negative-definite matrix, while V(k) is assumed to be a symmetric positive-definite matrix and thus the following 'square roots', or Choleski factors, which are usually taken to be positive-definite and which can be either upper or lower triangular, are defined as

$$V = (-V^{1/2})(-V^{T/2}), \quad W = (-W^{1/2})(-W^{T/2})$$

The Kalman filter estimates the state of the system from a sequence of measurements; the predicted state estimate $\hat{x}(k + 1/k)$ is the estimate of the state at time $k + 1$ given measurements up until time k, the filtered state estimate $\hat{x}(k/k)$ is the estimate at the time k given measurements up until time k. The corresponding error covariance matrices are

$$P(k + 1/k) = E\{(\underline{x}(k + 1) - \hat{x}(k + 1/k))(\underline{x}(k + 1) - \hat{x}(k+1/k))^T\}$$

$$P(k/k) = E\{(\underline{x}(k) - \hat{x}(k/k))(\underline{x}(k) - \hat{x}(k/k))^T\}$$

The regular covariance Kalman filter [8] is defined by the following set of equations:

$$\hat{x}(k + 1/k) = A(k)\hat{x}(k/k) + B(k)\underline{u}(k) \tag{1}$$

$$P(k + 1/k) = A(k)P(k/k)A^T(k) + W(k) \tag{2}$$

$$V_e(k + 1) = C(k)P(k + 1/k)C^T(k) + V(k) \tag{3}$$

$$K(k + 1) = P(k + 1/k)C^T(k + 1)V_e^{-1}(k + 1) \tag{4}$$

$$\hat{x}(k + 1/k + 1) = \hat{x}(k + 1/k) + K(k+1)[\underline{z}(k + 1) - C(k + 1)\hat{x}(k + 1/k)] \tag{5}$$

$$P(k + 1/k + 1) = P(k + 1/k) - K(k + 1)C(k + 1)P(k + 1/k) \tag{6}$$

3.2 Square Root Covariance Filter

Rounding errors can produce numerical problems for the regular filter by destroying the symmetry of the error covariance matrices which eventually are no longer positive semidefinite, as required by the theory. This can cause the state estimates to diverge completely from the optimal estimates and in the worse case the filter 'blows up'. Square root forms of the filter force the error covariance matrices to remain symmetric by propogating their triangular square roots, or Choleski factors, rather than the matrices themselves.

The square root covariance Kalman filter, given in equations (7) and (8), was proposed by Morf and Kailath [9].

$$Q(k)\begin{bmatrix} P^{T/2}(k/k-1)C^T(k) & P^{T/2}(k/k-1)A^T(k) \\ V^{T/2}(k) & 0 \\ 0 & W^{T/2}(k) \end{bmatrix} = \begin{bmatrix} V_e^{T/2}(k) & V_e^{-1/2}(k)C(k)P(k/k-1)A^T(k) \\ 0 & P^{T/2}(k+1/k) \\ 0 & 0 \end{bmatrix}$$

(7)

$$\hat{x}(k+1/k)=A(k)\hat{x}(k/k-1)+B(k)\underline{u}(k)+A(k)P(k/k-1)C^T(k)V_e^{-1}(k)[\underline{z}(k)-C(k)\hat{x}(k/k-1)] \quad (8)$$

The relative complexity of the two filter algorithms may be assessed in terms of the number of mathematical operations performed within each iteration; the regular covariance filter, equations (6)-(11), requires $0(2n^3)$ multi-plications/divisions while the square root covariance filter, equations (12), (13) involves $0(5/2n^3)$ multiplications/divisions and $0(n^2)$ square roots. The additional complexity of the square root filter is offset by its significantly improved numerical stability but clearly, in both cases, the computational load is a problem for real-time applications and increasing state vector dimension n.

4. TWO SYSTOLIC KALMAN FILTERS

4.1 Regular Covariance Filter

The systolic filter which follows is derived using the Fadeev algorithm [10], as proposed by Yeh [11] for a mixed form of the regular filter. The Fadeev algorithm, for finding the Schur complement of a matrix A, can be introduced in terms of calculating

$$WB + D \quad (9)$$

given

$$WA = C \quad (10)$$

where A,B,C,D are known matrices and W is unknown. Clearly equation (10) gives W as

$$W = CA^{-1}$$

which can be substituted in (9) to produce the required result

$$CA^{-1}B + D$$

The same answer can be arrived at, without the necessity of finding W explicitly, as follows. The matrices A,B,C and D are loaded into the compound matrix of equation (11),

$$\begin{bmatrix} A & B \\ -C & D \end{bmatrix} \tag{11}$$

and a linear combination of the first row is then added to the second to produce

$$\begin{bmatrix} A & B \\ -C + WA & D + WB \end{bmatrix} \tag{12}$$

If W, which specifies the linear combination, is chosen such that

$$-C + WA = 0$$

then the matrix in equation (12) becomes

$$\begin{bmatrix} A & B \\ 0 & D + CA^{-1}B \end{bmatrix}$$

and the lower right quadrant now contains the answer. The Fadeev algorithm then involves reducing C to zero, in the compound matrix of equation (11), by row manipulation. The required matrix calculation is produced in the bottom right quadrant.

In practice a modified Fadeev algorithm, where the matrix A is changed to triangular form prior to annulment of C, is more suitable for systolic array processing. This is summarised as

$$\begin{bmatrix} A & B \\ -C & D \end{bmatrix} \rightarrow \begin{bmatrix} TA & TB \\ -C & D \end{bmatrix} \rightarrow \begin{bmatrix} TA & TB \\ 0 & D+CA^{-1}B \end{bmatrix}$$

where TA is an upper triangular matrix.

The trapezoidal systolic array for the Fadeev algorithm is shown in Fig. 1. Here all the cells have two modes of operation, changing from one mode

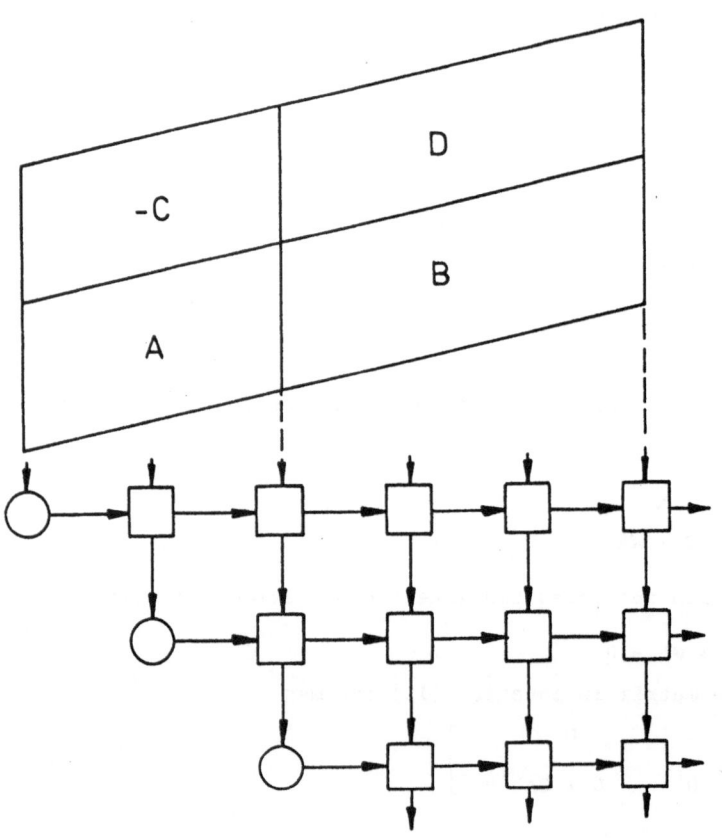

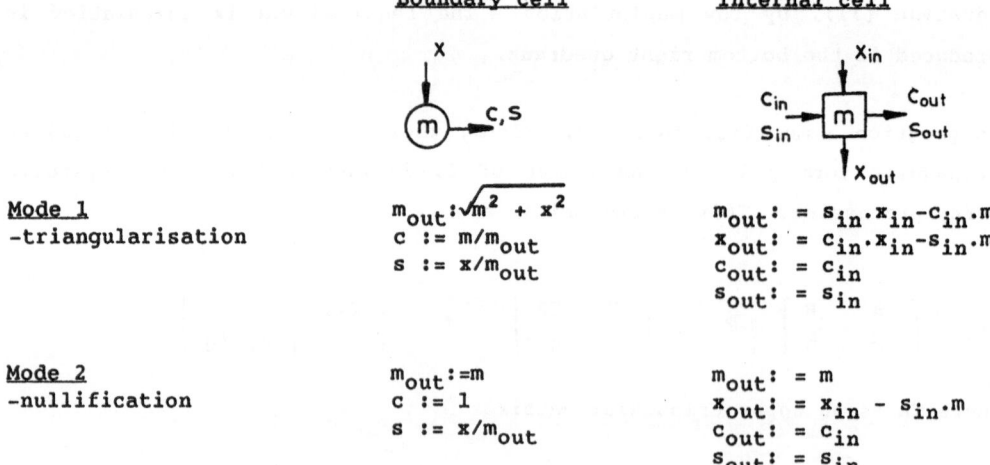

	Boundary cell	Internal cell

Mode 1
-triangularisation

$m_{out} := \sqrt{m^2 + x^2}$
$c := m/m_{out}$
$s := x/m_{out}$

$m_{out} := s_{in} \cdot x_{in} - c_{in} \cdot m$
$x_{out} := c_{in} \cdot x_{in} - s_{in} \cdot m$
$c_{out} := c_{in}$
$s_{out} := s_{in}$

Mode 2
-nullification

$m_{out} := m$
$c := 1$
$s := x/m_{out}$

$m_{out} := m$
$x_{out} := x_{in} - s_{in} \cdot m$
$c_{out} := c_{in}$
$s_{out} := s_{in}$

Fig. 1.   Trapezoidal array for the regular covariance Kalman filter.

to the next as required by the data. Givens rotations [12] are used to reduce the 'A' matrix in equation (11) to upper triangular form through successive premultiplication by rotation matrices. The computation is performed column by column, starting with the first element below the diagonal. These rotations can be mapped onto the array for mode 1 of the cell operation. The boundary cells calculate the rotation parameters and implicitly zero the elements in their respective columns below the diagonal, while the internal cells complete the corresponding row calculations.

The power of the algorithm lies in the fact that a range of matrix computations can be performed by proper choice of A,B,C and D; in a sense the compound matrix in equation (11) can be programmed. Thus for example, if

$$B = I, \ C = I, \ D = 0$$

then A^{-1} will be generated. In particular, the matrix equations (1) - (6) of the covariance Kalman filter can be produced in 9 successive passes of the Fadeev algorithm, as shown in Table 1. The time update of the state vector is completed in pass 1, the predicted error covariance matrix in pass 3, the gain matrix in pass 6, the filtered error covariance matrix in pass 7 and the measurement update of the state vector in pass 9.

4.2 Square root covariance filter

An orthogonal decomposition is required in equation (7) to triangularise the prearray. This is shown schematically in Fig.2 which illustrates how the data is fed into the triangular array and where the results are stored in memory. Since the order in which the rows of data are fed into the systolic array has no effect on the final result, the rows containing $V^{T/2}(k)$ and $W^{T/2}(k)$ can be preloaded and then only the rows containing $P^{T/2}(k/k-1)C^T(k)$ and $P^{T/2}(k/k-1)A^T(k)$ need to be fed in.

Step	'A'	'B'	'C'	'D'	Result
1	I	$\tilde{\underline{x}}(k/k)$	$-A(k)$	$B(k)\underline{u}(k)$	$\tilde{\underline{x}}(k+1/k)$
2	I	$P(k/k)$	$-A(k)$	0	$A(k)P(k/k)$
3	I	$A^T(k)$	$-A(k)P(k/k)$	$W(k)$	$P(k+1/k)$
4	I	$C^T(k)$	$-P(k+1/k)$	0	$P(k+1/k)C^T$
5	I	$P(k+1/k)C^T(k)$	$-C(k)$	$V(k)$	$V_e(k+1)$
6	$V_e(k+1)$	I	$-P(k+1/k)C^T(k+1)$	0	$K(k+1)$
7	I	$[P(k+1/k)C^T(k+1)]^T$	$-K(k+1)$	$P(k+1/k)$	$P(k+1/k+1)$
8	I	$\tilde{\underline{x}}(k+1/k)$	$C(k+1)$	$\underline{z}(k+1)$	$\Delta\underline{z}(k+1)$
9	I	$\Delta\underline{z}(k+1)$	$-K(k+1)$	$\tilde{\underline{x}}(k+1/k)$	$\hat{\underline{x}}(k+1/k+1)$

Table 1.   Regular covariance filter using the Fadeev algorithm
$[\Delta z(k+1) = \underline{z}(k+1) - C(k+1)\,\tilde{\underline{x}}(k+1/k)]$

The results, stored in the memory of the top m rows of the systolic array, can now be used to obtain the next predicted state estimate based on the Fadeev algorithm, as follows. If equation (8) is transposed, then the right hand side is in the form of a Schur complement, as follows

$$\underbrace{[A(k)\tilde{\underline{x}}(k/k-1)+B(k)u(k)]^T}_{'D'}+\underbrace{[z(k)-C(k)\tilde{\underline{x}}(k/k-1)]^T}_{'C'}\underbrace{[V_e^{T/2}]^{-1}}_{'A'^{-1}}\underbrace{[A(k)P(k/k-1)C^T(k)V_e^{-1/2}]^T}_{'B'}$$

$$(13)$$

Therefore, since the $V_e^{T/2}$ term is already triangular and loaded into the memory along with the 'B' matrix in equation (13), only the second or nullification mode of the Fadeev architecture is required. Fig. 3 shows how the Schur complement, which corresponds to the state estimate $\tilde{\underline{x}}(k+1/k)$, can be obtained.

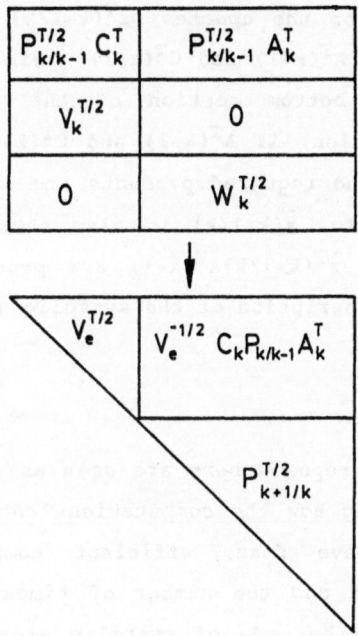

Fig 2. Data flow for orthogonal decomposition of the prearray in equation (7).

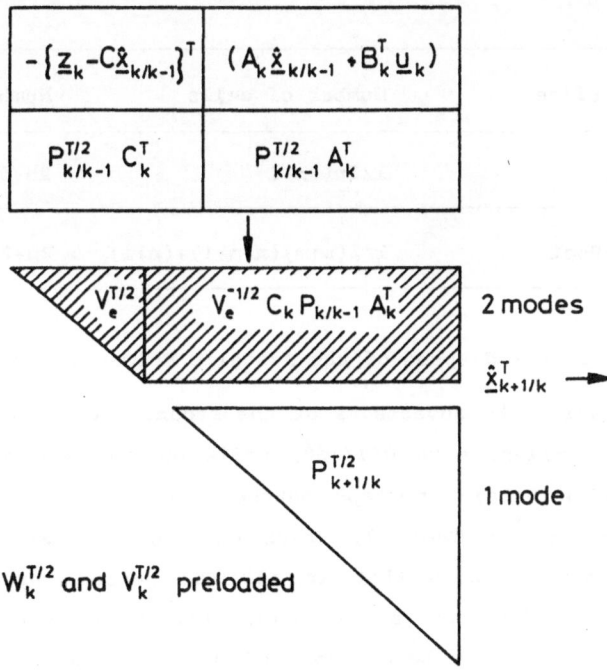

Fig. 3. Split architecture to include the state update of equation (8).

For the next iteration the updates, $\tilde{x}^T(k+1/k)$ and $P^{T/2}(k+1/k)$, must both
be postmultiplied by $A^T(k+1)$ and $C^T(k+1)$. Since $P^{T/2}(k+1/k)$ is stored in
the memory of the bottom section of the triangular array from the
orthogonal decomposition, if $A^T(k+1)$ and $C^T(k+1)$ are entered as shown in
Fig. 4 then two of the required products are obtained. Similarly, if the
updated state estimate, $\tilde{x}(k+1/k)$ is also stored in a linear array then
$\tilde{x}^T(k+1/k)C^T(k+1)$ and $\tilde{x}^T(k+1/k)A^T(k+1)$ are produced as shown in Fig. 4.
This completes the description of the systolic square root Kalman filter.

4.3 Discussion

The systolic arrays proposed here are seen as two-dimensional diagrams of
the algorithms showing how the computations can be organised and the data
manipulated to achieve fast, efficient computation. The number of
processing cells used and the number of timesteps between updates allow
simple comparisons to be made of systolic architectures. An analysis of
the two systolic Kalman filters given above produced the results contained
in Table 2.

Kalman Filter	Number of cells	Number of Timesteps
Regular	1/2n(3n+1)	9n+6m+3
Square Root	1/2(n+m)(n+m+1)+(n+1)	2n+2m+1

Table 2. Size and speed of the systolic Kalman filters.

On this basis, the advantages of the second architecture are clear; the
array is smaller, with $O(1/2n^2)$ cells as compared with $O(3/2n^2)$, and
faster, with $O(2n)$ timesteps between updates as compared with $O(9n)$.
An examination of Table 1, which defined the data required for the
systolic array, shows that in all the passes, except number 6, an
identity matrix is entered as matrix 'A'. This means that the identity
matrix is implicitly being inverted during these passes. Also a null
matrix is being added during passes 2, 4 and 6, because of the null 'D'

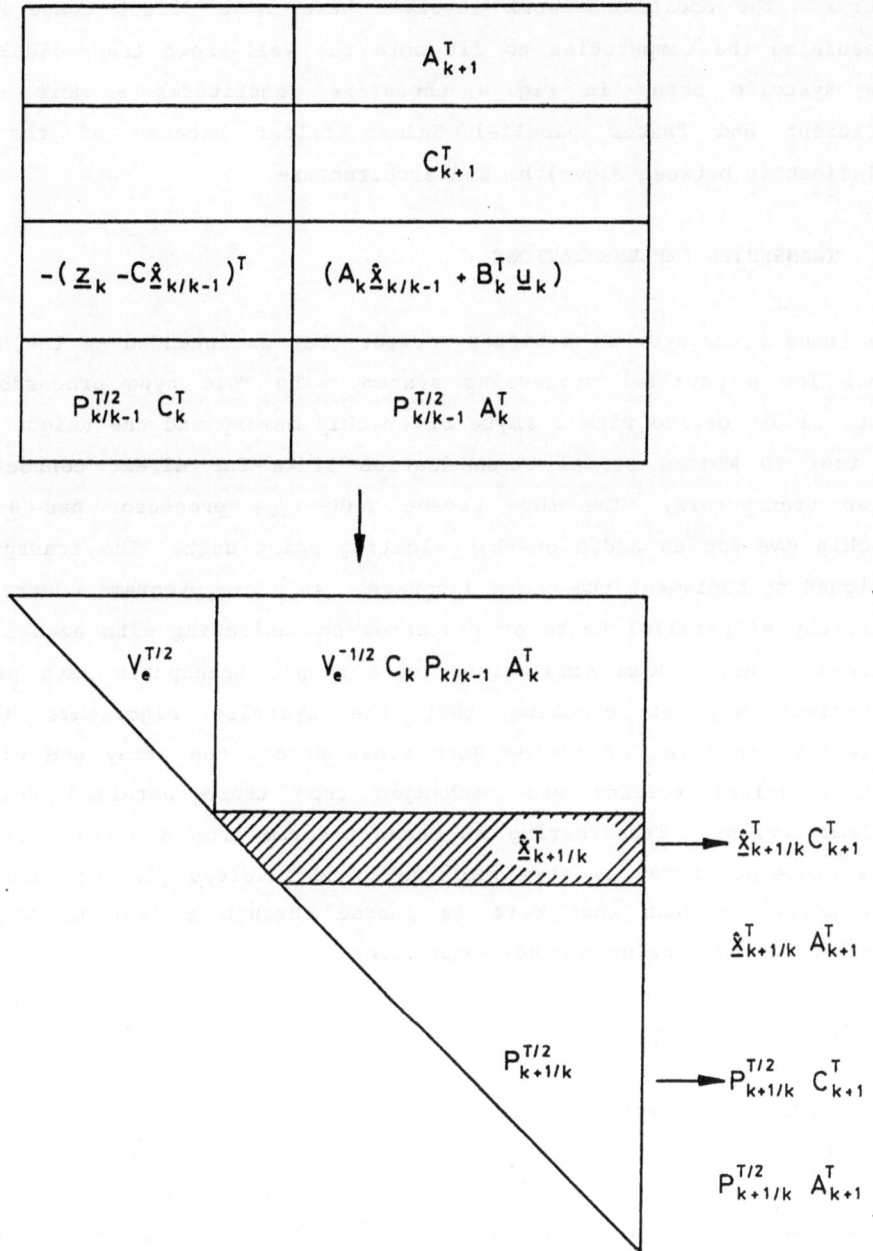

Fig. 4. Systolic array for the square root filter, incorporating a linear section for matrix products.

matrix. The additional work involved here is a direct consequence of organising the computation to fit onto the well-known trapezoidal array. The systolic array in Fig. 4 therefore constitutes a more compact, efficient and faster parallel Kalman filter because of the closer relationship between algorithm and architecture.

5.   TRANSPUTER IMPLEMENTATIONS

The Inmos transputer is a microprocessor that is intended as the building block for a parallel processing system. The T414 type processor is a fast, 32-bit device with 2 kbyte of on-chip memory and the unique feature of four 20 Mbit/s serial communication links for direct connection to other transputers. The more recent T800 type processor has 4 kbytes on-chip RAM and an added on-chip floating point unit. The transputer is designed to implement the occam language. An occam programe consists of a hierarchy of parallel tasks or processes communicating with each other by channels. Here occam simulations on a single transputer have proved a convenient way of checking that the systolic algorithms function correctly, in terms of timing data flows across the array and verifying that numerical results are unchanged from those obtained from ACSL implementations. Transferring an occam program from a single transputer to a processor array is straightforward and involves placing channels at link addresses such that data is passed through a link to a process attached to the link on another transputer.

5.1 Mapping Strategy

If parallel processing hardware, like transputers, is to be fully exploited in control, it is necessary that systematic mapping procedures emerge from the applications driven, heuristic methods of partitioning calculations which are used at present. The work described in this section was aimed at investigating the potential role of systolic algorithms in such a mapping process. The idea behind this was that systolic arrays provide a highly parallel description of an algorithm, suitable not only for a fine-grained VLSI realisation but also for coarser grained hardware like transputers.

A suitable mapping strategy must consider factors such as load balancing, communication requirements and the efficient utilisation of each transputer. A simple, but impractical, strategy would be to assign each cell of the systolic array to one transputer. However, a more efficient concurrent system is obtained when cells are grouped into coarser processes during the mapping procedure. Specifically, the cell computations in each row of the systolic array are grouped into a single row-process. This effectively projects the array structure into a collection of nearest-neighbour, row-processes communicating vectors of data which can be realised efficiently on a transputer pipeline. Each processor operates most effectively if the row-process is now carried out sequentially, since overheads are incurred in running parallel computational tasks on a single-processor. Assignment of row-processes to transputers must be done with care to ensure load balancing, especially with non-square systolic arrays, because the row-processes in this case do not require equal amounts of computation. Fig. 5 illustrates the mapping onto a three transputer pipeline, with a fourth processor to control the input and output of data and feedback results required for the next iteration.

5.2 Results

The application employed involved state-feedback control of a missile moving in the vertical plane against a stationary target, with the requirement to track a ramp trajectory [13]. The problem was to estimate the four missile states from a single noisy measurement of height error. Higher order systems were derived by taking multiples of the original one. The aim was to measure the performance of a transputer realisation of the square root covariance Kalman filter for varying system orders.

The graphs in Fig. 6 illustrate the average time required to estimate the state vector from a sequence of 200 measurements on four transputers (curve C). For comparison, results are also included for the single transputer implementations of both the systolic (curve A) and the sequential Kalman filter (curve B).

Systolic Array   Transputer Network

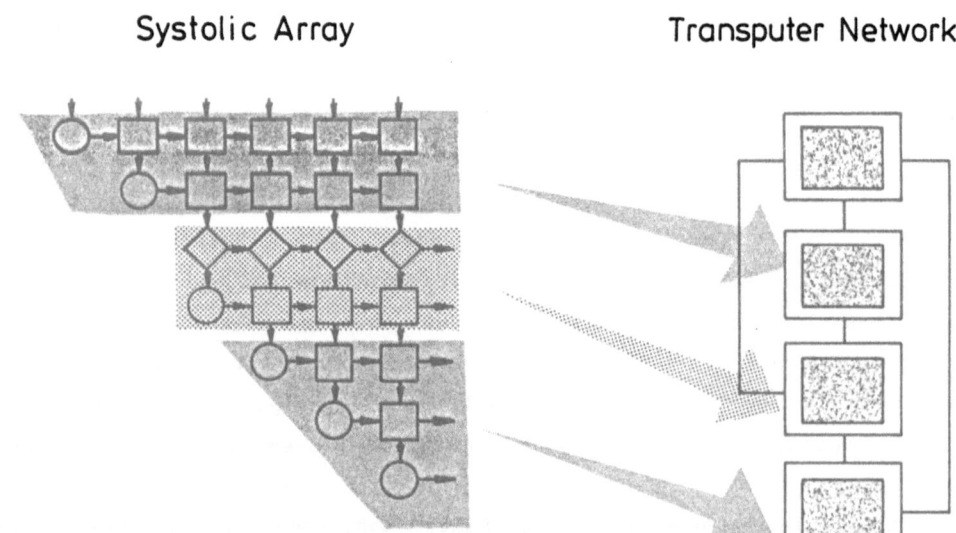

Fig. 5   Mapping a triangular systolic array onto a transputer pipeline.

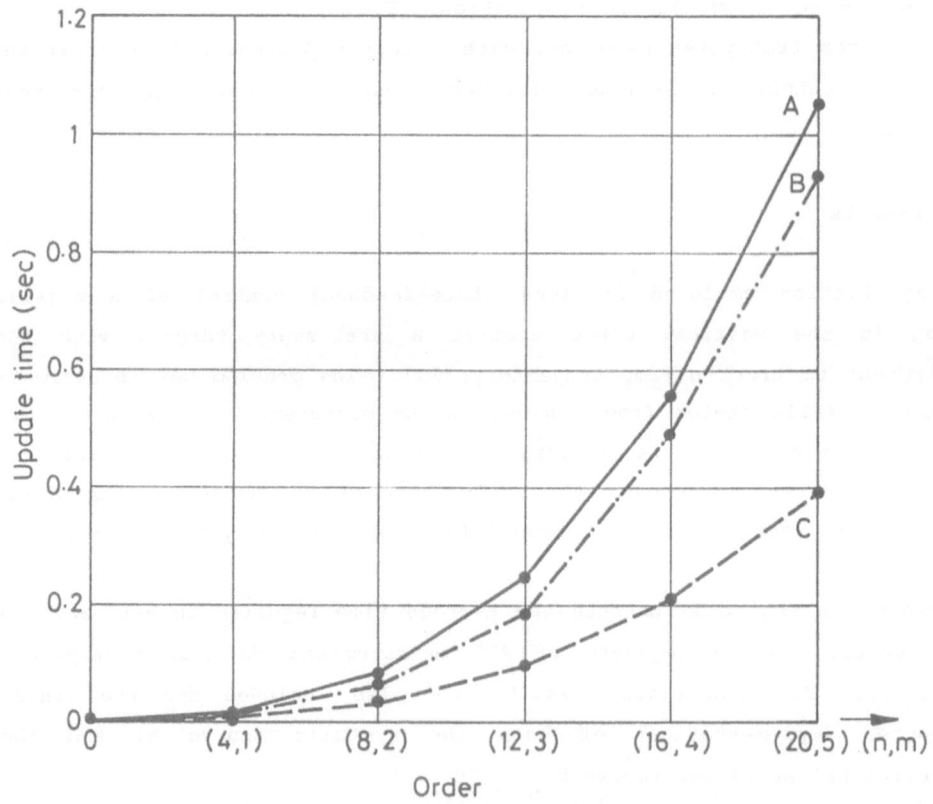

Fig. 6   Variation in state update times with system order for transputer
implementations of the squre root covariance filter.

The rapid increase in the update time on one processor, with increasing order of system, is clearly shown for both the systolic and regular filter algorithms. Note also that the systolic filter is slower than the regular one on a single transputer, although the cell processes are executed sequentially. However, the latter facilitates mapping onto multiple transputers and the use of a four processor array reduces the update times significantly.

Two measures of particular interest here are the speedup and the efficiency of the parallel filter. Speedup (S_p) is defined as the execution time on a single processor (T_1) to the execution time on p processors (T_p). Thus

$S_p = T_1/T_p$

Efficiency (E_p) is defined as the average utilisation of the processors, expressed as a percentage. Hence

$E_p = S_p/p \times 100\%$

There are two issues to highlight here, the success of the new mapping procedure and the performance of a transputer based Kalman filter. Comparing the single and multiple transputer implementations of the systolic filter (curves A and C) produced values of E_p in the range 55% to 68%, with better efficiencies for higher system orders. These are good figures and illustrate the success of mapping the systolic filter onto a fixed transputer array. However, to truly quantify the speed advantage it is necessary to compare the results from the four transputer array (curve C) with the fastest implementation on a single transputer (curve B). Here the speedings ranged from 1.6 to 2.4, with the better performances again with increased values of n and m.

6.   CONCLUDING REMARK

The matrix based nature of much of modern control theory makes it as natural area of application for parallel processing and in particular for systolic arrays. Although this chapter has concentrated on Kalman filtering, the ideas presented are considered to be of more general significance; to linear quadratic (LQ) optimal control through the Duality Principle [14], to least squares plant parameter estimation and hence to adaptive control [15].

ACKNOWLEDGEMENTS

The contributions of my former research student Dr. Fiona Gaston and Liam Maguire of the Institute of Advanced Microelectronics are gratefully acknowledged.

REFERENCES

1. Moden, P.E., "Self-tuning control in Novatune", Asea Innovation, KX 20-010E, 1983.

2. Tinham, B., "A new era of multi-algorithm control?", Control and Instrumentation, Morgan-Grampian (Process Press) Ltd, pp. 127-130, May 1990.

3. Stein, R.M., "T800 and counting", Byte, McGraw Hill, pp. 287-296, 1988.

4. Menzilcioglu, O., Kung, H.T. and Song, S.W., "A highly configurable architecture for systolic arrays of powerful processors" in Systolic Array Processors, eds. J.V. McCanny, et al., Prentice Hall International, pp. 156-165, 1989.

5. "Parallel Processing for Real-time Control", Special Issue, IEE Proc., Pt. D, Vol. 137, No. 4, 1990.

6. Kung, H.T. and Leiserson, E.C., "Algorithms for VLSI array processors", in Introduction to VLSI Systems, C. Mead and L. Conway, Addison-Wesley, Reading, Mass., 1980.

7. "Systolic Arrays", Special Issue, Computer, Computer Society of the IEEE, 1987.

8. Kalman, R.E., "A new approach to linear filtering and prediction problems", Trans. ASME, J. Basic Eng., Vol. 82D, pp. 34-45, 1960.

PROCESS ESTIMATION:
LINEAR ADAPTIVE ALGORITHMS AND NEURAL NETWORKS

Ming. T. Tham, Gary A. Montague and A. Julian Morris

Department of Chemical and Process Engineering
University of Newcastle upon Tyne
Newcastle upon Tyne, NE1 7RU, England

ABSTRACT

Two different approaches that can provide frequent and accurate estimates of process outputs which are subject to large measurement delays are outlined. The first is based upon linear adaptive techniques whilst the other makes use of a fixed parameter neural network model. The results of applications to industrial data are used to discuss and contrast the performance capabilities of the two techniques.

INTRODUCTION

In many process situations, limitations in instrumentation technology mean that outputs of interest can only be obtained infrequently and with long time delays. Nevertheless, these 'primary' outputs are often employed as feedback signals for process control. Typical examples are in the product quality control of distillation columns; chemical and biochemical reactors. Their use, however, can cause major and prolonged deviations from set-points since disturbance effects remain undetected in between the long sample periods. Even the implementation of advanced control algorithms may not provide adequate solutions to this problem of large measurement delays.

Traditionally, the control of primary outputs subject to large measurement delays has been achieved by the control of another (secondary) process output which can be sampled more frequently. Such strategies implicitly estimate/infer primary output behaviour from secondary output responses (inferential control). The underlying assumption is that those disturbances affecting the primary output will also affect the secondary output. Since control is executed more frequently, improved performances result. The complex and often time varying relationship between process variables can, however, cause implementation difficulties. As an example, consider the composition control of multi-component distillation columns. Single temperature feedback control is often used to regulate a product composition to set-point. This approach is not always effective since a constant tray temperature does not necessarily mean that product composition will remain at set-point. Hence, such inferential control loops usually have their set-points 'trimmed', either manually based upon off-line product quality analyses, or by an outer loop utilizing the delayed measurements of the controlled variable, ie. a parallel cascade strategy (Luyben, 1973).

However, it has been shown that by combining the basic concepts intrinsic to current inferential control practice, state and parameter estimation, a more practicable methodology results. This amalgamation of ideas culminates in the formulation of a family of adaptive inferential estimators (Guilandoust et al, 1987, 1988). The adaptation is necessary in order to track the changing nature of the relationship between primary and secondary outputs, in applications to non-linear systems. Alternatively, neural network based models, which are nonlinear representations of the process, can be used for the on-line estimation of process variables. This was also considered by Montague et al (1989). In this contribution, both the adaptive and the neural network approaches to inferential estimator design will be discussed, and the performances of the resulting algorithms compared, by applications to industrial data.

ADAPTIVE INFERENTIAL ESTIMATION

Adaptive inferential estimators can be designed based upon either a state-space or an input-output model of the process (Guilandoust et al, 1987, 1988). The development of the state-space approach is outlined here. The derivation of the estimator assumes that the process is described by the following discrete state space model:

$$x(t+1) = Ax(t) + Bu(t-m) + Lw(t) \tag{1}$$

$$v(t) = Hx(t) + v_s(t) \tag{2}$$

$$y(t) = Dx(t-d) + v_p(t) \tag{3}$$

where 't' is the time index, while $y(t)$ and $v(t)$ are the primary and secondary output measurements respectively. The corresponding measurement noises, $v_p(t)$ and $v_s(t)$, are assumed to be independent sequences with zero means and finite variances. The vector of process states is denoted by $x(t)$, while $w(t)$ is a vector of random disturbances. 'd' is the measurement delay associated with $y(t)$, and 'm' is the smallest of the delays in primary and secondary output responses to changes in the control signal $u(t)$. Any difference between the two delays can be included in the model by extending the state vector. It should be noted that the model is discretised with respect to the (faster) sampling time of the secondary output, $v(t)$.

If the matrices A, B, L, H, D and the statistics of the disturbances are known, a Kalman filter can be used to estimate $x(t)$, $\hat{x}(t)$, from $v(t)$. Substitution of $x(t)$ into Eq.(3) yields the estimates of $y(t+d)$ as:

$$\hat{y}(t+d) = D\hat{x}(t) \tag{4}$$

The Kalman filter for Eqs.(1) and (2) is:

$$\hat{x}(t+1) = A\hat{x}(t) + Bu(t-m) + K(t)\epsilon(t) \tag{5}$$

where $K(t)$ is the Kalman filter gain; $\hat{v}(t)=H\hat{x}(t)$ is the filtered value of $v(t)$, and $\epsilon(t) = v(t) - \hat{v}(t)$ is the innovations sequence. If the process is unknown, then $\hat{x}(t)$ can only be obtained via combined state and parameter estimation. In this case, the equations:

$$\hat{x}(t+1) = A\hat{x}(t) + Bu(t-m) + K\epsilon(t) \tag{6}$$

$$v(t) = H\hat{x}(t) + \epsilon(t) \tag{7}$$

$$y(t) = D\hat{x}(t-d) + v_p(t) \tag{8}$$

are used instead of Eqs.(1) to (3). K is the time invariant Kalman gain and $\hat{x}$ is the optimal estimate of x. Although Eqs.(6) and (7) are equivalent to Eqs.(1) and (2) (eg. see Anderson and Moore, 1979), the former set is more practical since the state estimates can then be obtained without having to compute $\hat{v}$. In addition, Eqs.(6) and (7) are parameterised in the following form:

$$\hat{x}(t+1) = \begin{bmatrix} -a_1 & 1 & 0 & \cdots & 0 \\ -a_2 & 0 & 1 & \cdots & 0 \\ & & \cdot & & \cdot \\ -a_{n-1} & 0 & & \cdots & 1 \\ -a_n & 0 & & \cdots & 0 \end{bmatrix} \hat{x}(t) + \begin{bmatrix} b_1 \\ \cdot \\ \cdot \\ b_n \end{bmatrix} u(t-m) + \begin{bmatrix} k_1 \\ \cdot \\ \cdot \\ k_n \end{bmatrix} \epsilon(t) \tag{9}$$

$$v(t) = [1, 0,..., 0]\hat{x}(t) + \epsilon(t) \tag{10}$$

so that the identified parameters would directly correspond to those of an input-output model (Ljung and Soderstrom, 1983). From the definition, $\epsilon(t) = v(t) - \hat{v}(t)$, Eq.(10) becomes:

$$\hat{v}(t) = \hat{x}_1(t) \tag{11}$$

Elimination of $\hat{x}_1(t)$ to $\hat{x}_n(t)$ from the RHS of Eq.(9) yields:

$$\hat{x}_1(t+1) = \hat{v}(t+1)$$

$$\hat{x}_2(t+1) = - a_2\hat{v}(t) - .. - a_n\hat{v}(t-n+2) + b_2u(t-m) + .. + b_nu(t-m-n+2)$$

$$+ k_2\epsilon(t) + .. + k_n\epsilon(t-n+2) \tag{12}$$

$$\cdot \qquad \cdot \quad \cdot \quad \cdot \quad \cdot \quad \cdot$$

$$\hat{x}_n(t+1) = - a_n\hat{v}(t) + b_nu(t-m) + k_n\epsilon(t)$$

Eq.(12) is then substituted into Eq.(8) leading to an expression with the following form:

$$y(t) = \beta_1u(t-m-d-1) + .. + \beta_{n-1}u(t-m-n-d+1) + \tau_0\hat{v}(t-d) + .. + \tau_{n-1}\hat{v}(t-n-d+1)$$

$$+ \delta_1\epsilon(\text{t-d-1}) + .. + \delta_{n-1}\epsilon(\text{t-n-d+1}) + v_1(t) \qquad (13)$$

which can be expressed more compactly as:

$$y(t) = \Theta^T\phi(\text{t-d}) + e(t) \qquad (14)$$

$$\Theta^T = [\beta_1, .. , \beta_{n-1}, \tau_0, .. ,\tau_{n-1}, \delta_1, .. , \delta_{n-1}]$$

$$\phi(\text{t-d})^T = [u(\text{t-m-d-1}), .. , \hat{v}(\text{t-d}), .. , \epsilon(\text{t-d-1}), ..]$$

and e(t) is the equation error. Each time a primary measurement becomes available, the parameter vector Θ can be estimated. Since Eq.(14) is independent of state estimates, this should not present any difficulties. After updating Θ at time 't', it is used in:

$$y(t+d) = \Theta^T\phi(t)$$

$$= \beta_1 u(\text{t-m-1}) + .. + \beta_{n-1}u(\text{t-m-n+1}) + \tau_0\hat{v}(t) + .. + \tau_{n-1}\hat{v}(\text{t-n+1})$$

$$+ \delta_1\epsilon(\text{t-1}) + .. + \delta_{n-1}\epsilon(\text{t-n+1}) \qquad (15)$$

to provide estimates of the controlled output. The values of $\epsilon(t)$ in Eqs.(13) and (15) can be obtained from Eqs.(9) and (10) once the parameters of Eq.(9) have been determined. Θ is again updated at time 't+d' when a new measurement of the controlled output becomes available. The relationship which is used to provide estimates of the controlled output at the faster secondary output sample rate is thus given by Eq.(15). Figure 1 shows a schematic of the algorithm, and the manner in which the estimates may be used for control. Details of this estimator are described in Guilandoust *et al* (1987).

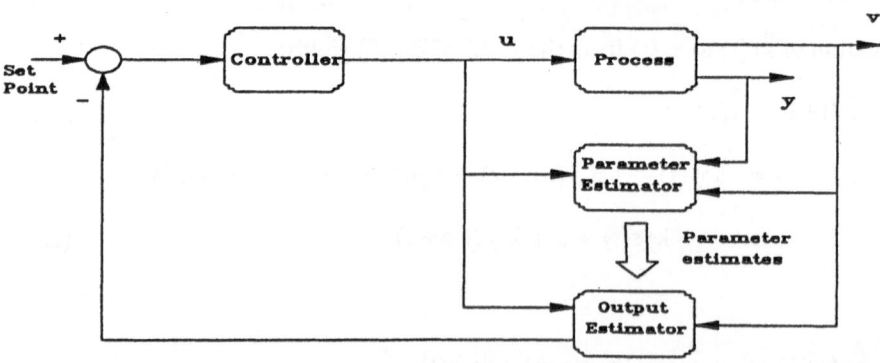

Figure 1. Schematic of Adaptive Inferential Estimation Scheme

A multirate estimator of similar form can be derived based on an input-output representation of the process (Guilandoust *et al*, 1988). As a result of adopting a different process representation, however, the implementation is slightly different. Although the input-output model based algorithm does not require the determination of an innovations sequence, there are more parameters associated in the final formulation.

INFERENTIAL ESTIMATION VIA ARTIFICIAL NEURAL NETWORKS

Rather than formulate an estimator based on a linear model, and incorporating an adaptive mechanism to deal with expected variations in process behaviour, a fixed parameter nonlinear model could be used instead. Nonlinear models can be developed from equations of continuity based upon a knowledge of the process. This approach is notoriously expensive in terms of model development time. Moreover, both the parameters and structure of the model would have to be determined. The utility of the resulting algorithm would therefore be limited as different models would have to be developed for different applications. A methodology offering a higher degree of flexibility, in terms of generality of model structure, and one which facilitates the identification of model parameters, would be preferred. Models based on artificial neural networks appear to meet these specifications.

Artificial neural networks (ANN's) are made up of highly inter-connected layers of simple 'neuron' like nodes. The neurons act as nonlinear processing elements within the network. For a given neural processing function, the accuracy of the neural network model may be influenced by altering the topology (structure) of the network. Although a number of network architectures have been proposed (see Lippmann, 1987), the 'feedforward' artificial neural network is by far the most widely applied. Indeed, Cybenko (1989) has recently shown that any continuous function can be approximated arbitrarily well on a compact set by a feedforward network, comprising two hidden layers and a fixed continuous nonlinearity. The implication of Cybenko's work is far reaching. It essentially states that a feedforward network is capable of characterising a large class of continuous nonlinear functional relationships. Furthermore, the structure of the resulting neural network models may be considered generic, since little prior formal knowledge is required in its specification. The methodology therefore promises to be a cost efficient and useful process modelling technique. In view of this, subsequent discussions will therefore concentrate on feedforward artificial neural networks (FANN's).

Feedforward Artificial Neural Networks

The architecture of a typical FANN is shown in Fig.2. The nodes in the different layers of the network represent 'neuron-like' processing elements. There is always an input and an output layer. The number of neurons in both these layers depends on the respective number of inputs and outputs being considered. In contrast, hidden layers may vary from zero to any finite number, depending on specification. The number of neurons in each hidden layer is also user specified. It is the structure of the hidden layers which essentially defines the topology of a FANN.

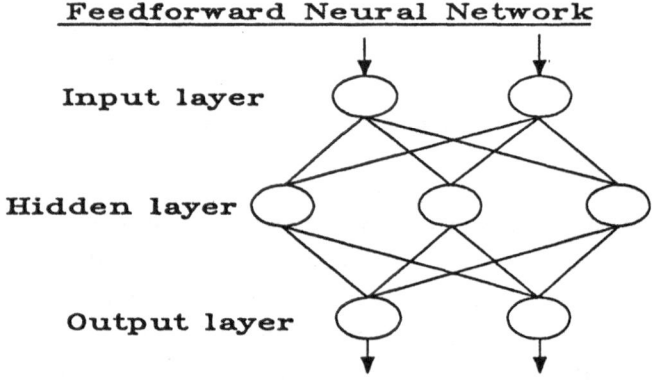

Figure 2. Architecture of Typical Feedforward Artificial Neural Network

The neurons in the input layer merely provide a means by which scaled data is introduced into the network, and do not perform data processing functions. These signals are then 'fed forward' through the network via the connections, through hidden layers, and eventually to the final output layer. Each interconnection has associated with it, a weight and a bias term which modify the strength of the signal flowing along that path. Thus, with the exception of the neurons in the input layer, inputs to each neuron is a weighted sum of the outputs from neurons in the previous layer. For example, if the information from the ith neuron in the j-1th layer, to the kth neuron in the jth layer is $I_{j-1,i}$, then the total input to the kth neuron in the jth layer is given by:

$$\alpha_{j,k} = d_{j,k} + \sum_{i=1}^{n} w_{j-1,i,k} I_{j-1,i} \qquad (16)$$

where $w_{j-1,i,k}$ is the weighting and $d_{j,k}$ is a bias term which is associated with each interconnection. The output of each node is obtained by passing the weighted sum, $\alpha_{j,k}$, through a nonlinear operator. This is typically a sigmoidal function, the simplest of which has the mathematical description:

$$I_{j,k} = 1/(1+\exp(-\alpha_{j,k})) \qquad (17)$$

and response characteristics shown in Fig. 3. Although the function given by Eq.(17) has been widely adopted, in principle, any function with a bounded derivative could be employed (Rumelhart *et al*, 1986). Nevertheless, it is interesting to note that human neuron behaviour also exhibits sigmoidal response characteristics, (Holden, 1976). Within an ANN, this function provides the network with the ability to represent nonlinear relationships. Additionally, the magnitude of bias term in Eq.(16) effectively determines the co-ordinate space of the nonlinearity. This implies that the network is also capable of characterising the structure of the nonlinearities: a highly desirable feature.

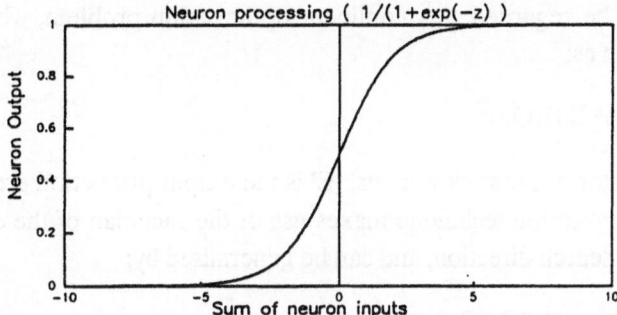

Figure 3. The Sigmoidal Function

Artificial Neural Network Model Development

In developing a process model using the neural network approach, the topology of the network must first be declared. The convention used in referring to a network with a specific topology follows that adopted by Bremmerman and Anderson (1989). For example, a FANN with 3 input neurons, 2 hidden layers with 5 and 9 neurons respectively, and 2 neurons in the output layer will referred to as a (3-5-9-2) network. A (2-10-1) network, thus refers to a FANN with 2 neurons in the input layer, 10 neurons in 1 hidden layer, and 1 neuron in the output layer.

Having specified the network topology, a set of input-output data is used to 'train' the network, ie. determine appropriate values for the weights (including the bias terms) associated with each interconnection. The data is propagated forward through the network to produce an output which is compared with the corresponding output in the data set. The resulting error is minimised by making changes to the weights and may involve many passes through the training data set. When no further decrease in error is possible, the network is assumed to have 'converged', and the latest set of weights retained as the parameters of the neural network model (NNM). Process modelling using ANN's is therefore very similar to identifying the coefficients of a parametric model of specified order. Loosely speaking, specifying the topology of an ANN is similar to specifying the 'order' of the process model. For a given topology, the magnitudes of the weights define the characteristics of the network. However, unlike conventional parametric model forms, which have an *a priori* assigned structure, the weights of an ANN also define the structural properties of the model. Thus, an ANN is capable of representing complex systems whose structural properties are unknown.

Algorithms for Network Training

As the task of weight adjustments is usually not amenable to analytical solutions, numerical search techniques are usually employed. Clearly, determining the weights of the network can be regarded as a nonlinear optimisation problem, where the objective function is written as:

$$V(\Theta,t) = \tfrac{1}{2} \Sigma \, E(\Theta,t)^2 \qquad (18)$$

Here, 'Θ' is a vector of network weights, 'E' is the output prediction error and 't' is time. The simplest optimisation technique makes use of the Jacobian of the objective function to determine the search direction, and can be generalised by:

$$\Theta^{t+1} = \Theta^t + \delta^t S \, V(\Theta,t) \qquad (19)$$

where δ^t, the 'learn rate', is a specified parameter which controls the rate of weight adjustments, and S is the identity matrix. Eq.(19) was used by Rumelhart and McClelland (1986) as the basis for their 'back-error propagation' algorithm: a distributed gradient descent search technique. In this approach, weights in the jth layer are adjusted by making use of locally available information, and a quantity which is 'back-propagated' from neurons in the $j+1^{th}$ layer. However, it is well known that steepest descent methods may be inefficient, especially when the search is in the vicinity of a minimum. Therefore, in most neural network applications, Eq.(19) is modified to include a 'momentum' term:

$$\Theta^{t+1} = \Theta^t + \delta^t S \, V(\Theta,t) + \beta(\Theta^t\text{-}\Theta^{t-1}) \quad 0 < \beta < 1 \qquad (20)$$

Thus the current change in weight is forced to be dependent upon the previous weight change, and 'ß' is a factor which is used to influence the degree of this dependence. Although the modification does yield improved performances, in training networks where there are numerous weights, gradient methods have to perform exhaustive searches and are also prone to failure. A potentially more appealing method would be a Newton-like algorithm which includes second derivative information. In this case, the matrix S in Eq.(19) would be the inverse of the Hessian. An alternative approach was proposed by Bremermann and Anderson (1989). Postulating that weight adjustments occur in a random manner, and that weight changes follow a zero-mean multivariate Gaussian distribution, their algorithm adjusts weights by adding Gaussian distributed random values to old weights. The new weights are accepted if the resulting prediction error is smaller than that recorded using the previous set of weights. This procedure is repeated until the reduction in error is negligible. They claimed that the proposed technique was 'neurobiologically more plausible'. Indeed, parallels have been drawn to bacterial motor-functions, and thus the procedure is referred to as the 'chemotaxis' algorithm.

Dynamic Neural Networks

The ANN's discussed above, merely perform a non-linear mapping between inputs and outputs, and dynamics are not inherently included within their structures. However, in many practical situations, dynamic relationships exist between inputs and outputs and thus the ANN's will fail to capture the essential characteristics of the system. Although, dynamics can be introduced by making use of time histories of the data, a rather more elegant approach is inspired by analogies with biological systems. Studies by Holden (1976) suggest that dynamic behaviour is an essential element of the neural processing function. It has also been suggested that a first-order low-pass filter may provide the appropriate representation of the dynamic characteristics (Terzuolo *et al*, 1969). The introduction of these filters is relatively straightforward, with the output of the neuron being transformed in the following manner:

$$y^f(t) = \Omega y^f(t-1) + (1-\Omega)y(t) \qquad 0 \leq \Omega \leq 1 \qquad (21)$$

where $y^f(t)$ is the filtered value of $y(t)$. Suitable values of filter time constants cannot be specified *a priori*, and thus the problem becomes one of determining 'Ω' in conjunction with the network weights. Another appealing feature of the chemotaxis approach is that the algorithm does not require modification to enable incorporation of filter parameters: the filter time constants are determined in the same manner as network weights. A particular instance where the use of a time history would still be appropriate is when uncertainty can exist over system time delays. The use of input data over the range of expected dead-times could serve to compensate for this uncertainty. In this situation a 'limited' time history together with neuron dynamics may be appropriate.

RESULTS

The potential of adaptive inferential estimators has been demonstrated using linear and non-linear simulation studies (Guilandoust *et al*, 1987,1988), and in applications to industrial processes (Tham *et al*, 1989). In this article, a selection of results from applications to industrial data will be presented to highlight the performance capabilities of both estimator design approaches. Due to commercial confidentiality, the ordinates the graphs will be presented without annotations.

Biomass Estimation in Continuous Mycelial Fermentation

Linear adaptive estimation: In this application, biomass concentration is the primary control variable. Due to limitations in current sensor technology, on-line measurements are not possible and biomass concentrations are obtained via laboratory analysis. The assays are only made available to the process operators at four hourly intervals, which is inadequate for effective control. This problem would be alleviated if the process operators were provided with more frequent information. Fortunately, a number of other process measurements such as dilution rate, carbon dioxide evolution rate (CER), oxygen uptake rate (OUR), and alkali addition rate, provide useful information pertaining to biomass concentration. If the complex relationships between these variables and biomass concentration can be established, then an estimator can be designed and used to infer biomass concentrations at a frequency suitable for control purposes. Although other variables, eg. pH, temperature etc., affect the nonlinear relationship, tight environmental regulation maintains these at an approximate steady state.

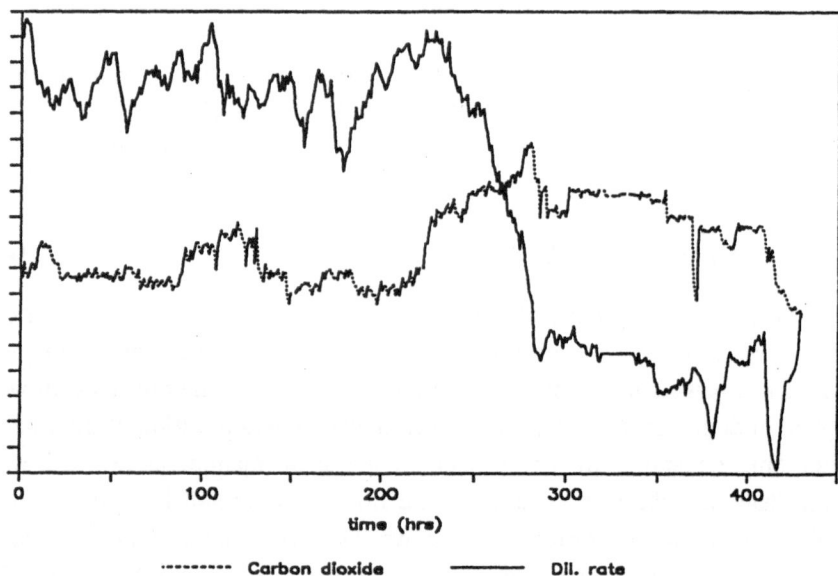

Figure 4. Dilution Rate and CO_2 Evolution Rate

The first set of results were obtained by applying the adaptive linear estimator, making use of CER and dilution rate, as the secondary output and manipulative input respectively. Estimator parameters were obtained on-line using the UD-factored recursive least-squares algorithm (eg. see Ljung and Soderstrom, 1983). CER and dilution rate (shown in Fig.4), were both sampled at 1 hr. intervals. Note that a step change in the fermenter dilution rate has been applied at approximately 200 hrs. The estimates of biomass, calculated every hour, together with off-line analysis results (step-like responses), are shown in Fig.5.

As can be observed, the linear adaptive estimator was able to provide good estimates of biomass concentrations. Notice in addition, that the estimates 'lead' the off-line biomass assays, demonstrating the predictive nature of the algorithm. At approximately 240 hrs., there was a sharp deviation from the downward trend in biomass assay. However, corresponding changes in the trends of either the CER or dilution rate responses were not observed. It is encouraging to note that the estimates were not influenced by this probable error in biomass assay. At approximately 280 hrs., there was a noticeable 'undershoot' in the estimates. This can be attributed to the estimator having to 'tune-in' to the new operating conditions.

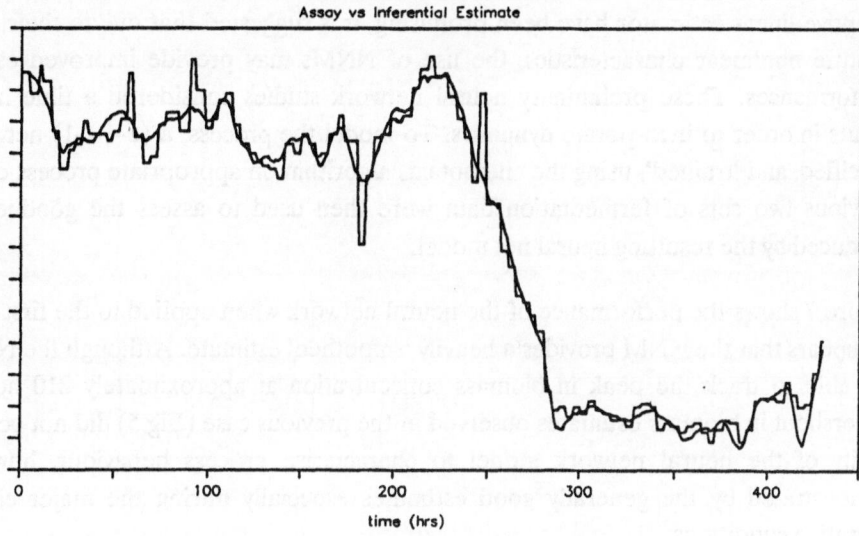

Figure 5. Measured and Estimated Biomass using the Linear
Adaptive Estimator (Mycelial Fermentation Data Set 1)

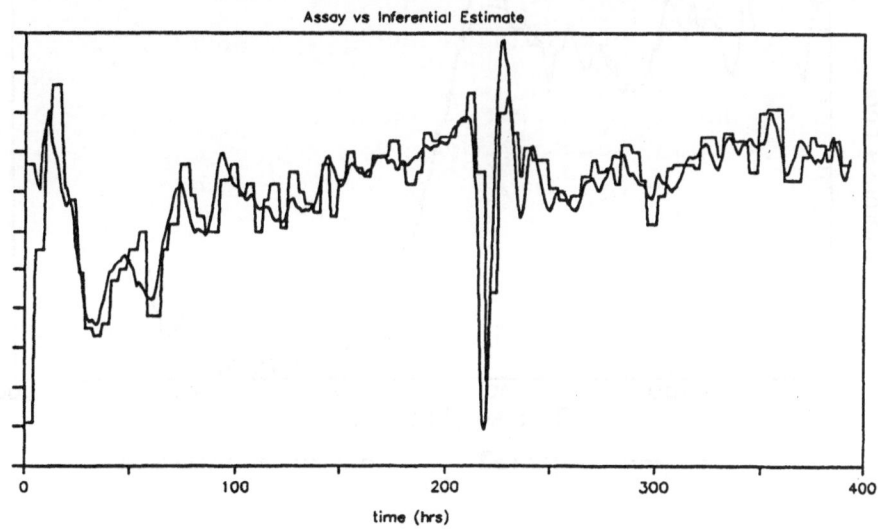

Figure 6. Measured and Estimated Biomass using the Linear
Adaptive Estimator (Mycelial Fermentation Data Set 2)

Figure 6 shows another application to the same process. Once more, good estimator performance was achieved. In particular, note that the estimator was able to predict the large deviation in biomass response at approximately 220 hrs. An overshoot in the estimate occurred when the biomass returned to near its normal operating region. Due to the tuning effect, overshoots in estimates are commonly encountered in the application of adaptive estimators, especially after the onset of large process transients.

<u>**Estimation using neural network based estimators**</u>: Although results obtained using the adaptive linear estimator have been promising, it is suggested that due to their ability to capture nonlinear characteristics, the use of NNMs may provide improved estimation performances. These preliminary neural network studies considered a time history of inputs in order to incorporate dynamics. To model the process, a (6-4-4-1) network was specified, and 'trained', using the chemotaxis algorithm on appropriate process data. The previous two sets of fermentation data were then used to assess the goodness of fit produced by the resulting neural net model.

Figure 7 shows the performance of the neural network when applied to the first data set. It appears that the NNM provides a heavily 'smoothed' estimate. Although the NNM was not able to track the peak in biomass concentration at approximately 210 hours, the undershoot in biomass estimates observed in the previous case (Fig.5) did not occur. The ability of the neural network model to characterise process behaviour, however, is demonstrated by the generally good estimates especially during the major change in operating conditions.

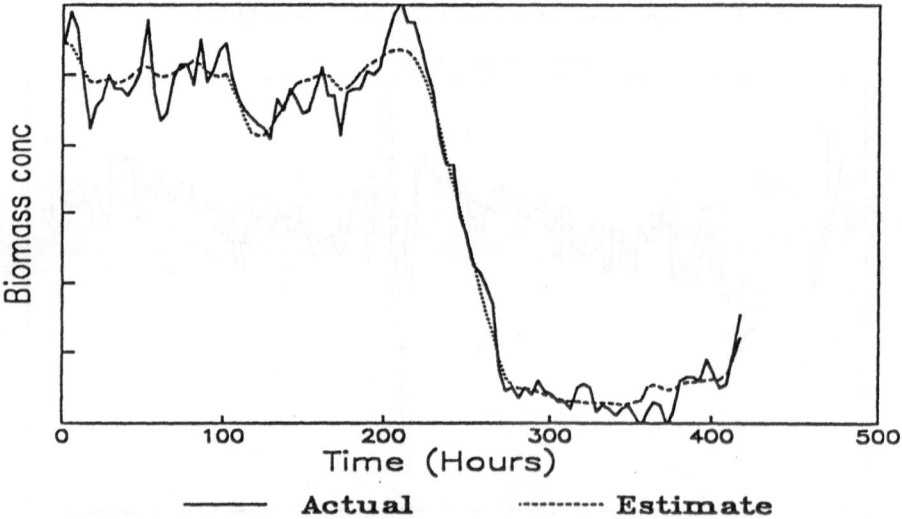

Figure 7. Performance of Neural Network Model Based Estimator (Mycelial Fermentation Data Set 1)

The performance of the NNM when applied to the second data set is shown in Fig.8. The overshoot observed when the linear adaptive estimator was applied, did not occur.

However, the transients during the early part of the response were slightly poorer when compared to that obtained using the linear adaptive estimator (see Fig.6). Nevertheless, good overall biomass estimation was again achieved.

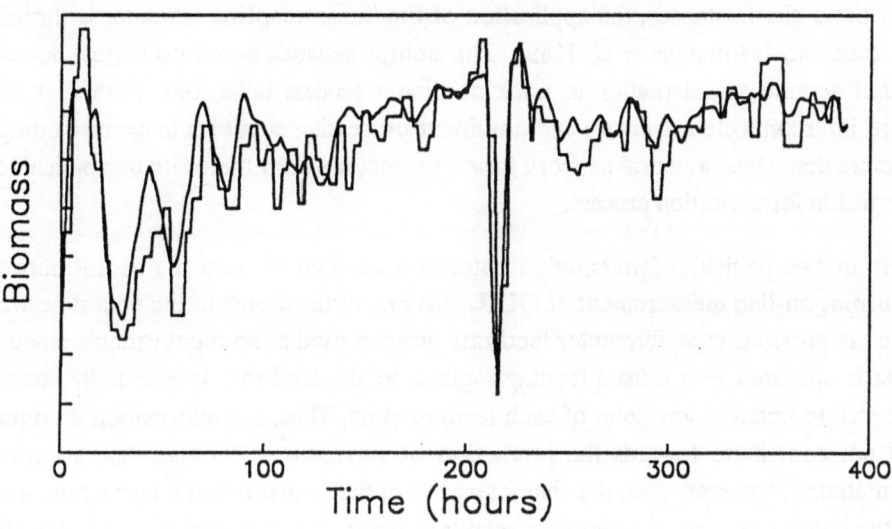

Figure 8. Performance of Neural Network Model Based Estimator (Mycelial Fermentation Data Set 2)

Biomass Estimation in Fed-batch Penicillin Fermentation

The measurement and control difficulties encountered in the production of penicillin are similar to those encountered in the continuous fermentation system described above. The growth rate of the biomass has to be controlled to a low level in order to optimise penicillin production. Although the growth rate can be influenced by manipulating the rate of substrate (feed) addition, it is not possible to measure the rate of growth on-line. The aim is, therefore, to develop a model which relates an on-line measurement to the rate of biomass growth. In this case, the OUR was used as the on-line measured variable, although the CER could have provided the same information. This is because the ratio of CER to OUR, known as the respiratory quotient (RQ), remains constant after approximately 20 hrs. into the fermentation.

The continuous fermentation process considered previously normally operates around a steady state. Thus an adaptive linear estimator may prove sufficient in the vicinity of normal process operating conditions. In contrast, the fed-batch nature of penicillin fermentation presents a more difficult modelling problem since the system passes through a wide spectrum of dynamic characteristics, never achieving a steady state. Moreover, the assays of biomass are available only at 24 hourly intervals. Thus, due to the limited number of primary output assays during the course of a fermentation run, an adaptive algorithm would not have sufficient time adjust to changing process behaviour. Under these circumstances, the application of the linear adaptive estimator is therefore not practicable (Montague *et al*, 1989). The neural network based estimator, however, does not depend on adaptation to track nonlinear process behaviour. Rather, it relies upon its interconnected structure and nonlinear processing elements to capture complex characteristics. Thus, a neural network based estimator would therefore be applicable to the penicillin fermentation process.

Data from two penicillin fermentation batches were used to train the neural network. The current on-line measurement of OUR, was one of the inputs to the neural network. Unlike the previous case, fermenter feed rate was not used as an input variable since the process is operated with a fixed feeding regime, ie. the feed rate is essentially constant from batch to batch at any point of each fermentation. Thus, the information it contains would not contribute towards the prediction of variations in biomass levels between fermentations. However, since the characteristics of the fermentation is also a function of time, the batch time was considered a pertinent input. The specified network therefore had a (2-3-1) topology. Again the chemotaxis algorithm was used for determining network weights. Figures 9 and 10 show the performance of the neural network estimator when applied to the 2 training data sets. As expected, good estimates of biomass were achieved.

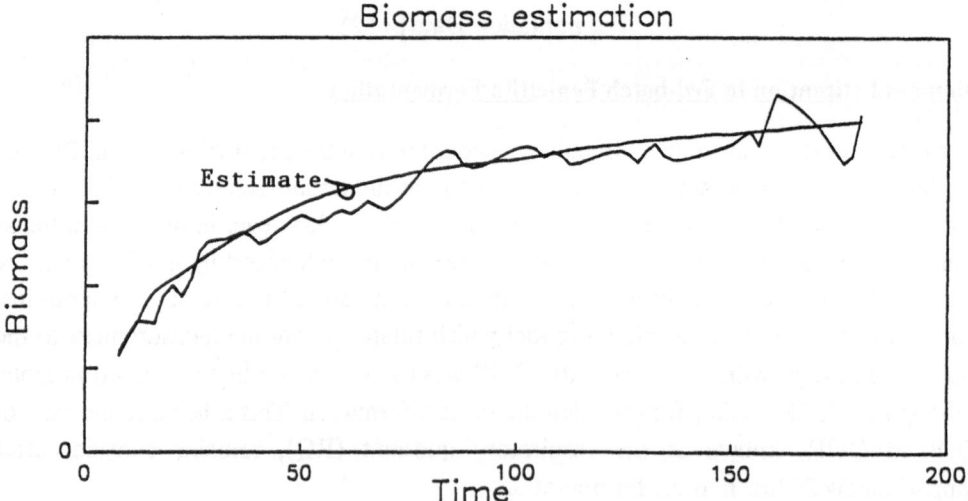

Figure 9. Neural Network Estimation of Biomass in Penicillin Fermentation
(Training Data Set 1)

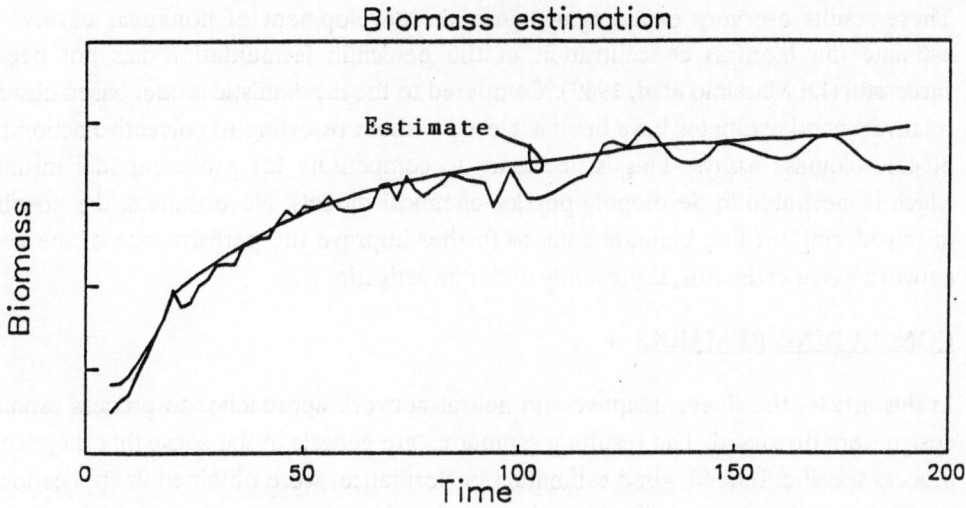

Figure 10. Neural Network Estimation of Biomass in Penicillin Fermentation
(Training Data Set 2)

OUR data from another fermentation batch was then introduced to the NNM, resulting in the estimates shown in Fig.11. It can be observed that the estimates produced by the NNM is very acceptable, and may be considered to be almost as good as those observed in Figs.9 and 10.

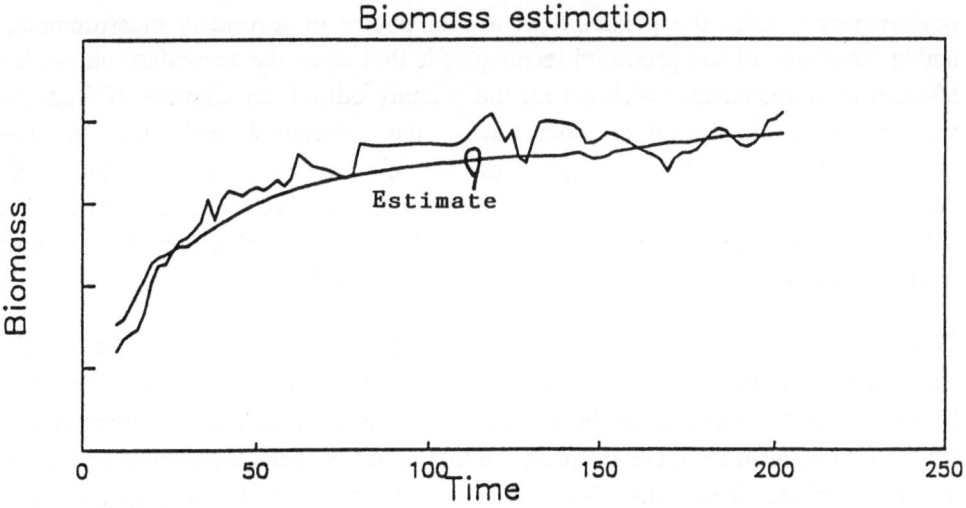

Figure 11. Neural Network Estimation of Biomass in Penicillin Fermentation
(Data Set 3)

These results are very encouraging, since the development of nonlinear observers to estimate the biomass concentration of the penicillin fermentation has not been as successful (Di Massimo *et al*, 1989). Compared to the mechanistic model based observer, relatively good estimates have been achieved without resorting to corrective action from off-line biomass assays. This is necessary to compensate for process-model mismatch which is inevitable in developing physico-chemical models. Nevertheless, the possibility of introducing off-line biomass data, to further improve the performance of the neural network based estimator, is presently under investigation.

CONCLUDING REMARKS

In this article, the linear adaptive and neural network approaches to process estimator design were discussed. The resulting estimators are generic in the sense that they are not process specific. Indeed, good estimation performances were obtained in applications to processes as diverse as distillation columns, chemical reactors, polymerisation processes and bioreactors (Montague *et al*, 1989; Tham *et al*, 1989; Willis *et al*, 1990). Both techniques therefore exhibit potential as 'soft-sensors', ie. sensors based on software rather than hardware.

In the previous references, it was also shown that the use of the inferred estimates for feedback control yielded significant improvements in closed loop performances. The ability to estimate the primary output accurately at the (faster) secondary output sampling frequency, means that disturbance effects can be quickly detected. Using these estimates for control will therefore result in significant improvements in control performance despite the possibility of a poor choice of secondary measurement. An added advantage of the proposed techniques is that since the secondary output is also affected by disturbances which act on the primary output, an element of feedforward information is incorporated. Thus, using the inferential estimates for control automatically incorporates a degree of feedforward compensation: an especially desirable property for efficient disturbance rejection. Moreover, if an accurate neural network model is available, it can be used directly within a predictive control framework, to effect nonlinear model based control (Willis *et al*, 1990).

These results indicate that artificial neural network technology can be a suitable alternative to current adaptive approaches to both estimator and controller designs. However, it is stressed that the field is still very much in its infancy and many questions still have to be answered. Determining the 'optimum' network topology is one example. Whilst two hidden layers have been claimed to be sufficient to approximate any non-linear function, for some data sets, one hidden layer may be adequate to describe the system behaviour. Therefore the approach adopted in the applications described above, was to commence with a relatively small network and progressively increase both the number of hidden layers, and the neurons within these layers, until no significant improvement in fit was observed on the training data sets. Admittedly, this is an *ad hoc* procedure. Although present theory does not allow a more rigourous mathematical approach, recent work by Wang (1990) has indicated that it is possible to develop procedures which aid in specification of neural network topologies.

The mechanisms of the linear adaptive technique and that based on neural networks are clearly different. With the adaptive algorithm, the underlying model is linear. This approach therefore depends on the on-line identification of model parameters to track nonlinear and slowly time-varying characteristics. The neural network approach, on the other hand, relies on the ability of the network topology to capture complex process characteristics. Thus, the latter technique may be considered to be more robust in that its implementation is not subject to those problems normally associated with on-line parameter estimation. However, if the neural network model has not been trained on a representative set of data, then the resulting estimator will not perform well. Although both algorithms have shown themselves to be resilient to noisy and uncertain process data, it is possible that steady state estimation errors can occur with the neural network based estimator. Due to constant adaptation, the linear adaptive approach does not suffer from this deficiency. An on-line network training procedure would overcome this problem, provided the observed robustness properties of neural networks are not compromised.

There also appears to be no established methodology for determining the stability of artificial neural networks. This is perhaps the most important issue that has to be addressed before the full potentials of the methodology can be realised on-line. Nevertheless, given the resources and effort that are currently being infused into both academic and commercial research in this area, it is predicted that within the decade, neural networks will have established itself as a valuable tool for solving many estimation and control problems.

ACKNOWLEDGEMENTS

The authors gratefully acknowledge the support of the Department of Chemical and Process Engnrg., Uni. of Newcastle upon Tyne; SmithKline-Beecham, Irvine; ICI Engnrg.and ICI Bioproducts.

REFERENCES

Anderson, B.D.O. and Moore, J.B., (1979). Optimal Filtering, Prentice-Hall.

Bremermann, H.J. and Anderson, R.W. (1989). 'An alternative to Back-Propagation: a simple rule for synaptic modification for neural net training and memory', Internal Report, Dept. of Mathematics, Uni. of California, Berkeley.

Cybenko, G. (1989). 'Continuous value neural networks with two hidden layers are sufficient, Internal report, Dept. of Comp. Sci. Tufts Univ. Medford.

Di Massimo, C., Saunders, A.C.G., Morris, A.J. and Montague, G.A. (1989). 'Non-linear estimation and control of Mycelial fermentations', ACC, Pittsburgh, USA, pp.1994-1999.

Guilandoust, M.T., Morris, A.J. and Tham, M.T. (1987). 'Adaptive Inferential Control'. Proc.IEE, Vol 134, Pt.D.

Guilandoust, M.T., Morris, A.J. and Tham, M.T. (1988). 'An adaptive estimation algorithm for inferential control', Ind. Eng.Chem. and Res., 27, pp 1658-1664.

Holden, A.V. (1976). Models of the Stochastic Activity of Neurones, Springer Verlag.

Lippmann, R.P. (1987). 'An Introduction to Computing with Neural Nets', IEEE ASSP Magazine, April.

Ljung, L. and Soderstrom, T. (1983). Theory and Practice of Recursive Identification', MIT Press.

Luyben, W.L. (1973). 'Parallel Cascade Control', Ind. & Eng. Chem. Fundam., 12, 463.

Montague, G.A., Hofland, A.G., Lant, P.A., Di Massimo, C., Saunders, A., Tham, M.T. and Morris, A.J. (1989). 'Model based estimation and control: Adaptive filtering, Nonlinear observers and Neural Networks', Proc. 3rd Int. Symp. 'Control for Profit', Newcastle-upon-Tyne.

Rumelhart, D.E. and McClelland, J.L. (1986). Parallel Distributed Processing: Explorations in the Microstructure of Cognition. Vol.1: Foundations, MIT Press, Cambridge.

Rumelhart, D.E., Hinton, G.E. and Williams, R.J. (1986). 'Learning representations by back-propagating errors', Nature, 323, pp 533-536.

Terzuolo, C.A., McKeen, T.A., Poppele, R.E. and Rosenthal, N.P. (1969). 'Impulse trains, coding and decoding'. In Systems analysis to neurophysiological problems, Ed. Terzuolo, University of Minnesota, Minneapolis, pp86-91.

Tham, M.T., Morris, A.J. and Montague, G.A. (1989). 'Soft sensing: a solution to the problem of measurement delays', Chem. Eng. Res. and Des., 67, 6, pp 547-554.

Wang, J. (1990). 'Multi-layered neural networks:approximated canonical decomposition of non-linearity', Internal report, Dept. of Chemical and Process Eng., University of Newcastle-upon-Tyne.

Willis, M.J., Di Massimo, C., Montague, G.A., Tham, M.T. and Morris, A.J. (1990). 'On the applicability of neural networks in chemical process control', AIChE Annual Meeting, Chicago, Nov.

TOWARDS MORE CREDIBLE SELF-TUNERS ...

Josef Böhm and Rudolf Kulhavý

Institute of Information Theory and Automation,
Czechoslovak Academy of Sciences
182 08 Prague 8, Czechoslovakia

1 Introduction

After more than two decades of intensive research in *self-tuning control theory*, one can say that the discipline has matured. Theory evolved from simple minimum variance algorithms to more sophisticated GPC and LQG ones. A lot of qualitative features have been understood. Both steady-state and transient behaviours have been analysed in detail. Nevertheless, theoretical achievements have not penetrated the target market to the extent expected originally. Academic self-tuners are still recognized by practicioners as *little credible* for various reasons. The number of full-scale applications does not correspond to the possibilities and the effort made.

The purpose of this contribution is to recapitulate a complete line (by far not a unique one) of LQG self-tuning control design, evaluating critically its *questionable points*. These points are of very different nature — they are related mainly to the following demands:

- *adequateness* of the used theoretical framework that should be powerful enough to formulate most of the design demands in mathematical terms and explicit to the extent not to hide limits of its applicability;

- *justifiability* of the assumptions specializing the general theory and of necessary approximations of the ideal solution, both motivated by feasibility constraints;

- *effectiveness* of robustifying measures necessary to cope with violations of the theoretical assumptions, namely with the presence of outliers, nonlinearities, higher-order unmodelled dynamics and input constraints;

- *availability* of numerically safe and efficient algorithms for identification and control;

- *simplicity* of the resulting self-tuner that can be affected by minimizing the number of user choices ("tuning knobs"), by automating the prior-adjustment process as well as by systematic education of new control engineers.

More than in making a sufficiently complete survey of the present state of the art we are interested in thinking about future. What are the crucial bottle-necks on the road between theory and practice? Which problems should be attacked in the near future horizon (say, 3–5 years) to achieve essential progress in approaching the practical needs? Answers or at least discussion to these questions should help to direct next research.

Needless to add, this paper brings a quite subjective view of the problem. We support our evaluations and conclusions by extensive simulation experience, frequent consultations with practising engineers and collaboration in a couple of pilot as well as full-scale applications.

2   From General Theory to Feasible Algorithms

This section sketches a possible line of design of self-tuning control. The aim of this reminder is to support our analysis of questionable and worth-attacking points. For this reason, the explanation is intentionally very concise. We are omitting technical details and generalities. Especially we do not treat the most general versions of the ARX model including external disturbances, time delays and/or absolute term because generalization of the following derivations to such models is more or less straightforward and resolved — as such it is not the topic of this account.

2.1   Conceptual Solution

We shall consider a stochastic system on which input $u(\cdot)$ and output $y(\cdot)$ (both possibly multivariate) are measured at discrete time instants labelled $t = 1, 2, \ldots$ Let us emphasize that we adopt the notation convention that $u_{(t)}$ precedes $y_{(t)}$. We denote the pair of both quantities representing *data* measured at the time t through

$$d_{(t)} = (u_{(t)}, y_{(t)}) \ .$$

The collection of all data within an interval $(t+1, t+N)$ is denoted by

$$d_{(t+1)}^{(t+N)} = (d_{(t+1)}, \ldots, d_{(t+N)}) \ .$$

Moreover, we shall use the notation $d^{(t)}$ as a short-cut for $d_{(1)}^{(t)}$. By definition, $d^{(0)}$ is treated as empty.

Choice of a Loss Function. The optimal control design requires to quantify the designer's idea of the desired closed-loop behaviour within the current planning interval $(t+1, t+N)$. One way to do it is to define the "ideal" inputs $u_{0(t+1)}, \ldots, u_{0(t+N)}$ and outputs $y_{0(t+1)}, \ldots, y_{0(t+N)}$ and then to assign to any possible values $u_{(t+1)}, \ldots, u_{(t+N)}$ and $y_{(t+1)}, \ldots, y_{(t+N)}$ a scalar loss $J_{(t)}(1, N)$ that will measure the "distance" of the actual signals from the ideal ones. We restrict ourselves to the loss function in the form

$$J_{(t)}(1, N) = J_{(t)}(d_{(t+1)}^{(t+N)}) = \sum_{k=1}^{N} \left[q_{y(t+k)}(y_{(t+k)} - y_{0(t+k)}) + q_{u(t+k)}(u_{(t+k)} - u_{0(t+k)}) \right] \quad (1)$$

where the functions $q_{y(t+k)}(\cdot)$, $q_{u(t+k)}(\cdot)$ assign specific, generally time-varying, losses to output and input deviations, respectively. The additive form is motivated mainly by mathematical tractability but it is flexible enough to express the designer's ordering.

Handling of Uncertainty. Comparing the loss function values is not sufficient to order control strategies *a priori* in the design stage: random influences and/or incomplete knowledge of the system description admit to evaluate the loss *ex post* only. Therefore, we have to specify the optimal $u_{(t+k)}$, $k = 1, \ldots, N$ as a function of past available data (here the feedback property appears).

The resulting control strategy is described by the mapping

$$L_{(t)} : d^{(t+k-1)} \mapsto u_{(t+k)}, \quad k = 1, 2, \ldots, N .$$

A rationally justified way to make different control strategies a priori comparable is to adopt the expectation of the loss function (see Kárný (1985) for a more detailed discussion)

$$\mathsf{E}[J_{(t)}(1, N)] = \int J_{(t)}(d_{(t+1)}^{(t+N)}) \, p(d_{(t+1)}^{(t+N)} | d^{(t)}) \, \mathrm{d} d_{(t+1)}^{(t+N)} . \tag{2}$$

Here $p(d_{(t+1)}^{(t+N)} | d^{(t)})$ denotes the conditional probability density function (c.p.d.f.) of data $d_{(t+1)}^{(t+N)}$ supposing previous data $d^{(t)}$ to be known. The design problem then takes the form

$$L_{(t)}^*(\cdot) = \arg \min_{L_{(t)}(\cdot)} \mathsf{E}[J_{(t)}(1, N)] . \tag{3}$$

Solution via Dynamic Programming. It can be proved [see Ullrich (1964)] that even in the stochastic case the optimal strategy can be determined by deterministic functions $L_{(t)}(\cdot)$. The application of the principle of dynamic programming leads to the recursion

$$K_{(t)}^*(k, N | k - 1) = \min_{u_{(t)}} \mathsf{E} \left[q_{y(t+k)}(y_{(t+k)} - y_{0(t+k)}) + q_{u(t+k)}(u_{(t+k)} - u_{0(t+k)}) \right.$$
$$\left. + K_{(t)}^*(k + 1, N | k) \mid u_{(t+k)}, d^{(t+k-1)} \right]$$
$$\text{for } k = N, N - 1, \ldots, 1 \tag{4}$$

starting with

$$K_{(t)}^*(N + 1, N | N) = 0 . \tag{5}$$

The notation $\mathsf{E}[\cdot | u_{(t+k)}, d^{(t+k-1)}]$ means the conditional expectation. $K_{(t)}^*(k, N | k - 1)$ denotes the optimal cost-to-go. The minimal value of the criterion coincides with the last cost-to-go

$$\min \mathsf{E}[J_{(t)}(1, N)] = K_{(t)}^*(1, N | 0) .$$

Needed Information about System. The "expectation step" of the dynamic programming requires to specify the corresponding *predictive* c.p.d.f.'s

$$p(y_{(t+k)} | u_{(t+k)}, d^{(t+k-1)}) \quad \text{for } k = 1, 2, \ldots, N . \tag{6}$$

Such a degree of knowledge of the controlled system is, of course, exceptional in practice. To express the designer's uncertainty about the appropriate choice of (6), we consider a set of models θ. Notice that the *subjective* uncertainty becomes at the moment a part of the stochastic nature of the system.

Thus, a family of models $\{\theta\}$, possibly varying slowly in time, that define

$$p(y_{(t)} | u_{(t)}, d^{(t-1)}, \theta_{(t)}) \quad \text{for all } t \tag{7}$$

is selected. The controller is designed using the current state of knowledge about possible θ.

Bayes Parametric Inference. The self-tuning controller has a single source of knowledge about models θ — observed data. In probabilistic terms, it fulfils the "natural conditions of control" [Peterka (1981)]

$$p(u_{(t)}|d^{(t-1)}, \theta_{(t)}) = p(u_{(t)}|d^{(t-1)}) . \tag{8}$$

This fact makes Bayes inference substantially easier. The needed predictive c.p.d.f. is related to a "parametrized" c.p.d.f. by the formula

$$p(y_{(t)}|u_{(t)}, d^{(t-1)}) = \int p(y_{(t)}|u_{(t)}, d^{(t-1)}, \theta_{(t)}) \, p(\theta_{(t)}|d^{(t-1)}) \, d\theta_{(t)} . \tag{9}$$

The c.p.d.f. $p(\theta_{(t)}|d^{(t-1)})$, describing the designer's uncertainty about $\theta_{(t)}$ with data $d^{(t-1)}$ at disposal, evolves according to the formula

$$p(\theta_{(t)}|d^{(t)}) = \frac{p(y_{(t)}|u_{(t)}, d^{(t-1)}, \theta_{(t)}) \, p(\theta_{(t)}|d^{(t-1)})}{p(y_{(t)}|u_{(t)}, d^{(t-1)})} \tag{10}$$

that follows directly from the Bayes rule and the natural conditions of control (8).

To complete the recursion, the designer has to supply a probabilistic model of parameter variations $p(\theta_{(t+1)}|d^{(t)}, \theta_{(t)})$

$$p(\theta_{(t+1)}|d^{(t)}) = \int p(\theta_{(t+1)}|d^{(t)}, \theta_{(t)}) \, p(\theta_{(t)}|d^{(t)}) \, d\theta_{(t)} . \tag{11}$$

Finally, the designer has to specify his prior uncertainty about possible models through the p.d.f. $p(\theta_{(1)}|d^{(0)}) = p(\theta_{(1)})$.

2.2 Specializing Assumptions

Designing the control strategy we have to respect the finite computational speed and finite memory of our computational facilities. We have to cope with *feasibility constraints* done by the need

1. to realize the Bayes parametric inference in a closed manner using data statistics of a finite and fixed dimension;

2. to execute each step of the dynamic programming by the same algorithm of a fixed complexity.

The choice of the underlying system model and loss function influences complexity of the resulting self-tuner dramatically.

Choice of Model. To fulfil the first constraint, we assume the dependence of the system output on available data described by a *linear-in-parameters regression of a finite order* with a *normally distributed white-noise stochastic component*. These assumptions imply the following form of the system model

$$y_{(t)} = \theta' \phi_{(t)} + e_{(t)}, \quad e_{(t)} \sim N(0, R_e) \tag{12}$$

where θ denotes the vector of regression coefficients, $\phi_{(t)}$ is a known vector function of $u_{(t)}$ and $d^{(t-1)}$, and $e_{(t)}$ stands for the stochastic (unpredictable) component. For simplicity, the covariance R_e is assumed known here (see Kárný *et al* (1985) for generalization to the case of unknown covariance).

Under the above assumptions, the Bayes parametric inference reduces to the evolution of self-reproducing normal c.p.d.f.'s described unambiguously by the mean value $\hat{\theta}$ and covariance P of the unknown parameters θ that are updated by data via notorically known *recursive least squares*

$$\hat{\theta}_{(t|t)} = \hat{\theta}_{(t|t-1)} + \frac{P_{(t|t-1)}\phi_{(t)}}{1 + \phi'_{(t)}P_{(t|t-1)}\phi_{(t)}} \hat{\varepsilon}'_{(t|t-1)} , \tag{13}$$

$$P^{-1}_{(t|t)} = P^{-1}_{(t|t-1)} + \phi_{(t)}\phi'_{(t)} . \tag{14}$$

Here $\hat{\varepsilon}$ denotes the prediction error that is defined by the formula

$$\hat{\varepsilon}_{(t|t-1)} = y_{(t)} - \hat{\theta}'_{(t|t-1)}\phi_{(t)} . \tag{15}$$

Assuming moreover the model of parameter variations e.g. of a random walk type

$$\theta_{(t+1)} = \theta_{(t)} + v_{(t+1)} , \quad v_{(t+1)} \sim \mathrm{N}(0, R_v) , \tag{16}$$

we could complete the recursion by the formulae

$$\hat{\theta}_{(t+1|t)} = \hat{\theta}_{(t|t)} , \tag{17}$$

$$P_{(t+1|t)} = P_{(t|t)} + R_v . \tag{18}$$

The above assumptions have good physical motivation. The linear-in-parameters model can be interpreted as a linearization of the actual system behaviour around the current operating point. Since the reality is always more complicated, we admit (relatively slow) parameter variations in time. The use of an explicit model of parameter variations ensures tracking of slow changes of dynamics. Finally, the recursive least squares have proved to be weakly sensitive to the assumption of normality of $e_{(t)}$. Many other probability distributions, including the uniform one, can be well approximated by the normal distribution.

Choice of Loss Function. Any sufficiently smooth loss function can be approximated (near its minimum) by a quadratic form. Thus, a *quadratic form of the loss function*

$$J(1, N) = \mathsf{E} \sum_{t=1}^{N} \left[(y_{(t)} - y_{0(t)})'Q_{y(t)}(y_{(t)} - y_{0(t)}) + (u_{(t)} - u_{0(t)})'Q_{u(t)}(u_{(t)} - u_{0(t)}) \right] \tag{19}$$

may be considered as the simplest choice from the mathematical point of view. High flexibility is ensured by admitting time-variable penalization matrices $Q_{y(t)}$ and $Q_{u(t)}$.

Note that to evaluate the expectation of a quadratic form we need to know just the first two moments of θ. This perfectly corresponds with the normality of the c.p.d.f.'s $p(\theta_{(t)}|d^{(t-1)})$ induced by the choice of the model (recall the fact that the normal distribution have the maximal Shannon entropy from all distributions with fixed first two moments).

Altogether, the above assumptions defines the simplest possible specialization of the adaptive control problem for dynamic systems [cf. explanation in Kumar(1985)].

2.3 Feasible Approximations

Unfortunately, this specialization is still insufficient to make the solution of control design feasible. The problem is to perform the computation of the c.p.d.f.'s (6) over the whole planning horizon $(t + 1, t + N)$. This task is so time and memory consuming that it is not realizable using current computational facilities. Thus, we are forced to accept some approximation of the ideal solution.

Suboptimal (Non-Dual) Strategies.   There are two main ways to get a feasible solution.

1. *Cautious Strategy:* We suppose that information about parameters $\theta_{(t+k)}$ and, consequently, the predictive c.p.d.f.'s $p(y_{(t+k)}|u_{(t+k)}, d^{(t+k-1)}))$ remains unchanged during the planning interval $(t + 1, t + N)$ in the sense

$$p(y_{(t+k+1)}|u_{(t+k+1)}, d^{(t+k)}) = p(y_{(t+1)}|u_{(t+1)}, d^{(t)}) \quad \text{for } k = 1, 2, \ldots, N - 1 . \quad (20)$$

   The advantage of this strategy is that uncertainty of estimated parameters is automatically respected. The uncertainty is projected onto the increased covariance of the predicted output that leads to an important damping of control actions especially in the initial stage of the self-tuner operation. Unfortunately, poor excitation conditions deteriorate the quality of identification and the strategy may produce an unsatisfactory behaviour for a long period.

2. *Certainty Equivalence Strategy:* An alternative is to neglect uncertainty of parameter estimates completely

$$p(\theta_{(t+1)}|d^{(t)}) = \delta(\theta_{(t+1)} - \hat{\theta}_{(t+1)}) \quad (21)$$

   that results in

$$p(y_{(t+k+1)}|u_{(t+k+1)}, d^{(t+k)}) = p(y_{(t+1)}|u_{(t+1)}, d^{(t)}, \hat{\theta}_{(t+1)}) \quad \text{for } k = 1, 2, \ldots, N - 1.$$
$$(22)$$

   Obviously, this simplification is justified in case of good system excitation when we can rely on the "point" estimates of θ.

Use of a Shorter Horizon.   Freezing the knowledge about the system for the whole planning interval reduces the credibility of the designed control strategy in far future. It is why the whole optimization should be repeated, optimally whenever new parameter estimates are obtained. This measure augments significantly the computational burden especially when the horizon N is large. Nore that our formulation does not guarantee stability of the closed loop explicitly. Theoretically, to ensure stability requires to go with N to infinity. There are two basic ways of avoiding the infinite-horizon solution.

1. *Receding Horizon Strategy:* It was proved that the stability can be guaranteed even for a finite-horizon criterion provided a special terminal penalization is applied [cf. Kárný *et al* (1985), Poubelle *et al* (1988), Bitmead *et al* (1989)].

2. *Iterations-Spread-in-Time Strategy:* When making an effort to reduce the computational burden, we can try to use the criterion even with $N = 1$. To ensure stable control, we have to select the terminal penalization carefully. A good choice is to use the kernel of the optimal cost-to-go $K^*_{(t)}$ from the previous sampling period. In such a way we spread iterations that would be performed normally in one shot over more sampling intervals. The following features have been observed:

- For time-invariant parameters there is no problem and the design converges to the steady-state solution.

- For time-varying parameters the stability condition must be ensured by modifying the kernel of the optimal cost-to-go $K^*_{(t)}$ — either by restarting from a stabilizing initial condition, or by incorporating an additional penalization.

- The strategy can be interpreted as one-step-ahead with all related advantages (especially simplicity and the possibility to use data dependent penalizations, or to respect signal constraints).

- In case of fast parameter changes a slower tracking of the optimum control must be expected.

2.4   Robustifying Tricks

The assumptions and approximations chosen with the main goal to make the solution feasible are far from being quite realistic and optimal properties of the controller can be guaranteed only for a rather limited class of problems. Any violation of the above assumptions leads to a deterioration of the overall behaviour and sometimes results in a quite unsatisfactory (even unstable) operation. To ensure an admissible behaviour even in this case, we have to use some robustifying tricks. Unfortunately, many of robust addenda commonly used today are based on witty tricks, *ad hoc* procedures, heuristic decomposition of a complex problem into a series of simpler ones, and pragmatic combination of very different approaches. This makes theoretical analysis of the self-tuner properties very difficult. These are some of the most frequent measures.

Coping with Unknown Model of Parameter Variations.   Although the explicit modelling of parameter variations is conceptually the only consistent way to achieve tracking of slowly-varying parameters, there is a common practice to use more or less heuristic methods of forgetting obsolete information instead. The main reason is the attempt to avoid specification of the model of parameter changes that is usually, at least partially, unknown to the designer. For tracking parameters of the model (12), we use the technique of *directional forgetting* [see Kulhavý and Kárný (1984), Kulhavý (1987), cf. Hägglund (1983)] that results in a modified evolution of the covariance matrix of regression coefficients

$$P^{-1}_{(t+1|t)} = P^{-1}_{(t|t)} + w_{(t)}\phi_{(t)}\phi'_{(t)}. \tag{23}$$

The weight on the regressor dyad $\phi_{(t)}\phi'_{(t)}$ is defined as

$$w_{(t)} = \lambda - \frac{1-\lambda}{\phi'_{(t)}P_{(t|t-1)}\phi_{(t)}} \tag{24}$$

using an explicit forgetting factor $\lambda \in (0,1)$. This method has proved to be extremely robust with respect to poor system excitation caused by linear feedback, rarely changing external disturbances, input saturation etc. Its maximal robustness is paid off by the loss of exponential convergence that may arise in some special cases in non-uniform quality of parameter tracking in various "directions" [Bittanti *et al* (1988)]. The technique may be further improved by respecting available prior information [see Kulhavý (1986)] or using a time-variable forgetting factor [Fortescue *et al* (1981), Kulhavý (1987)].

Coping with Abrupt Parameter Changes. Sudden changes of estimated parameters are usually detected using various statistical tests based on analysing the sequence of prediction errors $\hat{\varepsilon}_{(t)}$ or the sequence of parameter estimates (see Basseville and Benveniste (1988) for a nice survey). In the simplest case, when a substantial change of parameters is confirmed, parameter estimation is restarted from suitable initial conditions. When a multi-model appoach is applied, the results of the best-fitted filter are returned.

Coping with Nonlinearities. "Smooth" nonlinearities can be compensated by adaptation that is responsible for tracking of slow changes of the system dynamics. The other situation arises when hard functional nonlinearities are encountered. In such a case, it is inevitable to take into account specific features of the controlled process. As a rule, *ad hoc* modifications of the standard design are tailored to the process.

Coping with Outliers. The recursive least squares are known to be very sensitive to "outliers" in the prediction error sequence $\hat{\varepsilon}_{(t)}$ which drastically influence estimation of parameters. These are produced when we are faced with noises having distributions with heavy-tailed p.d.f.'s (such as the Cauchy distribution). Simple statistical tests to decide about the outlier rejection are often sufficient [see Kárný *et al* (1985)]. The results of robust statistics may be helpful in design of specific measures [cf. Huber (1981), Tsypkin (1984) and Hampel *et al* (1986)].

Coping with Higher-Order Unmodelled Dynamics. Output disturbance, unmodelled higher-order dynamics or measurement noises may violate the assumption of the white noise property of the stochastic component. More or less *ad hoc* filtering of data signals is the prevalent solution now (Hebký (1984) suggested a simple ARX filter to determine the trend of the signal within a sampling period, Kárný (1990a) required a smoother behaviour ot the signal between the neighbouring sampling periods).

Coping with Signal Constraints. When the distances of output or input from the "ideal" values are large, different shapes of the loss function may influence the control design substantially. There are situations when a fixed quadratic form of the loss function does not express the technological demands appropriately.

The hard input constraints $u_{(t)} \in \langle \underline{u}, \overline{u} \rangle$, quite frequent in practice, are a typical example that cannot be treated directly by the classical LQG theory. However, using the iterations-spread-in-time strategy, these bounds can be incorporated into the design by the following trick. Clearly, if $u_{(t+1)} > \overline{u}$ or $u_{(t+1)} < \underline{u}$, $L_{(t+1)}$ must be modified. With respect to the fact how $Q_{u(t+1)}$ influences $L_{(t+1)}$, it is always possible to choose $Q_{u(t+1)}$ in such a way that

$$u_{(t+1)} = \overline{u} \quad \text{or} \quad u_{(t+1)} = \underline{u}$$

Thus, we meet the constraints but remain in the linear region of the controller operation. The complexity of the resulting algorithm only slightly increases [see Böhm and Kárný (1982) and Böhm (1985)].

Coping with Non-Dual Control Effects. The use of non-dual control strategies may have two consequences.

1. The initial adaptation phase is quiet and damped, hence, the self-tuning control loop is poorly excited. As the self-tuner performance depends heavily on good excitation at the first steps of its operation, it is necessary to ensure favourable experimental conditions in another way. One of the simplest procedures is to inject an extra perturbation signal. It is suitable to modify the properties of the extra signal according to the results of the estimation process.

2. The initial adaptation phase is wild. The quality of control at the first steps is not satisfactory. It indicates that the model and prior conditions for parameter estimation have not been correctly chosen and the controller design is not robust enough. The re-design of the prior adjustment both of identification and control part is recommended. The start-up procedure has to be designed more carefully.

Coping with Mismodelling Effects. Not only identification but also control design must be made sufficiently robust with respect to unmodelled properties of the controlled system. Robustness of the resulting control is achieved at the cost of a loss of the optimal behaviour. Owing to permanent adaptation, robustness measures need not be so hard as in the case of a fixed model. The designer can e.g.

- use a higher penalization of inputs,

- modify the optimization (or directly the control law) to obtain larger robustness in stability (see Peterka (1990) for the application of a special choice of penalizations over the planning interval, for instance).

2.5   Algorithms

Even perfect and feasible theory can fail in practical application if not supported by numerically safe and effective routines.

Numerical Robustness. The main part of dynamic programming is the evolution of the symmetric positive definite matrix representing the kernel of the quadratic form $K^*_{(t)}(k, N|k - 1)$. The main part of recursive parameter estimation is the evolution of the symmetric positive definite covariance matrix $P_{(t|t-1)}$. In both cases, the numerically sensitive point is to ensure the properties of symmetry and positive definiteness. The standard tool to achieve it is to use either the Cholesky square-root factorization GG' of the mentioned matrices into the product of a lower triangular matrix G and its transposition

G' or the LD-factorization LDL' with a diagonal matrix D explicitly extracted. Instead of the original matrices, their factors are directly updated [the monograph of Bierman (1977) serves as a classical reference, see also Kárný *et al* (1985) and Peterka (1986)].

Another source of numerical troubles is connected with fast sampling of the controlled system. The system description in terms of the Delta operators has proved to enlarge the region of a successful operation (see Goodwin (1985) and Peterka (1986) for details).

Software Tools. To be able to test the research results experimentally as well as to accelerate the transfer of new algorithms into practice, the basic algorithms have to be really coded [cf. Kulhavý (1989)]. Moreover, a suitable supervisory environment with all needed system functions built in (simulation support, graphics, tools for interaction etc.) is needed for reasonable effectiveness of this stage of work [cf. Nedoma and Kulhavý (1988)].

2.6   Prior Tuning

The disadvantage of model-based adaptive control like the presented one is the fact that its behaviour is optimal only in the case that the model reflects well the reality. To avoid unnecessary deterioration of the control quality it is necessary to fit the model to the real system. Adaptivity of the self-tuning controller concerns only adaptation of model parameters. There remain a lot of parameters that must be selected before the controller is started. These parameters enable the designer

- to taylor the *model* to a given system by selecting

    - the model structure,
    - the model order,
    - the initial parameter estimates,
    - the type and value of forgetting,
    - the time delay,
    - the sampling period;

- to specify the *control design* by selecting

    - the type and value of penalization,
    - the horizon of the criterion,
    - the signal constraints;

- to choose the type and parameters of the *prefilter* which must mainly reject outliers and adjust the noise properties to conform to the model.

It is evident that to taylor the presented approach successfully to some practical application needs good knowledge and some experience to be able to incorporate prior knowledge into the model and to choose other parameters. Any computer support may be of great help at this stage.

Incorporating Prior Information. Except the parameters that are directly linked to technological requirements, such as the signal constraints, the translation of the designer's *a priori* information into the design terms is rather difficult.

A straightforward but rarely utilized possibility is to incorporate prior knowledge about the controlled system into parameter estimation. Both prior knowledge of global characteristics of the process [Kárný (1984)] and information about time variations of these characteristics [Kulhavý (1986)] can be included in a systematic way. The practical contribution of carefully expressing available prior knowledge (for instance, fixing the static gain in the case of unmodelled higher-order dynamics) may be striking.

Compact Self-Tuner. One way how to facilitate the implementation to the user is to design a maximally compact self-tuner with minimum of user's choices while the other parameters are fixed to values which have been found as suitable for most cases. On the contrary, such a compact controller has to possess maximum of robustifying measures to be able to work with the broad class of systems (see Böhm *et al* (1989) for a detailed description).

By experience, it is possible to reduce the set of tuning knobs to:

- the model order,

- the expected rate of static gain changes (relative to other parameters),

- the control period (a multiple of the sampling period),

- the time delay (a multiple of the control period),

- the range of admissible input changes.

Used robustifying measures comprise:

- directional forgetting endowed with the possibility to suppress information about the static gain more slowly;

- special local filter to get rid of a high frequency noise without adding substantial dynamics to the loop;

- respecting of hard restrictions on inputs or input increments in control design;

- handling of a preprogrammed command signal and compensation of a possible offset in the steady state [Böhm (1988)];

- extensions to models with external disturbance;

- the start-up procedure with bumpless transfer to the self-tuning operation ensured by the following modes of operation:

  - *Data Acquisition:* Actual measured data are fed into the appropriate vectors while the input signal is generated by another controller, or is determined by a suitable random-signal generator, or is man-operated.

- *Identification:* Data acquisition is completed by starting identification. The system remains in open loop (when stable), or the loop is closed by another controller. To achieve good parameter estimates requires a persistently exciting input signal. Often it is sufficient to inject an extra perturbation signal.

- *Self-Tuning Control:* The optimal input calculation is started and the self-tuning control loop is closed. To achieve a bumpless transfer, we have to choose a very high input penalization for a period of several first steps.

  If the controller is restarted, the data acquisition is followed directly by self-tuning control.

Computer Aided Adjustment. The prior adjustment of the self-tuner can be supported by a computer aided design. Using data collected beforehand (under manual or some simple regulator operation), computer analysis can help to find reasonable choices for a lot of model parameters and perform preliminary control design. There is a good theoretical support for this task (see e.g. Kárný (1980, 1983), Kárný and Kulhavý (1988) for structure determination, Kárný (1984) for initialization of parameter estimation, Kárný (1990b) for the choice of the sampling period). An example of such CAD support is the package DESIGNER, treated in detail by Kárný *et al* (1990).

Role of Common Sense. The drawback of adaptive controllers based on separate identification and control is their dependence on the goodness-of-fit of the model to reality. The planned control quality (measured through the optimal cost-to-go) need not correspond to the actual one. To overcome this discrepancy, some supervisor is needed. Some checks can now be automated. However, it is mainly the duty of the designer to avoid serious mistakes. Deep knowledge of the physical nature of the controlled process is highly recommendable and in nonstandard cases still inevitable. We are a bit pessimistic about the possibility of a fully automatic implementation of self-tuners especially to complex technological processes. On the contrary, we expect an increasing need for preparing control engineers who will be mathematically well-educated but still able to understand the technological side of the problem.

3   Discussion of Questionable Points.

Robustification tricks make the design of adaptive regulators problematic. It is the task for the future to provide the controller with extensions guaranteing needed properties without requiring excessive knowledge and experience of the designer.

Quadratic versus Non-quadratic Loss Functions. Many practicioners criticize the quadratic criterion as not being flexible enough to cope with different practical demands. It seems, however, that many failures can be attributed to little experience of users with expressing their demands in terms of penalization matrices $Q_{u(t)}$, $Q_{y(t)}$. For this reason, it seems reasonable not to attack optimal control design for non-quadratic loss functions but concentrate our effort to the problem how to automate the choice of the penalization matrices in order that they might reflect signal constraints, approximate "non-quadratic" criteria, increase stability robustness etc. One possible methodology was sketched in Böhm (1985).

Suboptimal versus Dual Control Strategies. The loss of duality when suboptimal control strategies are applied results in a poor balance between good system excitation and satisfactory control quality especially in the start-up phase. From the practical point of view, this problem does not appear crucial. First, the designer has usually some prior knowledge about the controlled process which can be translated into the probabilistic terms to initialize parameter estimation properly (cf. Section 2.6). A careful choice of initial values for estimation can make the transient behaviour of the self-tuner quite satisfactory. Hence, the problem is only to ensure sufficient system excitation. Nevertheless, there are cases when close-to-dual solutions would be of practical interest — if the start of control is to be optimal or when the prior-tuning role of the (human) designer is to be suppressed as much as possible.

Linear and Normal versus More Realistic Models. Two situations are of interest. First, we know that the system model is different from linear and/or normal one and are able to determine it. Second, we know only that the system model belongs to some class of models. In such a case we design the estimator with respect to a least-favourable model (e.g. the least-favourable noise distribution in a min-max sense). In both cases the problem is to design a recursive nonlinear estimator for the assumed or least-favourable model. The design requires a rationally based approximation of the optimal Bayes parametric inference. Much effort has been devoted to elaborate a systematic framework for solving this task (see Kárný and Hangos (1988) for a "point-estimation" procedure and Kulhavý (1990a,b,c) for a completely recursive approach).

Forgetting versus Modelling Time Variations. No doubt, it is much more systematic and consistent with the Bayes methodology to model parameter variations than to apply forgetting. The simplest case is to model the changes of the regression coefficients $\theta_{(t)}$ through a random walk process (16) and to estimate recursively the covariance matrix of the coefficients increments R_v. Needless to add, this task is inherently nonlinear and requires a rationally justified approximation again. Hopefully, the above mentioned approaches to approximation of Bayes inference will be extendable to this case, too.

ARX versus ARMAX Modelling. When a correlated noise is encountered, the natural solution is to model $e_{(t)}$ as a moving average stochastic process. It means to consider a more general and wide-spread ARMAX model instead of the ARX one. The design of self-tuning control is elaborated in detail now for the case that the moving average parameters are assumed known [Peterka (1986)]. Unfortunately, there is no satisfactory (at least from the Bayesian point of view) procedure to estimate the moving average parameters, safe enough for routine recursive application. This drawback disqualifies the approach so far.

Tailored versus Universal Self-Tuners. Both lines have a sense. Compact pre-adjusted self-tuners are demanded for routine repetitive implementations on chosen classes of processes. If used succesfully in several typical situations where pre-adjusting was carefully selected they can be widely used without much additional effort. The robustness is preferred to quality here and the prior tuning must reflect it (through higher penalization, stronger filtration etc.). The main advantage consists in substantial reduction of the designer's work. Special tailored self-tuners are attractive when meeting extreme requirements or mass production conditions when any improvement may realize great income. In this case the optimality is more important. A careful analysis is then inevitable to get the best knowledge of the process. The "adaptive" optimality close to the ideal case of known parameter changes is essential here.

4 Conclusions

We would like to conclude by guessing which directions of research suggest to bring the most important progress from the practical point of view (the academic viewpoint may be a bit different). Our conclusions reflect the results of the above discussion of questionable points of design. Clearly, they have to be very subjective.

We consider the following research projects the most promising:

1. *Theoretical and algorithmical support for prior tuning of LQG controllers:* It should complete the classical LQG control design to the extent that, at least for some classes of controlled processes, implementation of self-tuners would become a routine matter.

2. *Robustification of control design with respect to unmodelled dynamics:* A sophisticated choice of time-variable penalizations in the quadratic criterion, both over the planning horizon and the real time scale, can substantially enlarge the scope of applicability of LQG self-tuners.

3. *Robust parameter estimation with respect to heavy-tailed distributions and quantized data:* Nonlinear estimators able to provide the first two moments of unknown parameters for non-normal distributions may be directly combined with the LQ control design and can endow the resulting self-tuners with quite new properties.

4. *Identification of models of parameter variations:* A systematic nonlinear filtering approach to parameter tracking based on identifying an explicit model of parameter changes can be a good alternative to a bulk of heuristic techniques used today; it should offer better insight and more space for utilizing prior information.

5 References

Basseville, M. and A. Benveniste (1986). *Detection of Abrupt Changes in Signals and Dynamic Systems*. Springer-Verlag, Berlin.

Bierman, G.J. (1977). *Factorization Methods for Discrete Sequential Estimation*. Academic Press, New York.

Bitmead, R.R, M.R Gevers and V. Wertz (1989). *The Thinking Man's GPC*. Snofart Press.

Bittanti, S., P. Bolzern and M. Campi (1988). Convergence and exponential convergence of identification algorithms with directional forgetting factor. Report No. 88-002, University of Milano. To appear in *Automatica*.

Böhm, J., M. Kárný and R. Kulhavý (1989). Practically-oriented LQ self-tuners. *Preprints IFAC Workshop on Evaluation of Adaptive Control Strategies in Industrial Applications*, Tbilisi.

Böhm J. (1985). LQ selftuners with signal level constraints. *Preprints 7th IFAC/IFORS Symposium on Identification and System Parameter Estimation*, York, Vol. 1, pp. 131-135.

Böhm J. (1988). Set point control and offset compensation in discrete LQ adaptive control. *Problems of Control and Information Theory*, *17*, 125-136.

Böhm J., and M. Kárný (1982). Selftuning regulators with restricted input. *Kybernetika*, *18*, 529-544.

Fortescue, T.R., L.S. Kershenbaum and B.E. Ydstie (1981). Implementation of self-tuning regulators with variable forgetting factor. *Automatica*, *17*, 831-835.

Goodwin, G.C. (1985). Some observation on robust estimation and control. *Preprints 7th IFAC/IFORS Symp. on Identification and System Parameter Estimation*, York, Vol. 1, pp. 851-858.

Hägglund, T. (1983). The problem of forgetting old data in recursive estimation. *Proc. IFAC Workshop on Adaptive System in Control and Signal Processing*, San Francisco.

Hampel, F.R., E.M. Ronchetti, P.J. Rousseeuw and W.A. Stahel (1986). *Robust Statistics. The Approach Based on Influence Functions*. Wiley, New York.

Hebký, Z. (1984). Multiple sampling in digital control (in Czech). *Automatizace*, *27*, 142-145.

Huber, P.J. (1981). *Robust Statistics*. Wiley, New York.

Kárný, M. (1980). Bayesian estimation of model order. *Problems of Control and Information Theory*, *8*, 33-46.

Kárný, M. (1983). Algorithms for determining the model structure of a controlled system. *Kybernetika*, *19*, 164-178.

Kárný, M. (1984). Quantification of prior knowledge about global characteristics of linear model. *Kybernetika*, *20*, 376-385.

Kárný, M. (1990a). Local filter robust with respect to outlying data. Report No. 1644, Institute of Information Theory and Automation.

Kárný, M. (1990b). Estimation of sampling period for selftuners. Accepted for *11th IFAC World Congress*, Tallinn.

Kárný, M., A. Halousková, J. Böhm, R. Kulhavý and P. Nedoma (1985). Design of linear quadratic adaptive control: theory and algorithms for practice. *Kybernetika, 21*, Supplement to No. 3, 4, 5, 6.

Kárný, M. and K.M. Hangos (1987). One-sided approximation of Bayes rule: theoretical background. *Preprints 10th IFAC Congress*, Munich.

Kárný, M. and K.M. Hangos (1988). One-sided approximation of Bayes rule and its application to regression model with Cauchy noise. *Kybernetika, 24*, 321–339.

Kárný, M. and R. Kulhavý (1988). Structure determination of regression-type models for adaptive prediction and control. In J.C. Spall (ed.), *Bayesian Analysis of Time Series and Dynamic Models*, Chapter 12, pp. 313–345. Marcel Dekker, New York.

Kárný M. and A. Halousková (1990). Preliminary tuning of selftuners. This issue.

Kulhavý, R. (1986). Directional tracking of regression-type model parameters. *Preprints 2nd IFAC Workshop on Adaptive Systems in Control and Signal Processing*, Lund, pp. 97–102.

Kulhavý, R. (1987). Restricted exponential forgetting in real-time identification. *Automatica, 23*, 589–600.

Kulhavý, R. (1989). *SIC — Package for Simulation, Identification, and Adaptive Control of Stochastic Systems: User's Guide*. IBM Personal Computer Version 2. Report No. 1592, Institute of Information Theory and Automation.

Kulhavý, R. (1990a). Recursive Bayesian estimation under memory limitation. *Kybernetika, 26*, 1–16.

Kulhavý, R. (1990b). Recursive nonlinear estimation: a geometric approach. *Automatica, 26*, No. 3.

Kulhavý, R. (1990c). Differential geometry of recursive nonlinear estimation. Accepted for *11th IFAC World Congress*, Tallinn.

Kulhavý, R. and M. Kárný (1984). Tracking of slowly varying parameters by directional forgetting. *Preprints 9th IFAC World Congress*, Budapest, Vol. X, pp. 78–83.

Kumar, P. R. (1985). A survey of some results in stochastic adaptive control. *SIAM J. Control and Optimization, 23*, 329-380.

Nedoma, P. and R. Kulhavý (1988). *KOS — Conversational Monitor for Managing FORTRAN Libraries*. IBM PC Version 1.0. Report No. 1515, Institute of Information Theory and Automation.

Peterka, V. (1981). Bayesian approach to system identification. In P. Eykhoff (ed.), *Trends and Progress in System Identification*, Chapter 8, pp. 239–304. Pergamon Press, Oxford.

Peterka, V. (1986). Control of Uncertain Processes: Applied Theory and Algorithms. *Kybernetika*, *22*, Supplement to No. 3, 4, 5, 6.

Peterka V. (1990). Robustification of LQG optimum control design. This issue.

Poubelle M.A., R.R. Bitmead and M. Gevers (1988). Fake Algebraic Riccati Techniques and Stability. *IEEE Trans. Automatic Control*, *AC-33*, 379–381.

Tsypkin, Ya.Z. (1984). *Fundamentals of Information Theory of Identification* (in Russian). Nauka, Moscow. German translation (1987), VEB Verlag Technik, Berlin.

Ullrich, M. (1964). Optimum control of some stochastic systems. *Proc. 8th Conf. ETAN*, Beograd. pp. 291–298

THEORY AND IMPLEMENTATION OF PAPER CROSS PROFILE ADAPTIVE CONTROL

I. Nagy and F. Dušek

Institute of Information Theory and Automation
Institute for Paper and Board Industry Development
Prague, Czechoslovakia

1 Introduction

A constant or prescribed shape of the basis weight cross profile is an important property of produced paper. It takes part in determining the resulting paper quality. Up to now, only the machine direction profile has been controlled (see [1]). The innovation of the control taking into account and controlling also the cross profile properties of the paper is supposed to rise significantly its quality. Unlike the machine direction control where the time profile (the average basis weight taken across the paper sheet as a function of time) has been measured by a fixed measuring device, the moving gauge has to be used now. Due to mutual movements of the paper sheet and the traversing gauge, only a diagonal profile which is a mixture of the time and the cross profiles can be sensed. The task is to extract the cross profile from the measured mixture and to control it by adjusting screws shaping the slice through which the raw material solution flows on the net. The screws are manipulated by a special robot. As the measuring device can be placed only at the end of the whole paper machine, there is a big transport delay between applying the control action and sensing its influence to the profile. Also the application of control actions itself is relatively slow because there is only one robot supposed which adjusts sequentially one screw after another to values sequentially generated by the controller. The influence of the screws on the shape of the cross profile is supposed to be static. One screw influences symmetrically the effects of the other screws which are near to it. The range of influencing is smaller than the width of the paper sheet. During the cross profile control the control of time profile is performed by an independent regulator through changing the speed of the paper machine. This process is supposed to be almost static. The main effect of the cross control is expected at the initial tuning.

2 Pre-filtering of data

As the measuring device is able to give more data than it is possible to transfer to the control computer, a filtering of data is used. The filter is local. It gathers data between two screws interpolates them by a line and takes the representant (filtered variable) on the line in the prescribed point. Such filter can not only smooth the measured data but also to make them equidistant if the rough data are not and the equidistance is required. Even the particular points of "filtered data measuring" can be arbitrary chosen. For details concerning the filter see [3].

3   Modelling and identification

The paper machine is a process with distributed parameters. That is why the nonpara-metric continuous-time convolution model

$$Y(t,x) = C(x) + K(\dot{x}) * U(t - d, x) + J(t) * V(t - d) + E(t,x)$$

has been chosen as a starting point for modelling. Here:

Y is the modelled basis weight at time t and (cross) position x.

C is the initial cross profile which is supposed to be a smooth function.

K, J are cross and time kernels of the corresponding convolutions. The cross kernel is symmetric with a finite (small) support, the time kernel represents very quick time dynamics. Both of them are smooth.

U is a smooth function describing the shape of the slice.

V is a deviation of the machine speed from its initial value.

d is a transport delay.

E is a stochastic term of the model.

As all functions in the model are smooth and most of them have finite support, non-traditional first order spline approximation has been chosen for their parametrization. It means the supports of all functions of the model are divided by nodes into intervals and the functions are expressed in the spline basis in the following way

$$F(.) = f'q_f(.),$$

where F is the approximated function,

$$f' = (f_1, f_2, ..f_n)$$

is a vector of coefficients of the approximation and

$$q'_f = (q_{f,1}, q_{f,2}, ..q_{f,n})$$

are the base functions which form the space of approximation. The base functions are represented by so called (first order) fundamental splines. Each base function belongs to one node. It has value 1 in this node and is zero in all other nodes. Then the coefficients are directly the function values of the approximated functions at corresponding nodes.

Substituting the approximation formula to all functions of the model and making use of the fact that the coefficients of the approximations do not depend on the integration variable, the model reads

$$Y(t,x) = c'q_c(x) + k'H^{k,u}(x)u(t - d) + j'H^{j,v}(t)v(t - d) + E(t,x),$$

where matrices $H^{k,u}$ and $H^{j,v}$ are

$$H^{k,u}(x) = \int_0^L q_k(\varsigma)q'_u(x - \varsigma)d\varsigma,$$

$$H^{j,v}(t - d) = \int_0^t q_j(\tau)q_v'(t - \tau - d)d\tau,$$

L denoting the width of the paper sheet.

For the equidistant grids of the signal approximations and the constant structures of the kernels, the matrices depend only on the points of data sampling. Due to the data pre-filtering used, the data can be considered as if measured equidistantly in the apriori selected points, say corresponding to the screw positions. Then the matrices $H^{k,u}$ and $H^{j,v}$ are constant and can be pre-computed off-line before the control starts.

Introducing filtered variables $\tilde{u}$ and $\tilde{v}$ by

$$\tilde{u}(t) = H^{k,u}u(t - d), \qquad \tilde{v}(t) = H^{j,v}v(t - d)$$

the model is

$$Y(t, x) = c'q_c(x) + k'\tilde{u}(t) + j'\tilde{v}(t) + E(t, x)$$

Denoting x_t the position in which the measuring device is in time instant t, the above model can be writen in the form of a regression model for data available

$$\bar{y}(t) = P'z(t) + \varepsilon(t)$$

where $\bar{y}(t) = Y(t, x_t)$, $\varepsilon(t) = E(t, x_t)$, $P' = (c', k', j')$, $z'(t) = [q_c(x_t)', \tilde{u}(t)', \tilde{v}(t)']$. The parameters can be identified by the least-squares method [4]. The adaptive forgetting [5] is recommended.

4 Control design

For the purpose of control, the controlled variable is defined as a vector of derivatives of the cross profile

$$y'(t) = (\frac{\partial Y(t, x)}{\partial x}|x = x_1, \frac{\partial Y(t, x)}{\partial x}|x = x_2, ..., \frac{\partial Y(t, x)}{\partial x}|x = x_n)$$

in the prespecified fixed points $x_1, x_2, ...x_n$. The number of points n is equal to the number of the screws minus one.

Making use of the model for identification, the model for the controlled variable can be written

$$y(t) = a + Bu_d(t) + e(t),$$

where:

a, u_d, e are vectors of derivatives of C, U, E similarly as in the definition of the controlled variable y .

Matrix B is a band symmetric matrix given by its columns b^i

$$B = (b^1, b^2, ..., n^n).$$

The $j - th$ item of the $i - th$ column denoted by b_j^i is given by the formula,

$$b_j^i = \sum_l k_l \int_{x_j}^{x_{j+1}} q_{k,l}(x_i - x)dx.$$

This multi-input multi-output static model of the paper machine does not fulfill the requirement that only one screw can be adjusted at one time instant. To meet this condition it is suitable to define a dynamic variable (pointer) $i(t)$ indicating the $i - th$ screw which is to be adjusted at a current time instant t. Then, a new scalar control variable $u_m(t)$ which is the value to which the $i - th$ screw is to be adjusted at time t can be defined. Using these definitions and the state variable in the form

$$s(t) = a + Bu_d(t)$$

the state-space single-input model can be written

$$s(t) = s(t - 1) + (b^{i(t)+1} - b^{i(t)})u_m(t),$$

$$y(t) = s(t) + e(t).$$

The control is to be designed to minimize the quadratic criterion

$$\frac{1}{H}E\sum_{t=1}^{H}[s'(t)s(t) + \omega u_m^2],$$

where E denotes the mean value and $\omega \geq 0$ is a penalization of changes of the inputs.

The control synthesis leads to the LQ stochastic design with periodical solution given by the discrete time Riccati equation (cf. [2])

$$S(H - t + 1) = S(H - t)[I - (b^{i(t)+1} - b^{i(t)})L'(t)] + I,$$

where I denotes the unit matrix.

The linear optimal controller has the form

$$u_m(t) = -L'(t)s(t - 1),$$

where the control law is given by

$$L(t) = S(H - t)\frac{(b^{i(t)+1} - b^{i(t)})}{(b^{i(t)+1} - b^{i(t)})'S(H - t)(b^{i(t)+1} - b^{i(t)}) + \omega}.$$

5 Implementation aspects of the adaptive controller

The implementation of the controller consisted of two main tasks:

- Development of the complete application software for the control computer.

- Real time simulation experiments.

The control design had to respect some particularities of the whole system. The most important of them are:

1. Slow application (20 – 30 sec) of a control action which is performed by the robot traversing from one screw to another.

2. Condition that only the shape of the cross profile is to be influenced independently of its mean value which forms the time profile and is controlled separately.

3. Special demands on transfering the data from the measuring device to the control computer in reasonably big batches with respect to an acceptable transport delay added this way. (16 measurements 4 times during the way of the measuring device across the paper are transferred.)

4. Problems connected with a relatively small memory of the control computer used.

The control computer comunicates with both the control unit of the robot and the measuring device by means of two serial ports RS232. They are the only connection of the control computer with the controlled process. From the measuring device, the batch information about the first and the last measured position, the speed of the paper machine and the values of the measured diagonal profile come to the control computer. From the control computer, the commands are sent to the robot. In the implementation, the identification is completely separated from the control. It is connected with the availability of a new piece of information provided by the measuring device, it has higher priority and it is performed with other time period than control. With the same priority, the identified parameters are recomputed to the parameters of the model which is used for the control. The recomputation is connected with the information provided by robot that the previous control action has been finished. Computations of the control law and the value of a control action have lower priority. They are synchronized with finishing the application of the control action so that the computations may be performed while the screw is being adjusted. The identification can interrupt these computations.

To be able to check the implemented adaptive controller, a system simulating behaviour of the real paper machine has been developed and implemented on a separate computer. The basic properties of the paper machine which are considered in the simulation are:

1. (a) communication of the robot with the control computer on serial port RS232

   (b) constant period for shifting the robot from one screw to another (20 sec)

   (c) variable time interval for accomplishing the control action depending on its absolute value

2. (a) computation of an actual cross-profile (64 values) depending on the actual positions of the screws (15 values), the chosen initial cross-profile, the chosen influence of particular screws and the actual speed of the paper machine

   (b) computation of the actual speed of the paper machine depending on the value of the desired speed and the time constant representing the inertia of the machine (the desired value of the speed can be set in the arbitrary time instant)

   (c) generation of the cross-direction and the machine-direction noises with their specific statistical properties

3. (a) variable transport delay depending on the actual speed of the machine (about 60 sec)

   (b) collection of the measured values of the diagonal profile and sending data in batches to the control computer (each 8 sec 16 values are sent)

   (c) communication of the measuring device with the control computer on serial port RS232.

The computation of the actual values takes some 0.5 sec. The time necessary for the measuring device to traverse across the paper sheet is 32 sec. The values of the cross profile, the positions of screws and the actual speeds of the machine are stored each 30 sec. They can be currently evaluated. Other functions are performed immediately after their occurance.

6   Conclusions

The design of the adaptive controller for control of the cross-profile of the paper sheet and its implementation aspects have been presented. For modelling, the integral description parametrized by spline approximation with both the equidistant and the nonequidistant grids has been considered. For the least-squares identification, the data are filtered by constant matrices. To maintain the filter matrices constant, the rough measured data have to be pre-filtered by local filter which brings no additional closed-loop dynamic. The control is built on a dynamical single input model which is obtained from the multi-input static one. The approach of the dynamic programming resulting into the periodical solution of the discrete Riccatti equation is used.

7   References

1. Lízr, A.: Linear quadratic self-tuning regulators in paper machine control systems. 2nd IFAC Workshop on Adaptive Systems in Control and Signal processing, Lund, 109–111, 1986

2. Kárný et al.: Design of linear quadratic adaptive control – theory and algorithms for practice. Supplement to Kybernetika 21, No 3,4,5,6, 1985.

3. Hebký, Z.: Multiple sampling in digital control. Automatizace (in Czech), 27, 142–145, 1984.

4. Peterka, V.: Bayesian approach to system identification. In: "Trends and Progress in System Identification" (P.Eykhoff ed.), Pergamon Press, Oxford, 1981.

5. Kulhavý, R.: Restricted exponential forgetting in real time identification. Automatica, 23, 589–600, 1987.

6. Kárný at al.: Modelling, identification and adaptive control of cross-direction basis weight of paper sheets. Intern. conf. CONTROL 88, Oxford, 159–164, 1988.

CASE STUDIES IN APPLICATION OF STOCHASTIC CONTROL THEORY

by

O.L.R. Jacobs

Dept. of Engineering Science, University of Oxford, Parks Rd, Oxford OX1 3PJ, U.K.

ABSTRACT

Case studies are reported which show that ideas from stochastic control theory can be exploited to design useful practical controllers. The ideas indicate that the way to design a feedback controller is to implement two separate functions of estimation and control, each matched to some simple mathematical model capturing essential features of the controlled object. Good design does not usually require that either the estimator or the control law be exactly optimal, not even for the simplified model. Where the controlled object is essentially nonlinear, as in many real applications, this suboptimal stochastic design may give better performance than is achievable with classical, black-box linear controllers or their adaptive derivatives.

Two current collaborative case studies are summarised.

1.   Control of pressure and position in an air knife which regulates the thickness of zinc deposit on sheet steel galvanized by hot dip. Feedback in the form of a time-delayed measurement of deposited thickness drives a Kalman filter which estimates an uncertain multiplicative coefficient in a simplified model of the deposition. The updated model is then used to determine control action.

2.   Fishery management as a problem in feedback control. In collaboration with the Fisheries Laboratory of the UK Ministry of Agriculture Food and Fisheries, management of one particular harvested population (mackerel) has been considered from the point of view of control engineering. A suboptimal, certainty-equivalent control law is derived to regulate fluctuations about a specified equilibrium state. This control law includes as special cases strategies of constant effort and constant catch which are well known to fishery managers.

The overall conclusion is that the structural, stochastic, suboptimal approach can be recommended as a general procedure for designing feedback controllers to achieve desired performance of controlled objects which are subject to uncertainty.

1   INTRODUCTION

From the point of view of practical engineering, stochastic control theory can be valuable as a source of insight into the design of feedback controllers for systems such as that shown in Fig 1. In many real applications this is the type of system used to ensure

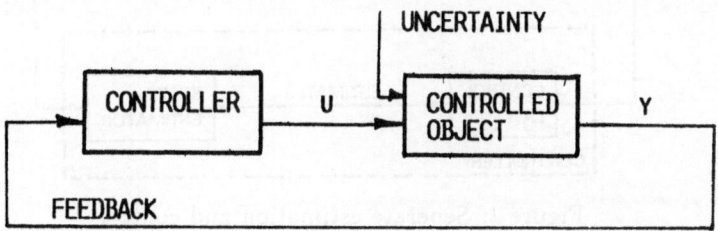

Figure 1: Feedback control system

that a controlled object, characterised by time-varying inputs u, internal states x and outputs y and subject to unavoidable uncertainty shall perform in some desired manner.

Important insights from stochastic control theory are:

1. Feedback, such as is shown in Fig 1, is the means by which control is achieved in the presence of uncertainty. Therefor any feedback system should benefit from a design approach which takes explicit account of uncertainty; that is from a stochastic approach. Classical control engineering design is deterministic: it concentrates on choosing dynamic feedback controllers to compromise between conflicting requirements for accuracy, closed-loop stability and robustness.

2. Two different types of uncertainty can affect controlled objects; uncertainty about current values of the internal states x and uncertainty about future values resulting from uncertainty about controlled object dynamics governing the evolution of future states.

3. A feedback controller must combine dual functions of estimation and of control (Fel'dbaum 1965). The estimation reduces uncertainty about current states whose current values are needed in order to determine current control action which could include probing to reduce uncertainties about future states (Jacobs and Patchell 1972).

4. Dynamics in feedback controllers serve the purpose of estimation (Jacobs 1974, 1982). Familiar controllers such as the lead-lag compensator and the PI controller can be regarded as the cascade of a dynamic linear observer and a certainty-equivalent control law, each of which is approximately matched to a reduced order linearised model of the controlled process (Friedland 1986).

The purpose of this paper is to show that the above insights can be exploited to design useful practical controllers in which two separate functions of estimation and control are implemented, as shown in Fig 2. Truly optimal control in specific applications is nor-

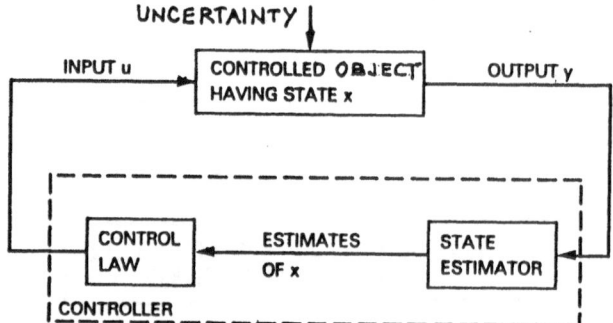

Figure 2: Seperate estimation and control

mally neither sought nor achieved; it being impracticable to obtain sufficiently accurate mathematical models or to solve the resulting equations characterising optimal estimation and optimal control. Useful results can be obtained by explicitly designing each of the two functions to match some simple mathematical model capturing essential features of the controlled object. This may need to be a structural, rather than behavioural or black box model. Good design does not usually require that either the estimator or the control law be exactly optimal, not even for the simplified model. Where the controlled object is essentially nonlinear, as in many real applications, this suboptimal stochastic design may give better performance than is achievable with classical linear controllers or their adaptive derivatives.

It is believed that the way to prove this approach is by case studies showing that good results can be achieved in specific real-world applications. Two such applications have already been described (Jacobs et al 1985, 1987). Two more current collaborative case studies are summarised here. Regulation of the thickness of zinc deposit on galvanized sheet steel is described in Section 2. Discussion of fishery management as a problem in feedback control is described in Section 3. Neither has yet proceeded beyond the stage of realistic simulation, but both show promising results.

2   FEEDBACK CONTROL OF AN AIR KNIFE

Production of galvanized sheet steel has for the past 50 years (Butler, Beam and Hawkins (1970); Wynne, Vickers and Williams (1987)) been based on the Sendzimir (1938) hot-dip process illustrated in Fig 3. Steel strip, of order 1 m wide by 1 mm thick, is preheated and passed at speeds of order 1 ms^{-1} through a bath of liquid zinc at temperature of order 500°C. A zinc film is entrained onto the strip as it emerges from the bath. This film solidifies whilst the strip runs vertically upwards, cooling as it goes, for a distance of order 50 m to the first of a set of rollers which accept the finished product.

Tight control of the amount of solidified zinc deposit to target values of order 100 gm^{-2} has commercial value. Overdeposition results in excessive use of zinc; un-

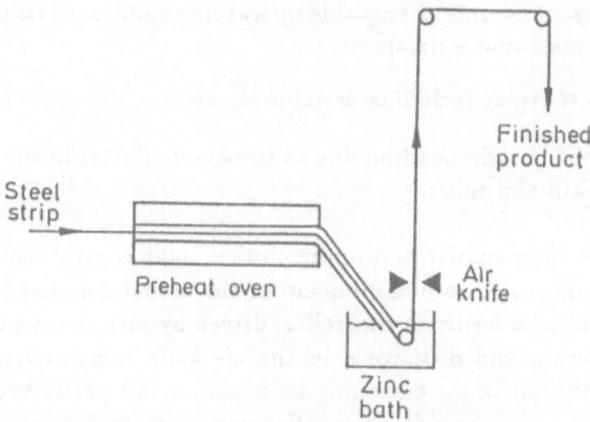

Figure 3: Continuous strip galvanizing

derdeposition results in an unsatisfactory product. Bertin and others (1978) estimated potential annual savings from a single production line to be of order $0.5M (1978).

Control is exercised by an "air knife" which consists of a pair of nozzles, located either side of the vertically rising strip, as shown in Fig 4. These nozzles deliver horizontally

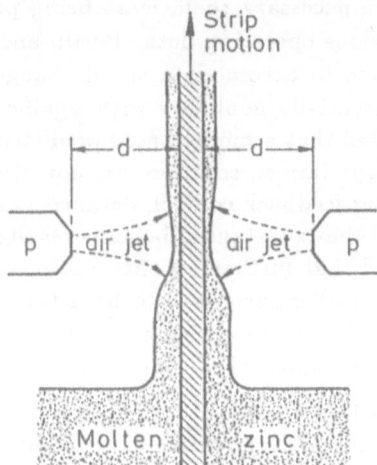

Figure 4: Air knife

extended air jets which augment the gravity-fed return flow of liquid zinc into the bath. Principal control variables are:

- Pressure p of air supplied to the nozzles (more pressure, less deposit)

- Displacement d between nozzles and strip (more displacement, more deposit)

The mass of zinc which eventually solidifies is governed by complex and nonlinear physics of fluid flow and heat exchange in the liquid film. It also depends on:

- More or less randomly changeable operating conditions including desired value of deposited mass and strip speed

- Strip imperfections including variable shape

- Fluctuations in strip position due to transverse flutter in the long vertical section between bath and roller.

Given the above complexities and uncertainties, tight control requires use of feedback driven by some downstream measurement of the actual deposited mass. The problem of designing a suitable feedback controller, driven by measurements of deposited mass, to adjust pressure p and distance d in the air knife is non-trivial partly because of nonlinearity in the equations governing knife action and partly because of unavoidable time-delay which arises because X-ray fluorescent sensors measuring deposited mass cannot function well at high temperatures and may need to be located well down-stream of the air-knives. Best accuracy is provided by a "cold" sensor which may need to be located beyond the next rollers, with consequent distance-velocity lag of order $100\ s$. "Hot" sensors are available which can be located near the air-knife, but they are an order of magnitude less accurate.

Butler, Beam and Hawkins (1970) proposed an open-loop control scheme in which pressure p and distance d would be made proportional to target coating weight and measured strip speed: the necessary coefficients being pre-determined from multiple regression analysis of previous operating data. Bertin and others (1978) reported suc-cesful use of such a scheme to accomodate speed changes. Wilhelm (1972) pointed out that the process is essentially nonlinear with significant dynamics in the form of the time-delay. He suggested that a model-based predictive controller might be needed to cope with the time-delay: implementation was not discussed. Edwards and others (1975) implemented a linear feedback control, detuned to sample slowly because of the time-delay. They reported that substantial benefits resulted from introduction of feed-back control and suggested that further benefits would result from use of an adaptive model-based predictive controller such as is to be described here. Noe and Schuerger (1975) reported succesful full-scale implementation of such a scheme, with adaptive gain updated according to a simple integral-of-control-error algorithm.

The study summarised here (Jacobs 1989) appears to be the first to exploit ideas from stochastic control theory with the problem explicitly formulated as one of control-ling a stochastic nonlinear process having dynamics in the form of time-delay. Feedback, in the form of time-delayed measurements of deposited mass, is used to drive an esti-mator, in the form of a Kalman filter, updating a conventional, simple two-dimensional mathematical model of knife action. The resulting model provides a basis for computing the next control input. Simulation indicates that the proposed controller gives reliable control, sufficiently promising that further work can be recommended to validate the model against full-scale plant data and subsequently to test a full-scale implementation of the scheme.

2.1 Mathematical model

Flows of molten zinc and incident air in the jet stripping process of an air-knife are governed by the Navier Stokes equations. The flows are complex and three-dimensional: they may subdivide into several regions according to whether viscosity or gravity dominates the zinc flow, according to the geometry of the air nozzles, according to the pressure gradient of incident air, etc. The model of knife action used here, both as a basis for controller design and for the simulation which tests the resulting controller, has the conventional simple form proposed by Harvey and Carlton (1974) and by Thornton and Graff (1976)

$$m = Kd\sqrt{\frac{v}{p}} + \xi_m \tag{1}$$

in which K is assumed to account for uncertainties about complex physics of knife action, and ξ_m accounts for unpredictable time-varying effects. This model is two-dimensional in that it neglects transverse variability accross the strip.

The model could be used to determine values for the controlling variables p and d so as to achieve some desired value m_0 of m. If values of K and v were known precisely, p and d could be chosen to satisfy

$$\frac{d}{\sqrt{p}} = \frac{m_0}{K\sqrt{v}} \tag{2}$$

giving

$$m = m_0 + \xi_m$$

and the resulting error in deposited mass would be only the unpredictable random term ξ_m.

In practice, good information about the strip speed v may well be available but the quantity K is unlikely to be known a priori. Nevertheless, information about K is available from measured plant behaviour, and could be extracted by a suitable estimation algorithm driven by known values of p, d and v together with the time-delayed measurements y of m. This estimation problem is simple and linear, but must correctly account for the time-delay between knife and gauge.

Assuming that control is to be implemented by an online digital computer, dynamic effects are represented here by difference equations, with independent variable i, representing values separated in time by a sampling interval Δ. Introducing notation

$$u \equiv d\sqrt{\frac{v}{p}} \tag{3}$$

to account for the known variables p, d and v, Eqn 1 becomes

$$m(i) = u(i)K(i) + \xi_m(i) \tag{4}$$

with measurements

$$y(i) = m(i - k(i)) + \xi_y(i) \tag{5}$$

where $\xi_y(i)$ represents unpredictable random noise in the gauge and

$$k(i) = \text{INT}(L/v(i)\Delta) \tag{6}$$

represents the dead-time as that needed to travel the distance L between knife and gauge at the current velocity $v(i)$, with INT() being the smallest integer greater than ().

To design a control law choosing a value for $u(i)$ in Eqn 4 should not be too difficult given that Eqn 4 is bilinear in u and in K, and that the measurement Eqn 5 is linear. Once $u(i)$ is determined, appropriate values for the pressure $p(i)$ and the distance $d(i)$ to keep p in mid-range can be found to satisfy Eqn 3 in the form

$$\frac{\sqrt{p(i)}}{d(i)} = \frac{\sqrt{v(i)}}{u(i)} \tag{7}$$

where $v(i)$ is the current value of strip speed.

Estimation of the uncertain coefficient $K(i)$ requires some prior model representing its dynamics. A general, first-order, linear stochastic difference equation is assumed here,

$$K(i+1) = a_K K(i) + \xi_K(i) + (1 - a_K)K_0 \tag{8}$$

This has no greater justification than that it can capture principal features of a variable coefficient. K_0 is the nominal value, a_K characterises detail of the dynamics, and $\xi_K(i)$ is an independent, zero-mean, random variable with variance $\text{var}(\xi_K)$ related to the variance $\text{var}(K)$ of excursions in $K(i)$ by

$$\text{var}(\xi_K) = (1 - a_K^2)\text{var}(K)$$

Fig 5 is a block diagram of the complete mathematical model specified by Eqns 1 to 8.

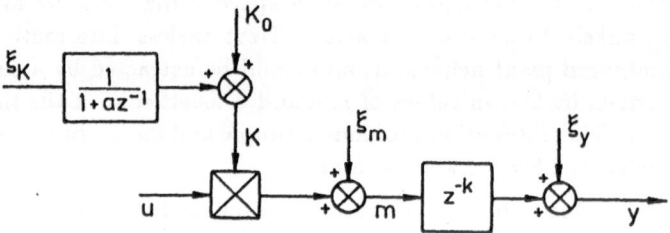

Figure 5: Block diagram of the mathematical model

2.2 Estimation and control

Estimation of $K(i)$ from the measurements $y(i)$ in Fig 5 is linear and can be formulated as a problem in Kalman filtering. Define N state-variables

$$
\begin{aligned}
x_1(i) &= K(i) - K_0 \\
x_j(i) &= x_{j-1}(i-1), \qquad j = 2, \cdots, N
\end{aligned}
$$

where N is larger than the largest possible delay k. Then Eqns 4 to 8 can be written form

$$x(i+1) = Ax(i) + b\xi_K(i)$$
$$y(i) = c(i)x(i) + u(i-k)K_0 + \eta(i)$$

where

$$A = \begin{bmatrix} a_K & & & & \\ 1 & 0 & & & \\ & 1 & 0 & & \\ & & \cdot & \cdot & \\ & & & 1 & 0 \end{bmatrix}, \qquad b = \begin{bmatrix} 1 \\ 0 \\ \cdot \\ \cdot \\ 0 \end{bmatrix}$$

with time-varying selector vector

$$c(i) = [0 \ \ . \ . \ \ u(i-k) \ \ . \ \ 0] \tag{9}$$

in which only the $(k(i)+1)$th element is non-zero, and with

$$\eta(i) \equiv \xi_m(i-k) + \xi_y(i)$$

This use of a time-varying selector to account for possibly time-varying time-delays may be original. Implementation of a matching Kalman filter, which generates estimates $\hat{K}(i)$ and reports estimated variances $\sigma_K(i)$, is routine (Bierman 1977) apart from the need to ensure that the number N of estimated states does exceed the greatest possible delay k.

The resulting estimates of $K(i)$ can be used in a control algorithm which chooses $u(i)$ so as to minimise the expected square of error $m_0 - m(i)$ in deposited mass. Eqn 4 gives this error as

$$m_0 - m(i) = m_0 - u(i)K(i) - \xi_m(i)$$

Assuming that the random variable ξ_m is independent of uncertainties about K, the minimising value is easily shown to be

$$u^*(i) = m_0 \frac{\hat{K}(i)}{\hat{K}^2(i) + \sigma_K(i)} \tag{10}$$

Once $u^*(i)$ has been obtained from this Eqn 10, values for the physical control variables $p(i)$ and $d(i)$ can be determined as at Eqn 7.

The above feedback control algorithm includes the two functions of estimation and control which theory indicates to be essential for optimal control in the presence of uncertainty. Although simple in principle, the resulting controller has several "advanced" features: it could be described as

- "Model-based" in its use of Eqns 7 and 10 to determine values for $p(i)$ and $d(i)$

- "Adaptive" (Åström and Wittenmark 1988) in its use of operating data to drive a recursive algorithm updating estimates of the coefficient K

- "Cautious" (Jacobs and Patchell 1972) in the way that Eqn 10 takes account of the uncertainty $\sigma_K(i)$ in the estimate $\hat{K}(i)$ of $K(i)$.

More important, is the recognition that this is a problem where the feedback measurement y must be used to drive an estimation algorithm, the Kalman filter here, and that y does not appear explicitly in the control algorithm, Eqn 10. In many conventional linear feedback control problems, good control can be achieved without this need to make explicit distinction between the two functions of estimation and control. The need arises here because of the combination of nonlinearity in knife action and time-delay in the feedback measurement.

2.3   Simulations

A simulation was implemented to test the above feedback controller. Lacking any more detailed mathematical model of knife action, the simulated controlled process had the same form as that used to design the controller, with modelling errors introduced to represent discrepancies between the real controlled object and the model. Fig 6 shows simulation results obtained using the feedback controller over 200 sampling intervals, a

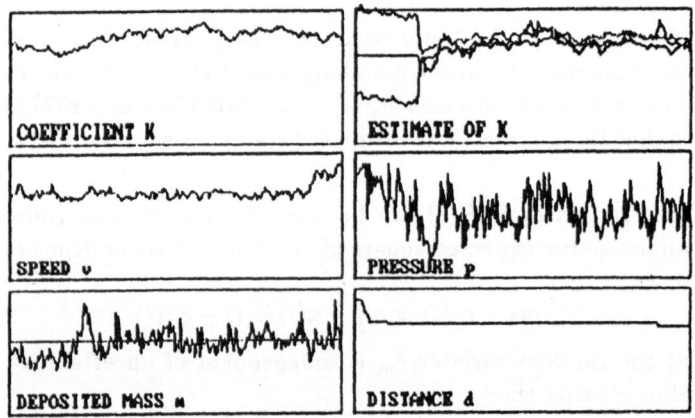

Figure 6: Suboptimal feedback control

short time in relation to the length of a real operating run, but long enough to show the controller in action. The graph labelled ESTIMATE OF K shows $\hat{K}(i)$, frozen at the nominal value K_0 for the first N samples, it also shows confidence limits of width one estimated standard deviation about the true value, $K(i) \pm \sigma_K(i)$. These limits nicely bracket the estimate showing that, in spite of significant fluctuations in $K(i)$ and in $v(i)$, good estimation of $K(i)$ is achieved. It can also be seen that $m(i)$ is succesfully regulated to the target value m_0, shown as a straight line in the graph labelled DEPOSITED MASS m.

Also simulated were constant controls u_0 having the initial nominal value

$$u_0 = m_0 \frac{K_0}{K_0^2 + \text{var}(K)}$$

with and without additional conventional PI control from measurement gauge y_m to air pressure p . The PI parameters had to be heavily detuned (Gain = 0.002, Integral time constant = $10^4 s$) to avoid instability due to the time-delay. Resulting rms errors in deposited mass over 5000 time intervals, for each of the three controls, are in Table 1 Sequences of random variables were identical to the extent that all three controllers

CONTROLLER	rms ERROR
Suboptimal stochastic control	18.49
Nominal control u_0	40.27
Nominal u_0 plus PI	37.72

Table 1: Simulated rms error in deposited mass

faced exactly the same task. Absolute numerical values of rms error here have little significance but the relative values show clearly that the suboptimal stochastic control gives significantly better regulation of $m(i)$ than do the other controls.

3   FISHERY MANAGEMENT

Management of a sea fishery is a complex process (May et al 1978) governed by a variety of political, economic, and biological factors. These include:

- Dynamics of recruitment to and mortality from the harvested population which are nonlinear, stochastic and poorly understood (Beddington & May 1977, Beverton & Holt 1957, Clark & Kirkwood 1974, Schaefer 1954).

- Uncertainty about size of the population (Lockwood & Shepherd 1984).

- Economic objectives which are ill-defined and subject to political considerations (Hansard 1988).

- Various and controversial methods for implementing regulation (Pope 1982a)

- Decision procedures which involve consultation within and between several interested nations and can be subject to significant dead-time between availability of data and its use in regulation.

Problems of managing fisheries could be formulated and solved in terms of control engineering according to the block diagram of Fig 1. The controlled object is a harvested population from which fishing effort f produces catch c and population size x. The relationship between these variables is governed by uncertain internal dynamics and renewal processes of the population as well as by exogenous disturbances. Values of the catch c and size x are not precisely known, but more or less accurate estimates $\hat{c}, \hat{x}$ are used as a basis for management decisions, represented by the control law in Fig 2.

A management objective of minimising fluctuations is reflected in notation which expresses variables as the sum of an equilibrium value and a perturbation; for example

$$c = C + \Delta c, \qquad\qquad f = F + \Delta f$$

Quality of control in any one year can then be measured by a dimensionless quadratic cost function

$$I(t) = (1 - \lambda)\left(\frac{\Delta f(t)}{F}\right)^2 + \lambda\left(\frac{\Delta c(t)}{C}\right)^2 \qquad 0 \le \lambda \le 1 \qquad (11)$$

where λ is a constant which weights the relative importance of fluctuations in catch c against fluctuations in effort f. Over an extended period T years, quality of control is measured by the average sum

$$J_T = \frac{1}{T}\sum_{t=1}^{T} I(t) \qquad (12)$$

Analytic work to date about fishery management, summarised in several books (Clark 1976, 1985, Goh 1980, Mangel 1985), includes discussions of feedback control and (Walters 1986) possible application of ideas from stochastic control theory, including the use of probing action to enhance adaptive control. Because there has been little opportunity to implement proposed solutions, with all the motivation that this can generate to complete a workable if suboptimal design, the work has an applied mathematical, rather than an engineering bias. Attention has focussed on obtaining truly optimal feedback control laws for idealised mathematical models, some of which account for uncertainty affecting future states of the controlled process. There has been little discussion of how a practical feedback control might be designed which would accomodate uncertainty.

Simulation studies have been reported in which the engineers' familiar linear feedback controllers were coupled to mathematical models of fish populations (Tanaka 1980, Ballance et al 1988a, 1988b, Muratori et al 1989). One conclusion (Jacobs 1987) has been that the problem of fishery management is too nonlinear to be well controlled by such simple linear feedback controllers.

The collaborative project summarised here (Jacobs et al 1990) explored fishery management problems from the point of view of control engineering and of stochastic control theory. The controlled object, at least for initial studies, could not be a real fishery. A mathematical model of one particular isolated population (the Western stock of mackerel) was used, in the form of a set of simultaneous difference equations, one to account for the size $x_j(i)$ of each year's age-group (Horwood & Shepherd 1981). Closed loop behaviour was investigated under feedback control driven by estimates $\hat{x}_j(i)$ of population size, such as are generated by standard processes of virtual population analysis (Pope & Shepherd 1982, Lockwood et al 1981) which are used to estimate uncertain fish populations. These estimates were assumed to relate to the true values according to

$$\hat{x}_j(i) = \eta_j(i)x_j(i). \qquad (13)$$

where the noises $\eta_j(i)$, are independent random variables having log-normal distributions with mean unity and given variances σ_j^{η}. Delays in making and implementing management decisions were modelled in the form of a one year dead-time in the estimation such that the management decision $u(i)$ affecting fishing in year i was based on a prediction $\hat{x}(i \mid i - 1)$. No attempt was made to design any other estimator.

Attention was focussed mainly on designing the control law in Fig 2 to minimise fluctuations measured by equation 12. The resulting control law was found to range between strategies of constant catch and of constant effort, both of which are well known to fishery managers (May et al 1978). Numerical data were obtained which could be used to tune a controller within the range of these two extremes. Stability analysis, not reported here, lead to known results (Horwood and Shepherd 1981) about instability of the population at high levels of exploitation.

3.1 Mathematical model

Controller design was based on a reduced-order model having one single aggregated age-group with population size $x(i)$ governed by a stochastic difference equation

$$x(i + 1) = \xi(i)R\left(x(i)\right) + x(i)e^{-m-f(i)} \tag{14}$$

The recruitment function $R(x)$ is (Shepherd 1982)

$$R(x) = \frac{\alpha x}{1 + (x/k)^\beta} \tag{15}$$

where α, β, k are fitting parameters which can be matched to observed data such as is shown in Fig 7. Large, apparently random, fluctuations in recruitment data are

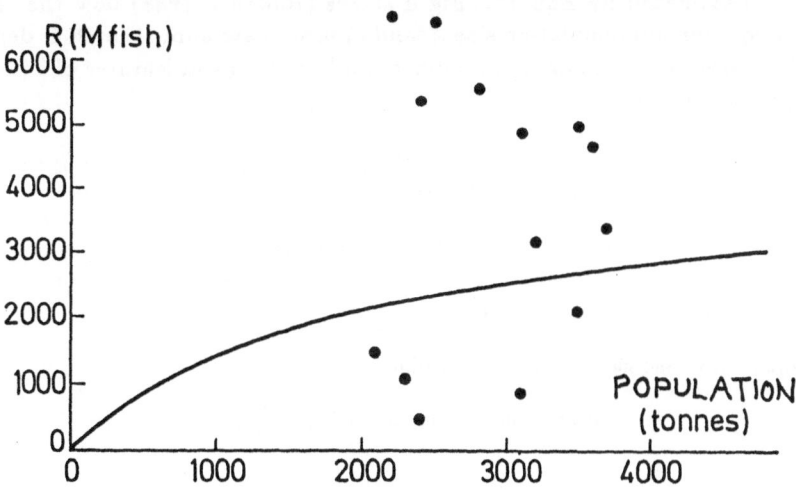

Figure 7: Observed recruitment data for Western mackerel

modelled (Garrod, 1983) by the log-normal [1] random variable $\xi(i)$ which is assumed to have mean value of unity and variance σ_ξ. The catch is

$$c(i) = \frac{f(i)}{f(i) + m} \left(1 - e^{-m-f(i)}\right) x(i) \tag{16}$$

and with low mortality ($| m + f | << 1$), as when most of the population survives from one year to the next, Eqn 16 can be approximated by

$$c(i) \approx f(i)x(i) \tag{17}$$

A convenient feature of this model is that, using Eqn 16 or 17 together with information about population size $x(i)$, the effort $f(i)$ in Eqn 14 can be replaced by an easily computed function of $x(i)$ and $c(i)$ so that catch $c(i)$ can be regarded as the actuating signal $u(i)$ for purposes of control. Thus a control law can be explicitly designed to determine annual catch quotas, corresponding to common administrative practice within the European Community of management by setting a "Total Allowable Catch "(TAC) (Pope 1982a).

Equilibrium of the model can be expressed as a function of an equilibrium value F of fishing effort $f(i)$. Ideally, equilibrium values of the states and the catch would be computed, for given F, as expected values $E[x]$ of $x(i)$ and $E[c]$ of $c(i)$, from the stochastic equations 14 to 17. Unfortunately, nonlinearities in the system make this an impracticable computation. A deterministic approximation is therefore used which is obtained by assigning to the random variable $\xi(i)$ its mean value of unity in Eqn 14 to give an implicit equation

$$X = R(X) + X e^{-m-qF} \tag{18}$$

A corresponding approximate equilibrium value C of catch can then be obtained from Eqn 16 or approximated by Eqn 17. Fig 8 shows (Ballance 1988) how the resulting approximate equilibrium population size X and approximate annual catch C depend on effort F. The curves in the figure are nonlinear and have typical features of a harvested population, in particular:

- Population X decreases as effort F is increased. There is a limiting value F_e of sustained effort at and above which the population is extinguished.

---

[1] If x is a normally distributed random variable with probability density

$$p(x) = N(\mu_x, \sigma_x) = \frac{1}{\sqrt{2\pi\sigma_x}} exp\left(\frac{(x - \mu_x)^2}{2\sigma_x}\right)$$

$z = e^x$ is always positive and has a log-normal distribution with

Mean value $\qquad m_z = exp\left(\mu_x + \frac{\sigma_x}{2}\right)$

Variance $\qquad \sigma_z = m_z^2 \left(exp(\sigma_x) - 1\right)$

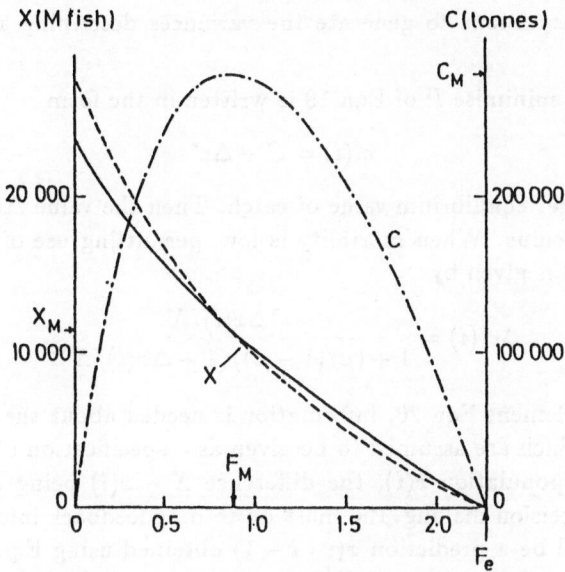

Figure 8: Equilibrium of fish population models: age-structured (full line) and simplified (dashed line). The two C-characteristics (chain-dotted) are indistiguishable

- Catch C increases with small F but decreases with large F. There is a value F_M ($F_M < F_e$) at which the catch takes a maximum value C_M, the "Maximum Sustainable Yield (MSY)". The corresponding population size is X_M.

3.2 Control

A control law was designed to minimise fluctuations about the desired equilibrium state of the simplified model. Ideally, control of fluctuations in both catch and effort might be achieved by designing a controller to minimise the expected value of the summed cost function J_T of Eqn 12 over some future time horizon T. Unfortunately, nonlinearities and uncertainties in fishery management, which cannot be neglected even in the simplified model, make optimal design impracticable. The design presented here is therefore suboptimal in two conventional respects:

1. The controller was designed to optimise, over the minimal time horizon $T = 1$, a one-step-ahead cost $I'(i)$ similar to that of Eqn 11

$$I'(i) = (1 - \mu) \left(\frac{\Delta f(i)}{F} \right)^2 + \mu \left(\frac{\Delta c(i)}{C} \right)^2 \qquad 0 \le \mu \le 1 \tag{19}$$

the weighting coefficient μ not necessarily having the same value as the coefficient λ in Eqn 11.

2. Uncertainty about the current state of the controlled process was accomodated by using certainty-equivalent control (Jacobs 1974) which proxies the uncertain value $x(i)$ in a control law with the best estimate available at time i, the prediction $\hat{x}(i \mid i - 1)$. To account for the magnitude of uncertainty, for example by using variance terms to introduce caution (Jacobs & Patchell 1972), would require introduction of

further estimator equations to generate the variances describing accuracy of current estimates.

The catch c^* to minimise I' of Eqn 19 is written in the form

$$c^*(i) = C + \Delta c^*$$

where C is the target equilibrium value of catch. Then the value Δc^* to minimise I' is easily found by calculus. When mortality is low, permitting use of Approximation 17 to eliminate f, Δc^* is given by

$$\Delta c^*(i) = \frac{C \Delta x(i)/X}{1 + (\mu/(1-\mu))(1 + \Delta x(i)/X)^2} \qquad (20)$$

In order to implement Eqn 20, information is needed about the target equilibrium values C and X, which are assumed to be given as a specification of desired state, and about the current population $x(i)$, the difference $X - x(i)$ being $\Delta x(i)$. Because of the dead-time in decision making, the most up-to-date feedback information about the population $x(i)$ will be a prediction $\hat{x}(i \mid i - 1)$ obtained using Eqn 14, and assuming that $\xi(i-1)$ took its expected value of unity, to make a certainty-equivalent prediction

$$\hat{x}(i \mid i - 1) = R(\hat{x}(i-1 \mid i-1)) + \hat{x}(i-1 \mid i-1)e^{-m- f(i-1)}$$

The estimate $\hat{x}(i-1 \mid i-1)$ of last year's population was taken to be a weighted (to allow for non-uniform biomass) sum $\hat{x}(i-1)$ of last year's estimates (Eqn 13). Substituting $\hat{x}(i-1)$ in place of $\hat{x}(i-1 \mid i-1)$ and using approximation 17 gives

$$\hat{x}(i \mid i - 1) \approx R(\hat{x}(i-1)) + (1-m)\hat{x}(i-1) - c(i-1) \qquad (21)$$

The catch $c^*(i) = C + \Delta c^*(i)$ to be implemented at time i can then be computed using $X - \hat{x}(i \mid i - 1)$ from Eqn 21 to proxy $\Delta x(i)$ in Eqn 20.

The control law of Eqn 20 is generally nonlinear as illustrated in Fig 9, where normalised catch c^*/C is plotted against normalised population x/X for different values of μ in the range $0 \le \mu \le 1$. Limiting cases at the extremes of the range in μ are:

- When $\mu = 0$, Eqn 20 becomes

$$\frac{\Delta c^*(i)}{\Delta x(i)} = \frac{C}{X}$$

  so that $c^*(i)/\dot{x}(i)$ remains constant and, with approximation 17, $f(i)$ remains constant: a **constant effort** control law. This control law, which passes through the origin and through the desired equilibrium point in Fig 9 corresponds to some of current practice (Pope, 1982b).

- When $\mu = 1$, Eqn 20 becomes $\Delta c^* = 0$ so that $c^*(i) = C$ remains constant: a **constant catch** control law. This is not a feedback control. It is well known in the fisheries literature (Clark 1985) and corresponds to what in the short run may seem convenient to the industry. It is sensitive to random fluctuations in recruitment, and could lead to extinction of the population.

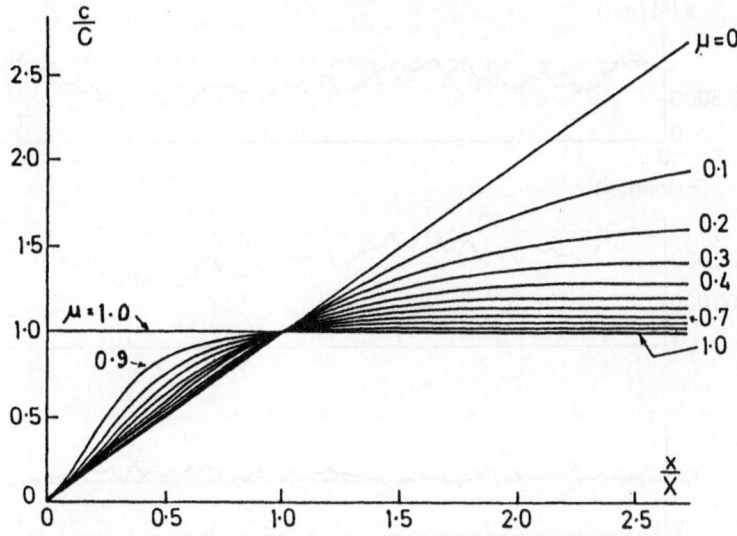

Figure 9: Suboptimal control for fishery management

At intermediate values of the tuning parameter μ, the control law lies between the extremes of constant effort and constant catch. For values of population less than the target equilibrium value ($x/X < 1$) the control law is "exploitive" in the sense that catch is greater than it would be with constant effort. For values of $x/X > 1$ the control law is "conservative" in that catch is less than it would be for constant effort. This can be interpreted as a strategy of moderation; it does not cut right back on fishing effort when the population is temporarily low, but holds back in times of temporary plenty so as to avoid large fluctuations in catch.

3.3   Simulations

Control of a full age-structured model was simulated using the control law of Eqn 20, designed for a matching simplified mathematical model. Free parameters in the simulations were:

- The desired equilibrium state, characterised by F (or X or C, all three being interelated as explained in Section 2.1.1)

- The coefficient μ in the control law of Eqn 20, which served as a tuning parameter.

Performances of different controllers were compared by using the same sequence of computer-generated pseudo-random numbers to give the same sequence of disturbances and noises in each of a set of comparative simulations. Quality of control was assessed using the average sum J_T of Eqn 12, with $\lambda = 0.5$ in Eqn 11, to represent equal weighting on fluctuations in catch c and in effort f. A standard simulation time T of 1000 years was found to give repeatable equilibrium values.

Fig 10 gives typical time-histories of aggregated population x, catch c, fishing effort f, and recruitment ξR, with desired equilibrium values shown dashed. Aggregated

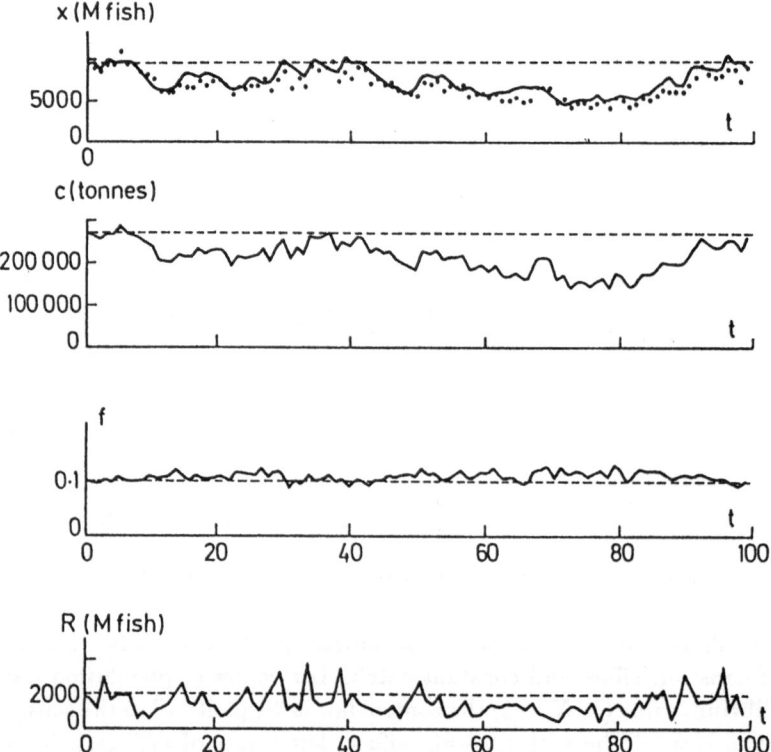

Figure 10: Time histories using suboptimal control on the age-structured fishery model

measurements y of population size are shown as dots with x. Fig 11 shows some typical average costs J_{1000} from simulating control over periods of 1000 years with different desired equilibrium operating points, specified in terms of F, and with different values of the controller tuning parameter μ over the range $0 \leq \mu \leq 1$. The point marked **A** locates the simulation of Fig 10.

It is concluded that the suboptimal feedback controller, designed to match the simplified model, can give satisfactory control of the more complex age-structured population. Costs escalate with increasing exploitation F and for each F there is an optimal value of μ giving minimal cost. At reasonably low levels of exploitation below MSY ($F < F_M$) the optimal value of μ is closer to unity (constant catch, nonfeedback control) but if exploitation is attempted above MSY ($F > F_M$), as may be necessary for reasons of short-term political expediency, the optimal μ approaches zero indicating that a constant effort, feedback control is required.

4 Conclusions

Two case studies have been summarised which show how insights from stochastic control theory can provide a general approach to designing controllers for real control systems. The approach is particularly suitable for nonlinear systems which may not be well controlled simply by using a conventional, black box, linear or adaptive controller to

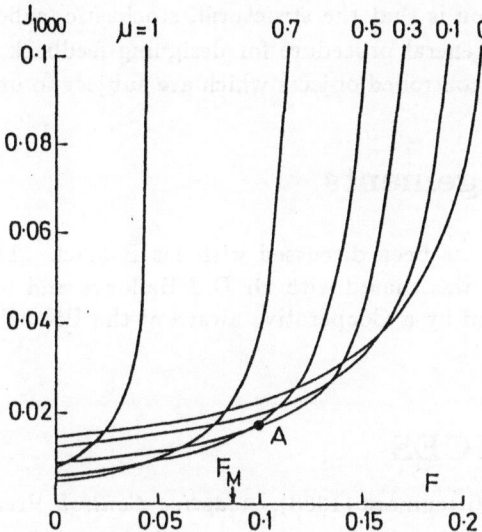

Figure 11: Average costs J_{1000} against F for various values of the tuning parameter μ. The point A locates Fig 10

close a feedback loop. The main feature of the approach is to give seperate consideration to each of the two functions of estimation and control which must be performed by a feedback controller. Neither function need be performed optimally; good control can be achieved with an estimator and a control law matched to a simple mathematical model capturing structural features of the controlled process.

Both case studies have been taken to the point where realistic simulations indicate that the structural approach gives significantly better control than would be achieved by using conventional black box control. Conclusions from the individual studies are:

Air knife regulating zinc deposition The main uncertainty here is about characteristics of the controlled process. If it were not for the time delay affecting measurement of deposited mass, good control might be achieved by closing the feedback loop with a conventional linear controller. Because of the time delay, the feedback should drive an estimator updating an adaptive parameter in a predictive model of knife action. This model can then used to compute control actions.

Management of an isolated fishery Fishery management requires exercise of control in the presence of large uncertainties about current and future states of a severely nonlinear controlled process. The estimation function is performed by standard processes of virtual population analysis (Pope & Shepherd 1982, Lockwood et al 1981) which have not been discussed here. A suboptimal, certainty-equivalent control law has been derived to regulate fluctuations about a specified equilibrium state. This control law includes as special cases strategies of constant effort and constant catch which are well known to fishery managers.

The overall conclusion is that the structural, stochastic, suboptimal approach can be recommended as a general procedure for designing feedback controllers to achieve desired performance of controlled objects which are subject to uncertainty.

5   Acknowledgements

The air knife problem has been discussed with Dr H Wick. The collaborative work on fishery management was shared with Dr D J Ballance and with Mr J Horwood of MAFF; it was supported by a Cooperative award of the UK Science and Engineering Council.

6   REFERENCES

Åström K.J. and B. Wittenmark (1988). *Adaptive Control.* Prentice Hall.

Ballance. D.J. (1988). *Control Methodology Applied to National Fishing Strategies.*, D. Phil Thesis, U. of Oxford.

Ballance, D.J., O.L.R. Jacobs, and J.W. Horwood (1988a). Application of CADCS Methods to a problem in Fishery Management. *4th IFAC Symp. on CADCS*, Beijing.

Ballance, D.J., O.L.R. Jacobs, and J.W. Horwood (1988b). State Estimation in Regulating a Harvested Population. *8th IFAC/IFORS Symp. Ident. and Par. Est.*, Beijing.

Beddington, J.R., and R.M. May (1977). Harvesting Natural Populations in a Randomly Fluctuating Environment. *Science*, **197**,463–465.

Bertin M.C., Brown L., Whitehead R.L. and E. Yrisarri (1978). Computer Control of Coating Weight on Galvanizing and Tinning Lines. *7th IFAC World Congress, Helsinki.* pp 199-206.

Beverton, R.J.H. and S.J. Holt (1957). On the dynamics of exploited fish populations. *Fishery Invest., Lond.*, Ser. 2, **19**.

Biermann G.J. (1977). *Factorisation Methods for Discrete Sequential Estimation.* Academic Press.

Butler J.J., Beam D.J. and J.C.Hawkins (1970). The Development of Air Coating Control for Continuous Strip Galvanizing. *Iron and Steel Engineer*, Feb, pp 77-86.

Clark, C.W. (1976). *Mathematical Bioeconomics: The Optimal management of Renewable Resources.* Wiley, New York.

Clark, C.W. (1985). *Bioeconomic Modelling and Fisheries Management.* Wiley, New York.

Clark, C.W. and G.P. Kirkwood (1974). On uncertain renewable resource stocks: optimal harvest policies and the value of stock surveys. *Tech. Rep. Inst. Appl. Math. Stats.*, **84-4**, University of British Columbia, Canada.

Edwards W.J., Carlton A.J., Harvey G.F., Evans R.F. and P.J. McKerrow (1975). Continuous Mass Control System Design for a Continuous Galvanizing Line. *6th IFAC World Congress, Boston*, **46.3**.

Ellen C.H. and C.V. Tu (1983). An Analysis of Jet Stripping of Molten Metallic Coatings. *Fluid Mechs Conf, Univ of Newcastle, N.S.W.*

Fel'dbaum A.A. (1965). *Optimal Control Systems*. Academic Press.

Friedland B. (1986). *Control System Design*. Prentice-Hall.

Garrod, D.J. (1983). On the Variability of Year-class Strength. *J. Cons. Int. Explor. Mer*, **41**, 63–66.

Goh, B.S. (1980). *Management and Analysis of Biological Populations*, Chap 4, Elsevier, Amsterdam.

Hansard (1988). HMSO Vol 143, 12 Dec, 664–676.

Harvey G.F. and A.J. Carlton (1974). Mathematical Modelling of Air Jet Coating Mass. *John Lysaght (Australia) Ltd Res. and Tech. Report*, **533**.

Horwood, J.W. (1982). The variance of population and yield from an age-structured stock, with application to the North Sea herring. *J. Cons. Int. Explor. Mer*, **40**, 237–244.

Horwood, J.W., and J.G. Shepherd (1981). The Sensitivity of Age-Structured Populations to Environmental Variability. *Math. Biosci.*, **57**, 59–82.

Jacobs O.L.R. (1974). *Introduction to Control Theory*. OUP.

Jacobs O.L.R. (1982). Integral control action and separated stochstic control. *IFAC Symp. Theory and Applications of Digital Control. Delhi.*

Jacobs, O.L.R. (1987). Management procedure and stochastic control theory. *IWC Comprehensive assessment workshop on management*, Section 6.3, Reykjavik.

Jacobs O.L.R. (1989). Adaptive cautious control of an air knife. OUEL internal report.

Jacobs O.L.R., Badran W.A., Proudfoot C.G. and C.While (1987). On Controlling pH. *Proc IEE D*, **134D**, 3, pp196-200.

Jacobs O.L.R., Ballance D.J. and J.W. Horwood (1990). Fishery Management as a Problem in Feedback Control. In press, *Automatica.*

Jacobs O.L.R., Bullingham R.E.S., Lammer P., McQuay H.J., O'Sullivan G. and M.P. Reasbeck (1985). Modelling Estimation and Control in the relief of Post-Operative Pain. *Automatica*, **21**, 4, pp 349-360.

Jacobs O.L.R. and J.W. Patchell (1972). Caution and Probing in Stochastic Control. *Int J. Cont.*, **16**, 1, pp189-199.

Lockwood, S.J., J.H. Nichols, and W.A. Dawson (1981). The Estimation of a Mackerel (Scomber scombrus L.) Spawning Stock Size by Plankton Survey. *J. Plankton Research*, **3**(2), 217–233.

Lockwood, S.J., and J.G. Shepherd (1984). An assessment of the Western stock of Mackerel. *J. Cons. Int. Explor. Mer*, **41**, 167–180.

Mangel, M. (1985). *Decision and Control in Uncertain Resource Systems*, Chap 6. Academic Press, New York.

May, R.M., J.R. Beddington, J.W. Horwood, and J.G. Shepherd (1978). Exploiting Natural Populations in an Uncertain World. *Math. Biosci.*, **42**, 219–252.

Muratori,S., S.Rinaldi, and B.Trincher (1989). Performance evaluation of positive regulators for population control. *Modelling, Identification and Control*, **10**, 125-134.

Noe T.G. and T.R. Schuerger (1975). Computer Control of Zinc-Coating Weight on the Fairfield No 4 Galvanizing Line. *6th IFAC World Congress, Boston*, **46.2**.

Pope, J.G. (1982a). Background to Scientific Advice on Fisheries Management. *Lab. Leafl., M.A.F.F. Direct. Fish. Res., Lowestoft*, (**54**).

Pope, J.G. (1982b). Analogies to the Status Quo T.A.C.'s: Their Nature and Variance. *Can. Spec. Publ. Fish. Aquat. Sci.*, **66**, 99–113.

Pope, J.G., and J.G. Shepherd (1982). A Simple Method for the Consistent Interpretation of Catch-at-Age Data. *J. Cons. Int. Explor. Mer*, **40**, 176–184.

Reid P., (1972). The Theory of Jet Stripping, Part 1. *John Lysaght (Australia) Ltd Res. and Tech. Report*, **465**.

Schaefer, M.B. (1954). Some aspects of the dynamics of populations important to the management of commercial marine fisheries. *Bull. Inter-Amer. Trop. Tuna Comm.*, **1**, 25–56.

Sendzimir T., (1938). Process for Coating Metallic Objects with Layers of Other Metals. *US Patent*, 2110893.

Shepherd, J.G. (1982). A Versatile New Stock-recruitment Relationship for Fisheries, and the Construction of Sustainable Yield Curves. *J. Cons. Int. Explor. Mer*, **40**(1), 67–75.

Tanaka, S. (1980). A Theoretical Consideration on the Management of a Stock-Fishery system by Catch Quota and its dynamical properties. *Bull. Japan. Soc. Sci. Fish*, **46**, 1477–1482.

Thornton J.A. and H.F. Graff (1976). An Analytical Description of the Jet Finishing Process for Hot-Dip Metallic Coatings on Strip. *Metallurgical Trans. B*, **7**b, pp 607-618.

Walters, C. (1986). *Adaptive Management of Renewable Resources*. MacMillan.

Wilhelm R.G. Jnr, (1972). *Control Eng.*, Nov, pp 44-47.

Wynne D.C., Vickers A.L. and F.Graham Williams (1987). Galvanizing Developments in the U.K. *Iron and Steel Eng.*, July, pp 40-47.

DESIGN OF A MULTIVARIABLE SELF-TUNING CONTROLLER FOR A CLASS OF DISTRIBUTED PARAMETER SYSTEMS

B.Rohal-Ilkiv,P.Zelinka,R.Richter,I.Sroka,Z.Orszaghova
Department of Automatic Control and Measurement
Slovak Technical University
Nam.Slobody c.17, 812 31 Bratislava,Czechoslovakia

May 18, 1990

Abstract. This paper presents a simple method for designing digital adaptive multivariable controllers intended for the control of a class of continuous-time linear distributed parameter systems. To cope with infinite-dimensionality nature of the systems the spline functions technique is used for the approximation of the spatially distributed output signals and the kernels of the operators describing the systems. In this way the various kinds of simplified finite-dimensional input-output models are obtained. The resulting models can be recursively identified by standard algorithms and easily implementable for the self-tuning controllers design. Based on the formulation of the integral quadratic cost functional and its approximation two types of suboptimal controllers are constructed: first one generating usual piecewise constant control input signals and the second one generating continuous time control input signals in the form of piecewise polynomial curves - spline functions of given order. The results of a simulation study are included to illustrate the feasibility of the derived controllers.

Keywords. Distributed parameter systems; spline approximation; finite-dimensional models; identification; self-tuning control.

1 INTRODUCTION

In many industrial applications it is necessary to stabilize or to control some important quantities which depend not only on time but also on spatial coordinates.These problems can be solved by a variety of techniques developed for the control of distributed parameter (infinite-dimensional) systems (DPSs).The detailed mathematical modeling of the DPSs leads to partial differential equations (PDEs).Generally, the DPSs are described by a set of coupled partial differential (integrodifferential) equations.It should be stressed that depending on the shape of spatial region and on the type of the PDEs (which may include nonlinearities) the problems encountered in the attempt on the self-tuning control of the DPSs are often very complex.Therefore for engineers a common way of treating these problems is to find an appropriate approximation of the system behaviour and to control law synthesis conserving the distributed nature of the problem on reasonable level.

In the past years a number of adaptive control schemes including self - tuning and model reference control ones have been developed and/or tested for the DPSs, see Hamza and Sheirah (1978);Vajta (1982); Watanabe,Yamamoto and Omatu (1986);Kobayashi (1988).The schemes are based on both continuous time and discrete time formulations using different input-output and the state-space models and mostly rest upon the given PDEs.In this paper we consider an alternative approach suitable for the DPSs,the dynamic of which is represented with linear nonparametric models given in the form of convolution integrals. To ensure a finite-dimensionality character of the solved problem the unified viewpoint to approximation of the space and time functions (involved in the model and cost functional) based on spline functions, is adopted throughout the paper.

The essence of the submitted procedure is the above spline approximation applied to the discrete time and/or space sampled signals (of lumped ,or distributed type) and/or to the kernels of the convolution-type nonparametric linear model of the system.This step constitutes a nontraditional way of emphasizing the continuous time and continuous space nature of the controlled DPSs and promises to achieve a higher quality of control.In the separate sampling time instants the spatially distributed output signal is approximated and expressed in the form of a finite sum of products of random variables (spline coefficients) and spline basis functions of defined properties.This decomposition of the distributed output signal forms a basis for the creation a finite dimensional and input-output representation of the DPS. The continuous input/output signals are dealt with in terms of spline coefficient vectors with the smallest possible dimensions. The overall dimension of the representation and also of the resulting controller first of all depends on desired level of approximation error. The deduced models and the quadratic cost functional under rational conditions can be parameterized and the routine techniques can be applied to model parameters estimation and to criterion minimization. The motivation for the work were some problems encountered in designing self-tuning control of a temperature fields in industrial conditions.

Some standard questions encountered in the implementation of DPSs computer control are now seen from the following point of view:

- <u>sensors location</u>,in most DPSs , the measurements are taken at a finite number of spatial locations; this raises the problem of defining a suitable number and locations of the sensors. Here the only criterion for solving the problem is a required level of the approximation error of the functions we are interested in.

- <u>sampling rate</u> is still a crucial point in the digital control systems design. The above attempt at continuous time and space modeling makes the final controller law less sensitive to the time frequency of sampling of the input/ output signals. The frequency of sampling is selected to meet the identification and approximation processes demands. The proper input control signals can be designed as spline functions with given break points (imitating the sampling instants) and produced in the such way to get closer to their continuous nature.

To construct the adaptive system utilizing the above ideas,the following assumptions are indispensable:

- the distributed parameter systems are linear (linearizable) and time invariant ,or slowly time varying

- the signals and kernels of the convolutions integrals belong to function spaces which can be approximated by properly chosen splines on reasonable level.

- the past history of the controlled process has only a limited time effect for future process behaviour

- the DPS spatial domain is a fix region in the Euclidean space

In the contribution,the one dimensional spatial domain is considered for making the presentation simpler. The generalization to multidimensional spatial domains is in principle the same but with a larger computational burden.

Basic notation. Throughout the paper the following notational convention is introduced:

$x = [x_1, x_2, \ldots, x_{dx}]^T$ vector x with dx entries

$X = [x_{i,j}](i = 1, \ldots, dx), (j = 1, ..ex), (dx \quad x \quad ex)$ matrix X

T time horizon of interest

$[0, L] \in CR^i$ one dimensional spatial domain

$fs(x), fp(x, t)$ - spline approximation of a function $f(x), f(x, t)$; all quantities γ introduced in subsequent chapter(scalars,vectors,functions) and necessary for spline $fs(x), fp(x, t)$ definition are designed as $\gamma f, \gamma fp$.

2 APPROXIMATION BY SPLINES

The functions (signals,kernels) involved in the problem and defined on intervals $[0, L], [0, T], [0, L]x[0, T]$ are approximated by univariate and bivariate splines. In the paper we make use of a B-spline basis to represent the spline (B-representation of spline functions).Such a basis permits splines of any order, uniform and nonuniform knots sets, simple and multiple knots ,and various orders of continuity to be used. Moreover,it provides a ready ability to incorporate boundary information, generally good numerical conditioning and straightforward extensions to other splines.

Each B-spline is a *bell-shaped* function and has compact supports: a B-spline of order r (degree $r - 1$) is nonzero only over an interval spanning $r + 1$ knots. It is strictly positive within the open interval spanned by these knots.As a result of this compact support, the system of linear algebraic equations defining the coefficients of a spline interpolant or approximant has a special structure (band form). The B-spline coefficients in a spline interpolant or approximant tend to mimic the measured values from which they were obtained. The definition of spline functions in the terms of B-splines is also advantageous since the number of the unknown coefficients to be determined is minimal.

Univariate splines Let $[a, b]$ is a closed interval of a real line. Consider the strictly increasing sequence of real numbers:

$$a = \lambda_0 < \lambda_i < \ldots < \lambda_q = b$$

representing the partition of $[a, b]$ into q subintervals. Then the function $s(x)$ is called a spline of degree r on $[a, b]$, with knots λ_i, $i = 1, \ldots q - 1$, if:

- in each interval $[\lambda_i, \lambda_{i+1}]$, $i = 0, 1, 2, \ldots q - 1$, $s(x)$ is given by a polynomial of degree r or less

- $s(x)$ and its derivatives of orders $1, 2, \ldots r - 1$ are continuous everywhere in $[a, b]$.

Every such spline has a unique representation of the form:

$$s(x) = \sum_{i=1}^{z} c_i M(x_i) \tag{1}$$

in which $M(x)$ is i-th B-spline of order r specified for the knots sequence $\lambda = [\lambda_0, \lambda_1, \ldots \lambda_q]$ and c_i, $i = 1, 2, \ldots z$ are spline coefficients. Dimension z of the spline equals :

$$z = \sum_{i=1}^{q-1} \alpha_i + r \tag{2}$$

where α_i is a multiplicity of knots λ_i (if all knots are simple i.e. $\alpha_i = 1$ for all i, then $s(x)$ belongs to $C^{r-2}[(a, b)]$). Let us order the coefficients c_i and B-splines $M_i(x)$ to z dimensional vectors:

$$c = [c_1, c_2, \ldots, c_z]^T \tag{3}$$

$$m(x) = [M_1(x), M_2(x), \ldots M_z(x)]^T \tag{4}$$

$$s(x) = c^T m(x) \tag{5}$$

Bivariate tensor product spline Bivariate splines can be represented in terms of a tensor product of two univariate B-splines bases. Let a closed rectangular domain $D = [a, b] x [0, T]$ be given. Then the strictly increasing sequences of real numbers:

$$a = \lambda_0 < \lambda_1 < \ldots < \lambda_q = b$$

$$0 = \mu_0 < \mu_1 < \ldots < \mu_h = T$$

define a rectangular grid on D, and spline $s(x, t)$ of degree r in x and l in t, with knots λ_i, $i = 1, 2, \ldots, q$ in the x direction and $\mu_j, j = 1, 2, \ldots, h$ in the t direction , has a unique representation of the form:

$$s(x, t) = \sum_{i=1}^{z} \sum_{j=1}^{v} c_{i,j} M_i(x) N_j(t) \tag{6}$$

where the $M_i(x)$ are the above B-spline and $N_j(t)$ denotes the B-splines in variable t of order l for knot sequence $\mu = [\mu_1, \mu_2, \ldots \mu_h]$. Suppose that $c_{i,j}$, and $(M_i(x)N_j(t))$ for $i = 1, 2, \ldots z$, $\quad j = 1, 2, \ldots v$ are ordered to zv-dimensional vectors c and $m(x,t)$, then for (6) it holds:

$$s(x,t) = c^T m(x,t) \tag{7}$$

For detailed discussion see De Boor (1978), Schumaker (1981) and Korneitschuk (1984).

3 THE DPS INPUT-OUTPUT REPRESENTATION

The DPS is supposed to be practically controlled through a finite number of actuators acting on $[0, L]$ domain (distributed control). The actuators are directly manipulated by variables of lumped character $u_j(t)$, $\quad j = 1, 2, \ldots du$, $\quad du$ is the number of the actuators. The adjustment of the variables is performed by the controller.In the following explanation we shall refer to du dimensional vector $u(t)$

$$u(t) = [u_1, (t), u_2(t), \ldots u_{du}(t)]^T \tag{8}$$

as to the input vector of the DPS.An optimal number of the actuators and their optimal location is assumed to enable us to achieve the required system behaviour.

Let us further consider that the infinite dimensional output signal $y(x,t), x \subset [0, L], t > 0$ is observed at dy discrete locations in the spatial domain with coordinates x_i, $x_i C[0, L], i = 1, \ldots dy$ and $x_1 < x_2, \ldots < x_{dy}$. If necessary, $x_i = 0, x_{dy} = L$. Let us arrange the values $y(x_i, t_k)$ measured at sampling time instant $t_k, k = 1, 2, \ldots$, tody dimensional vector $y(t_k)$:

$$y(t_k) = [y(1,k)y(2,k), \ldots, y(dy,k)]^T \tag{9}$$

where to short we use notation $y(x_i, t_k) = y(i, k)$ for all $i = 1, 2, \ldots dy$, $\quad k = 1, 2, \ldots$.

3.1 Discrete time,continuous space case

From the view point of an outer observer in each sampling time instant t_k the distributed output signal $y(x,t), x \subset [0, L], t = t_k$ of the given DPS seems to be realization of a stochastic function $\{y(x,t), x \subset [0,L], t = t_k\}$ defined on the parameter set $[0,L]$. This realization is conditioned by the values of the measurable and also unmeasurable variables acting on the system. To represent this stochastic function the families of finite-dimensional probability distributions are obviously needed. More practical representation can be found using properly chosen spline function (1) for the approximation of $y(x, t_k)$ the coefficient vector (3) of which forms a finite dimensional random vector. The probability distribution of the random vector then determines the be- haviour of the stochastic function.A large sub class of real stochastic function can be defined in similar way.(see Jazwinsky,1970) Let as formally denote the coefficient vector relating to time instant t as $cy(k)$. Then a sequence of conditional probability distributions describing the dependence of the output vector $cy(k)$ on the known past history of the system observed input-output data forms the discrete time, finite dimensional input-output representation of the DPS. Generally we can express this representation by the sequence of conditional probability distributions :

$$p_k(cy(k)/u(k), cy(k-1), u(k-1), cy(k-2), \ldots, \quad k = 1, 2, \ldots) \quad (10)$$

For lumped multivariable systems the input-output representation of system dynamics based on conditional probability distributions was introduced in Peterka(1981), where the general approach to Bayesian system identification is systematically developed.This approach can be adopted in the problems of parameterization of the representation (10). Arbitrary mathematical model defining the set (10) of conditional probability distributions by means of finite number of parameters can be considered as a DPS model.As shown in Karny (1985),to determine the suboptimal control strategy for quadratic cost functions it is sufficient to know only the first two conditional moments of the distribu- tions (10). Mostly they can be present in the form of linear regression models.

3.2   Continuous time and space case

For deterministic and time invariant linear DPSs the relation between input (8) and the output $y(x,t)$ is given by convolution integrals:

$$y(x,t) = \sum_{k=1}^{du} \int_0^t G(x,k,t-\tau)u_k(\tau)d\tau \quad (11)$$

$$t \subset [0,T] \quad xc[0,L]$$

where the unknown kernels $G(x, k, t - \tau), k = 1, 2, \ldots, du$ are Green's functions,or space-time impulse response functions.

Let us approximate the output signal $y(x, t)$ and the input signals $u_k(\tau)$ by spline $ys(x, t) = cy^T my(x)$ (univariate spline in x direction at time instant t) and by spline $us_k(\tau) = cu^{kT} mu^k(\tau)$ and substitute the tensor product spline approximants $Gp(x, k, t - \tau) = cGp^{kT} mGp^k(x, t - \tau), \quad k = 1, \ldots, du$, for the unknown kernels $G(x, k, t - \tau)$; consider that the relations among the approximants can be described by the same structure as (11); then (11) can be converted to the relation:

$$ys(x, t) = \sum_{k=1}^{du} \int_0^t Gp(x, k, t - \tau) us_k(\tau) d\tau \tag{12}$$

For $k - th$ member of the above sum holds:

$$\int_0^t Gp(x, k, t - \tau) us_k(\tau) d\tau =$$

$$= cGp^{kT} \int_0^t mGp^k(x, t - \tau) mu^{kT}(\tau) d\tau \quad cu^k =$$

$$= \sum_{j=1}^{zGp^k} \left[cGp_j^{kT} A^k(t) cu^k \right] MGp_j^k(x) \tag{13}$$

where $cGp_j^k = [cGp_{j,1}^k, cGp_{j,2}^k, \ldots cGp_{j,vGp^k}^k], j = 1, 2, \ldots zGp^k$, are subvectors of the vector cGp_j^k; $A^k(t) = (vGp^k \quad x \quad zu^k)$ -matrix with $i, j - th$ entry defined by integral:

$$A_{i,j}^k(t) =$$

$$= \int_0^t NGp_i^k(t - \tau) Mu_j^k(\tau) d\tau$$

$$i = 1, \ldots vGp^k \quad j = 1, 2, \ldots zu^k \tag{14}$$

A simple analysis of (13) and (12) shows that basis splines $My_i(x), i = 1, 2 \ldots zy$, in the x direction of the approximant $ys(x, t)$ are shaped by basis splines of approximants $Gp(x, k, t - \tau), k = 1, 2, \ldots du$, in the x direction.

Keeping in mind the reasonable complexity of the resulting model,suppose that the kernels $G(x, k, t - \tau), k = 1, 2 \ldots du$, belong to the same function space approximable by spline space defined on a properly chosen basis $MGp_i(x), NGp_j(t), i = 1, \ldots zGp, j = 1, \ldots vGp$, i.e:

$$zGp^k = zGp$$

$$\begin{aligned} Gp^k &= vGp \\ dMGp_i^k(x) &= MGp_i(x) \\ NGp_j^k(t-\tau) &= NGp_j(t-\tau) \quad k = 1, 2, ...du \end{aligned} \qquad (15)$$

and if we similarly suppose the same basis splines $Mu_j(\tau), j = 1, 2, ...zu$ for the input signals $u_k(\tau)$, $k = 1, 2...du,$.i.e.:

$$\begin{aligned} zu^k &= zu; \\ Mu_j^k(\tau) &= Mu_j(\tau) \\ j = 1, \ldots, zu \quad & k = 1, \ldots, du \end{aligned} \qquad (16)$$

then we can put the relation (12) for approximant $ys(x, t)$ into the form:

$$ys(x, t) = cy^T my(x) = [G \ F(t) \ cu(t)]^T mG(x) \qquad (17)$$

$$G = \begin{bmatrix} cGp_1^{1T} & cGp_1^{2T} & \ldots & cGp_1^{zuT} \\ cGp_2^{1T} & cGp_2^{2T} & \ldots & cGp_2^{zuT} \\ \ldots & \ldots & \ldots & \ldots \\ cGp_{zG}^{1T} & cGp_{zG}^{2T} & \ldots & cGp_{zG}^{zuT} \end{bmatrix}$$

$G - (dG \times eG)$ matrix of the unknown spline coefficients of the approximants $Gp(x, k, t - \tau), k = 1, ..du, dG = zGp = zy, eG = (vGp \ zu)$.

$$F(t) = \begin{bmatrix} A(t) & & \\ & A(t) & \\ & & A(t) \end{bmatrix}$$

$F(t) - (dFxeF)$ matrix, $dF = (vGp \ zu), eF = (zu \ du)$; the $A_{i,j}(t)$ entry of the matrix $A(t)$ is given by the integral (12):

$$\begin{aligned} A_{i,j}(t) &= \int_0^t NGp_i(t-\tau)Mu_j(\tau)d\tau \\ cu^T(t) &= [cu^{1T}(t), cu^{2T}(t), ...cu^{duT}(t)] \end{aligned} \qquad (18)$$

$cu(t)$-vector of spline coefficients of the approximants $us_k(t), k = 1, 2, ...du,$ argument t in $cu(t)$ means approximation up to time t

$$mG(x) = [MGp_1(x), MGp_2(x), ..., MGp_{zG}(x)]^T$$

$mG(x)$ - spline basis vector in the x direction.

At time instant $t, t \subset [0, T]$, the approximant $ys(x, t)$ is specified by time invariant basis in the x direction :

$$my(x) = mG(x) \qquad (19)$$

and by time - variable spline coefficient vector :

$$cy(t) = G \ F(t) \ cu(t) \qquad (20)$$

The obtained equation (20) represents a finite dimensional approximate model simply describing relations between the *input* $cu(t)$ and the *output* $cy(t)$. The known matrix $F(t)$ is working as a filter of the input vector $cu(t)$; adopting this point of view then:

$$cy(t) = G \ cu_f(t) \qquad (21)$$

with

$$cu_f(t) = F(t) \ cu(t)$$

and G is a matrix of unknown *model parameters* to be determined. Using the relation between B-representation of splines and piecewise polynomial representation (see De Boor ,1978) we can convert equality (20) to the following useful form :

$$cy(t) = PM \ cup(t) \qquad (22)$$

where vector $cup(t)$ contains coefficients of all individual polynomial pieces representing the inputs $u_i(\tau) \quad, i = 1, 2, \ldots, du$, over time interval $\tau \in [t, t - Tm]$ in which it is significant to approximate elapsed inputs :

$$cup(t) = [pu_i(t - 1), ..pu_{du}(t - 1), ..., pu_i(t - m), ..pu_{du}(t - m)]^T \qquad (23)$$

m - number of polynomial pieces in interval $[t, t - Tm]$
PM - matrix of unknown model parameters for formulation (22).
$pu_j(t - k)$ - vector of coefficients of the polynomial which illustrates the $j - th$ inputs $u_j(\tau)$ in the $k - th$ piece, $j = 1, 2, ..du, \ k = 1, 2...m$.
The process finite memory assumption guarantees us that the only finite length Tm of the inputs $u_i(t), i = 1, 2, ..du$, is important for approximation.
Comment: The selection of the number and positions of the above approximant knots (breakpoints) is an important and difficult problem. In principle, the positions of the knots are free parameters occurring in a nonlinear way. Strictly speaking, the optimal knot positions are time varying. Taking into account :

the desired level of the approximation error

a priori information and physical reasoning

experimental data from the system

the number and positions of the knots can be determined in a sub-optimal way.The approximants degrees can be processed in a similar fashion.Due to periodic time sampling of the input and output signals $u_k(\tau), k = 1, ..., du,$ $y(x, \tau)$ we will stress equidistantly located knots of the approximants in the τ direction on regions $[t, t - Tm], [0, L]x[t, t - Tm]$.

Identification :The number dy of the measurement points,locations x_i and the sampling rates are taken to create good preconditions for identification and to fulfil necessary conditions for solving the spline approximation problems involved.The spline supports in the t direction are of finite length and are shifted with time.Then for the time invariant rectangular grid on region $[0, L]x[t, t - Tm]$ the matrix $F(t) = F(t_k), (k = 0, 1, ..., t_k = kTI,$ TI - time step of identification) do not depen d on time, $F(t_k) = F$, and can be computed in advance .In each identification step tk two activities are carried out:coefficient vectors $cy(t_k)$ and $cu(t_k)$ or $cup(t_k)$ of approximants $ys(x, t_k)$ and $us_i(t_k)$,$i = 1, 2, ..., du$,are calculated utilizing the measured data.In order to carry out the above approximation,the spline interpolation for the $u_i(t)$ and least-square approximation for the $y(x, t)$ is preferred.Based on (20) or (22) the unknown parameter matrix is estimated by standard off-line or on-line algorithms.

4   APPROXIMATION OF OPTIMAL CONTROL

To determine (sub)optimal control input signal u(t) we shall minimize the following two quadratic cost functionals:

$$
\begin{aligned}
J(u) &= \frac{1}{H} \sum_{k=1}^{H} \Big[\int_0^L \big[y(x, k - yr(x, k) \big]^2 w(x, k) dx + \\
&+ (u(k) - ur(k))^T Q(k)(u(k) - ur(k)) \Big], \\
&\quad u(k) \subset U_{ad}(k)
\end{aligned}
\tag{24}
$$

where:

$yr(x, k)$ - output reference signal
$ur(k)$ - suitably chosen input reference signal
$w(x, k)$ - weighting function
$Q(k)$ - input penalty matrix $(du \times du)$
$Q_{ad}(k)$ - admissible class of input functions
H - control horizon

The control horizon H, penalty matrix $Q(k)$ and reference values are selected by user. The above criterion requires to specify precisely the spatially distributed signals $y(x, k), yr(x, k)$ in the separate sampling time instants $t_k, k = 1, 2, \ldots, H$. The criterion (24) so is more or less of a theoretical meaning.In order to obtain more practical criterion we replace the functions $y(x, k), yr(x, k)$ by spline approximations. After substitution to (24) and minor derivation we obtain the criterion in the modified form :

$$J(u) = \frac{1}{H} \sum_{k=1}^{H} \left[q(k) + (u(k) - ur(k))^T Q(k)(u(k) - ur(k)) \right] \qquad (25)$$

with $q(k)$ written as

$$q(k) = cy^T(k)Q_1(k)cy(k) - 2cy^T(k)Q_2(k)cyr(k) + cyr^T(k)Q_3cyr(k) \qquad (26)$$

For matrices $Q_1(k), Q_2(k), Q_3(k)$ it can be found:

$$Q_1(k) = \int_0^L my(x, k)my^T(x, k)w(x, k)dx \qquad (27)$$

$$Q_2(k) = \int_0^L my(x, k)myr^T(x, k)w(x, k)dx \qquad (28)$$

$$Q_3(k) = \int_0^L myr(x, k)myr^T(x, k)w(x, k)dx \qquad (29)$$

The matrices are generally time variable,depending on the spline knots locations.The minimization of the above criterion (25) leads to the complicated problem of nonlinear optimization.The situation becomes easier in the regulation problem,where the task is to stabilize the distributed system near given time nonvariable reference values.If we select for the spline approximation the same time nonvariable basis, then:

$$Q_1(k) = Q_2(k) = Q_3(k) = Q$$

and

$$q(k) = \Big[cy(k) - cy(k)\Big]^T Q\Big[cy(k) - cyr(k)\Big] \tag{30}$$

The matrix Q can be computed in advance and to minimize criterion (25) with $q(k)$ defined by (30) the standard techniques for lumped multivariable system can be employed. For detailed discussion the case with time variable knots see Zelinka,(1989).

4.1 Continuous time and space case

In order to grasp better the continuous nature of the controlled system we will assume following functional :

$$\begin{aligned}
J(u) \;=\;& \frac{1}{Th} \int_0^{Th} \Big[\int_0^L \big[y(x,t) - yr(x,t)\big]^2 w(x,t)dx \;+ \\
&+\; (u(t) - ur(t))^T Q(t)(u(t) - ur(t))\Big], \\
& u(t) \subset U_{ad}(t)
\end{aligned} \tag{31}$$

where: Th - control horizon; the other symbols are of the
same meaning as in criterion (25)
The minimization of the cost functional (31)
will be performed under the following assumptions:

the output signal $y(x,t)$ and reference signal $yr(x,t)$ are replaced by tensor product spline approximants $yp(x,t)$, $yrp(x,t)$ defined over the same basis used in the model approximation;the same holds for input signals $u(t), ur(t)$

the admissible class $U_{ad}(t)$ of the input functions $u(t)$ can be expressed by upper and lower bound vectors $cu^u(t), cu^u(t)$ for the spline coefficient vector $cu(t)$ of the approximant $us(t)$

the parameters matrix PM is completely known. Applying the above assumptions to the cost functional (31) we find that:

$$J(cu) = \left[cyp(Th) - cyp^r(Th) \right]^T Qy \left[cyp(Th) - cyp^r(Th) \right] +$$

$$+ \left[cu(Th) - cu^r(Th) \right]^T Q_2 \left[cu(Th) - \right.$$

$$- \left. cu^r(Th) \right] cu(Th) \subset \left[cu^l(Th) cu^u(Th) \right] \qquad (32)$$

$Qy - (dQy \times eQy)$ penalty matrix with $dQy = (zG \quad vyp)$, $eQy = (zG \quad vyp)$ defined

$$Qy = \frac{1}{Th} \int_0^{Th} \int_0^L myp(x,t) myr^T(x,t) w(x,t) dx dt \qquad (33)$$

$myp(x,t)$-vector of products $MGpi(x) Nyp_j(t)$, $i = 1, ...zG, j = 1, ..., vyp$ $Q_2 - (dQ_2 \times eQ_2)$ penalty matrix $dQ_2 = (du \quad zu), eQ2 = (du \quad zu)$ defined by relation:

$$Q_2 = \frac{1}{Th} \int_0^{Th} Q_m^T(t) Q(t) Q_m(t) dt \qquad (34)$$

with

$$Q_m(t) = \begin{bmatrix} mu^T(t) & & \\ & mu^T(t) & \\ & & mu^T(t) \end{bmatrix}$$

$Qm(t) - (dQ_u \times eQ_u)$ matrix, $dQ_u = du$, $eQ_u = (du \quad zu)$

$$mu(t) = \left[Mu_1(t), Mu_2(t), ..., Mu_{zu}(t) \right]^T$$

The $(zG \quad vyp)$ - dimensional vector $cyp^r(Th)$ of the spline coefficients is calculated for the prescribed reference profile $yr(x,t)$. Since the model (22) used for control synthesis is only able to describe the time development of the coefficient vector $cy(tk)$, of the spline-paths on the approximant $yp(x,t)$ in the x direction at time instants tk, $k = 1, 2, ...$, it is necessary to express the coefficient vectors $cyp(Th), cyp^r(Th)$ of the actual and reference tensor product spline approximants $yp(x,t)$, $yrp(x,t)$ in terms of time sequences of the vectors $cy(tk), cy(tk)$.

Let us further consider the time instants tk which are chosen to divide the time interval $[0, Th]$ to vyp subintervals (for $t_0 = 0, t_{vyp} = Th$), then (32) can be modified to the final form:

$$J(cu) = \left[cy(Th) - cy^r(Th) \right]^T Q_1 \left[cy(Th) - cy^r(Th) \right] +$$

$$+ \left[cu(Th) - cu^r(Th) \right]^T Q_2 \left[cu(Th) - cu^r(Th) \right] cu(Th) \subset$$

$$\subset \left[cu^l(Th) cu^u(Th) \right] \qquad (35)$$

where:

$$cy(Th) = \left[cy^T(t_1), cy^T(t_2)\ldots, cy^T(t_{vyp})\right]^T \tag{36}$$

$$Q_i = [N^{-1}]^T QyN^{-1}$$

$N - (dN \times eN)$ matrix, $dN = (zG\text{vyp})$, $\quad eN = (zG\text{vyp})$

$$N = \begin{bmatrix} Nyp^T(t_1) & & & & \\ & \ddots & & & \\ & & Nyp^T(t_1) & & \\ Nyp^T(t_2) & & & & \\ & \ddots & & & \\ & & Nyp^T(t_2) & & \\ \cdots & \cdots & & \cdots & \\ Nyp^T(t_{vyp}) & & & & \\ & \ddots & & & \\ & & & Nyp^T(t_{vyp}) & \end{bmatrix} \begin{matrix} i \\ \vdots \\ zG \\ i \\ \vdots \\ zG \\ \cdots \\ i \\ \vdots \\ zG \end{matrix}$$

$$Nyp(t_i) = \left[Nyp_1(t_i)Nyp_2(t_i),\ldots, Nyp_{vyp}(t_i)\right]^T$$

The vector $cy^T(Th)$ is recalculated from the generated vector $cyp^r(Th)$ by the expression:

$$cy^r(Th) = Ncyp^r(Th) \tag{37}$$

The matrices Q_1, Q_2 are nondiagonal matrices and can be computed in advance. In order to obtain future outputs $cy(t_i)$ in the vector $cy(Th)$ a predictions with $1, 2, \ldots$ and vyp input polynomial pieces can be constructed on the basis of the model (22). With the predictive model it is possible to use some well-known state space techniques to design a predic- tive type controller with moving horizon. The resulting constrained opti- mization problem can be then solved by efficient algorithms of the quad- ratic programming methods, see Powell (1985). The optimization is realized repeatedly at every breakpoint based on the actual output and the predicted future outputs over a fixed time horizon Th. The model (22) results directly from the impulse response function of the distributed systems. The outputs is predicted using only preceding inputs. To accele- rate the controller convergence it is convenient to extend the model of an appropriate number ncy of the preceding inputs $cy(t_k)$.

Using the optimal vector $cu(Th)$ the continuous inputs signals (as spline functions) are calculated through the relation (1).In the periods between break points the functions can be generate with high sampling rate.

5   SIMULATION RESULTS

5.1   Discrete time,continuous space case

To demonstrate the applicability of the proposed methods many prac- tical experiments were done with self-tuning control of the temperature field of the physical laboratory model - the electrically heated tube with movable liquid medium (simple scheme see on Fig.1). The powers of five heaters were manipulated by computer and the temperature was measured by thermo-couples equidistantly located along the whole active lenght ($L = 1.24m$) of the tube. For specified structure of models (regression or ARMAX type) the self-tuning algorithms were implemented. The convergence properties of the controllers were justified with help of total deviation et(k) defined by relation :

$$et(k) = \int_0^L \Big[ys(x,k) - yrs(x,k)\Big]^2 dx \qquad (38)$$

The typical course of $et(k)$ is drawn on Fig.2. The dash line represents control strategy with unrestricted minimization ,the unrestricted inputs were cut off whenever they exceed the boundary of the admissible range. The full line represents the conditional minimization of the control criterion.The spline approximation error $er(k)$ evaluated in accordance with relation:

$$er(k) = \max_j \Big[|ys(x_j,k) - y(x_j,k)|\Big] \qquad (39)$$

is depicted in Fig.3. The control signals are given in Fig.4-Fig.8. The results show the successful convergence properties of the controller and its ability to reduce the effects of disturbances caused by various speed of the movable medium (see 100-th and 200-th step in the experiment).

5.2   Continuous time and space case

The simulation experiments were done under PC-KOS Conversational Monitor using SIC library (for detailed information about the system,see Karny, (1985)) where the distributed process was simulated by heat equa- tion with 3 actuators with influence functions acting on $[0, L]$ interval. The characteristic simulation results for deterministic conditions are shown on Fig.9

(total deviation) and Fig.10 (spline approximation error). The input signals $u_1(t), u_2(t)$ and $u_3(t)$ were generated (between the breakpoints)in the form of parabolic polynomials $a + b.t + c.t^2$. The time development of the coefficients a, b, c belonging to the input $u_1(t)$ (see Fig.11) is drawn on Fig.12 - Fig.14. The above simulation was done for time horizon $Th = 90s$ with 3 breakpoints for the model with $Tm = 150s$ and $ncy = 2$. Step changes of the distributed output reference signal was made during the experiment. The similar simulation experiment was repeated with pseudo random Gaussian noise added to output measurements. The results (see Fig.15 - Fig.20) show satisfactory controller behaviour in the presence of the noise.

6 CONCLUDING REMARKS

In this paper a self-tuning control scheme for DPSs is outlined. The resulting control algorithms are suitable for industrial distributed parameter processes where the processes modeling are difficult and expensive and the parameters of the processes and its environment change frequently.

It is obvious that this study submits only the main ideas,which must be further elaborated to achieve an implementable stage of the problem.

References

[1] De Boor,C.(1978). Practical guide to splines. Springer-Verlag.

[2] Hamza,M.H.,Sheirah,M.A.(1978). A Self Tuning Regulator of Distributed Parameter Systems.Automatica.Vol.14.No.5.

[3] Jazwinski.A.H.(1970).Stochastic Processesand Filtering Theory. Academic Press New York.

[4] Ljung,L.,Soderstron,T.(1983).Theory and Practice of Recursive Identifi- cation. MIT Press,Cambridge,Mass.

[5]
 Karny,M.,Halouskova,A.,Bohm,J.,Kulhavy,R.,Nedoma,P.,(1989).Design of Linear Quadratic Control:Theory and Algorithms for Practice. Kybernetika.Vol.21. Academia.Prague.(A supplement).

[6] Kobayashi,T.(1988).Finite-dimensional adaptive control for infinite- dimensional systems.Int.J.Control,Vol.48.No.1.

[7] Korneitschuk,N.P.(1984). Splines in approximation theory.(in Russian). Moscow.Nauka.

[8] Peterka,V.,(1981).Bayesian Approach to System Identification. Trends and Progress in System Identification.Pergamon Press.

[9] Powell,M.J.D.(1985).On the quadratic programming algorithm of Gold-farb and Idnani. Math. Programming Study, Vol.25.

[10] Schumaker,L.(1981).Spline functions. Basic theory. John Wiley.

[11] Vajta,M.J.(1982).Self-Tuning Control of a Heat Conduction Processes. IFAC 3rd Symposium of Control of Distributed Parameter Systems.

[12] Watanabe,E.,Yamamoto,T.,Omatu,S.(1986).A Design Method For a Self- Tuning Regulator For a Distributed Parameter Systems. Int.J.Control. Vol.43.No.1.

[13] Zelinka,P.(1989).Adaptive Control of Distributed Parameter Systems. Ph.D.Thesis.Slovak Technical University.Bratislava.

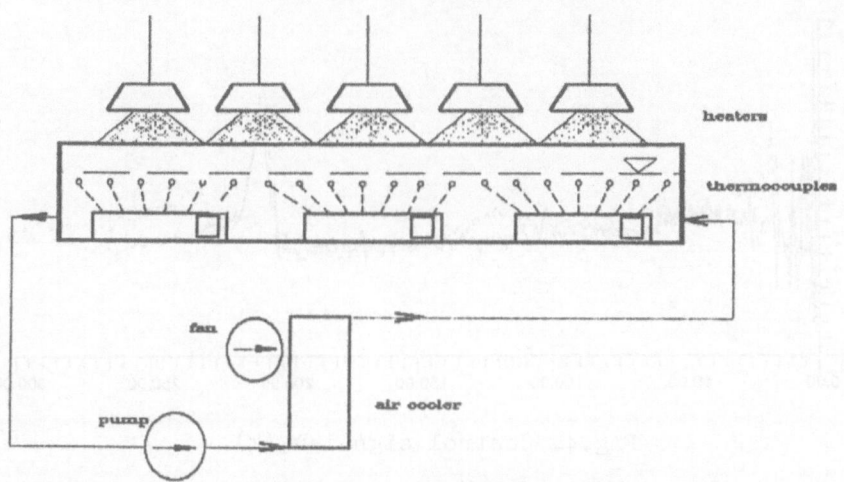

Fig.1. Laboratory model

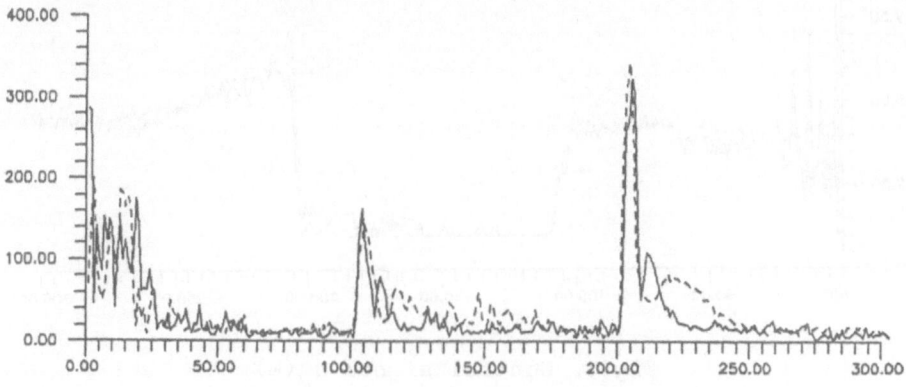

Fig.2. Total deviation

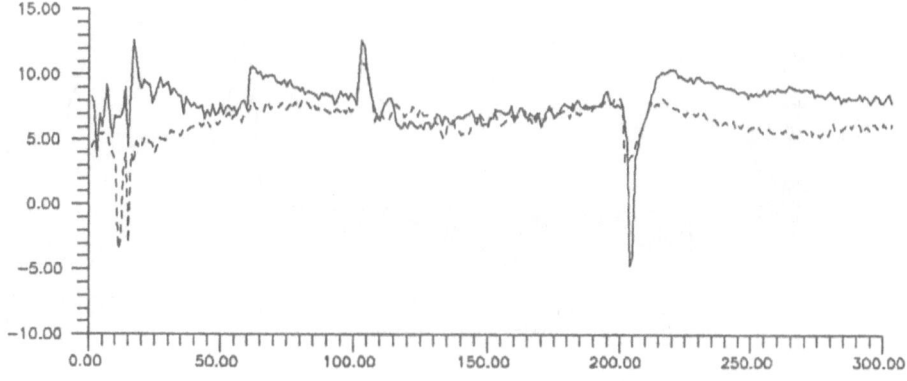

Fig.3. Spline approximation error

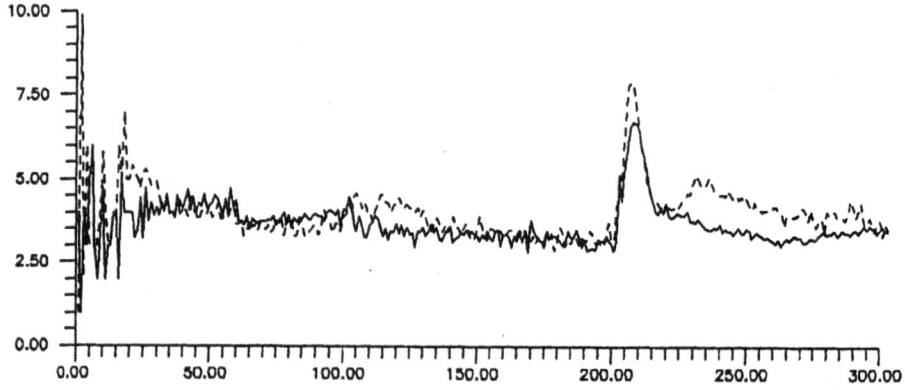

Fig.4. Control signal $u_1(k)$

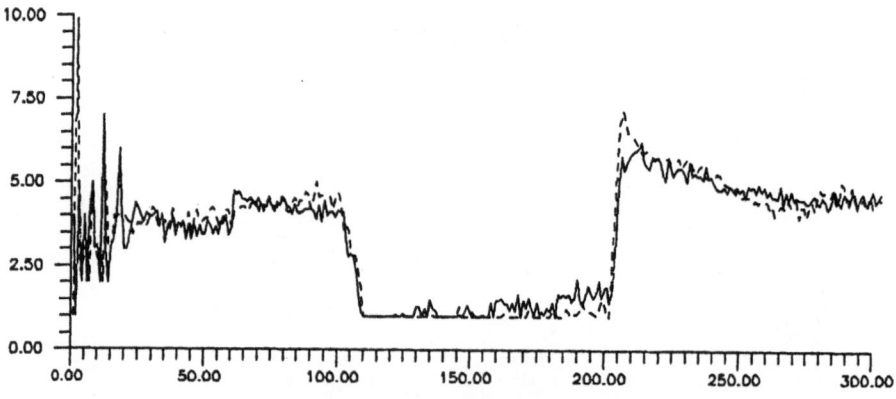

Fig.5. Control signal $u_2(k)$

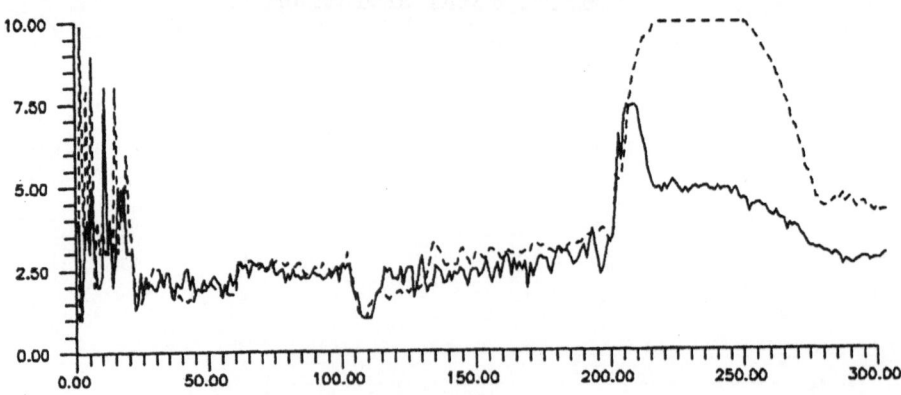

Fig.6. Control signal $u_3(k)$

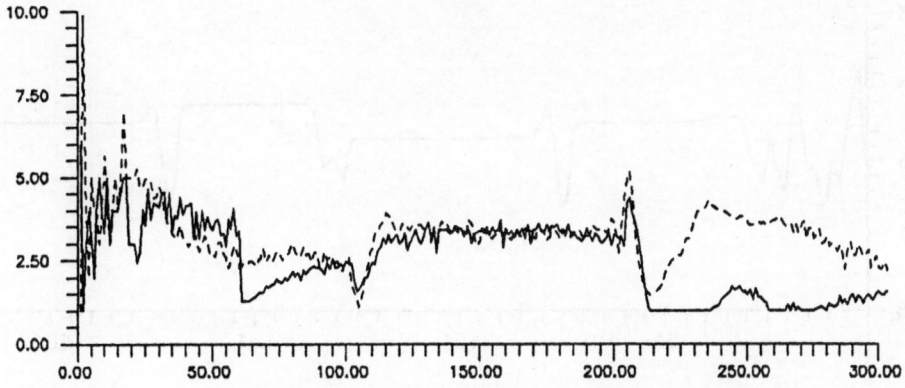

Fig.7. Control signal $u_4(k)$

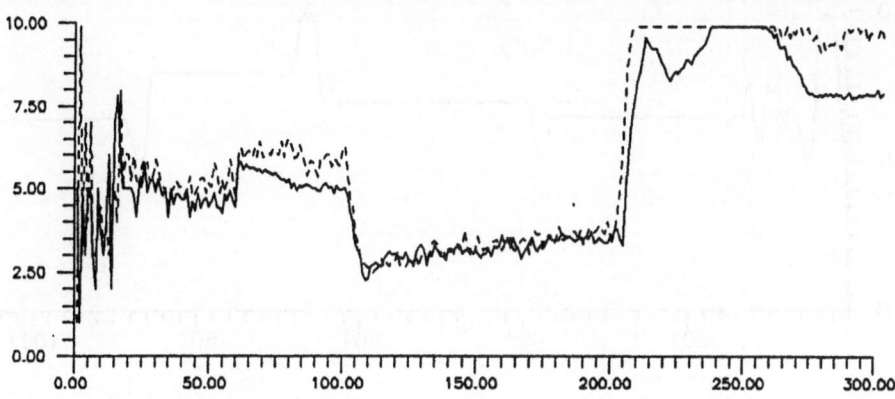

Fig.8. Control signal $u_5(k)$

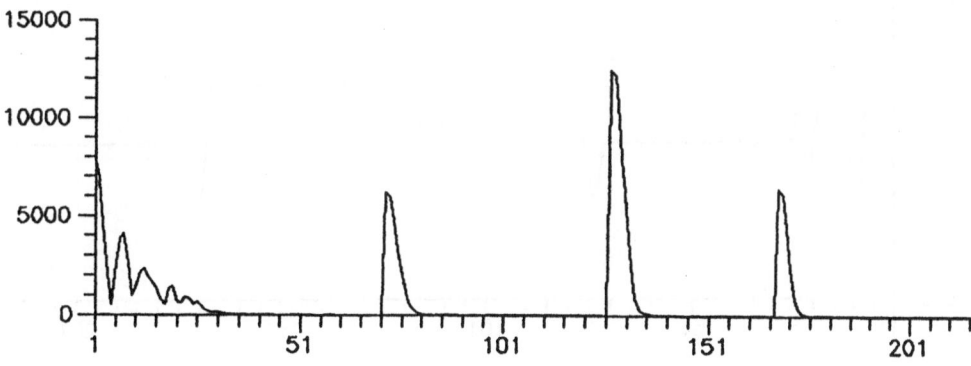

Fig.9. Total deviation

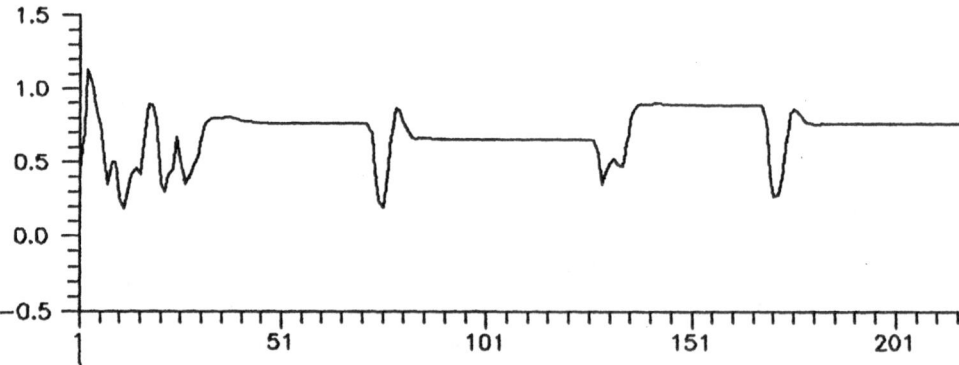

Fig.10. Spline approximation error

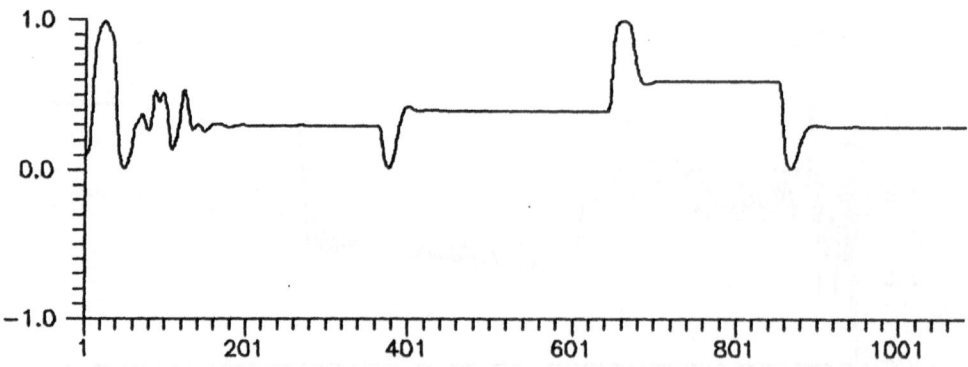

Fig.11. Input signal $u_1(t)$

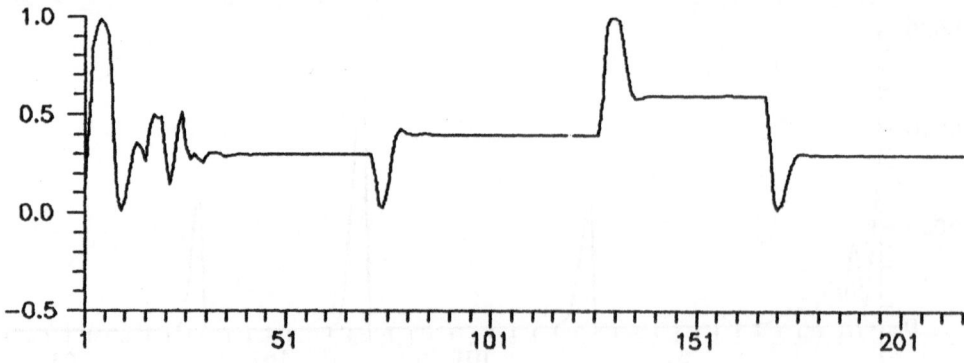

Fig.12. Coefficient "a" of the parabolic
polynomial

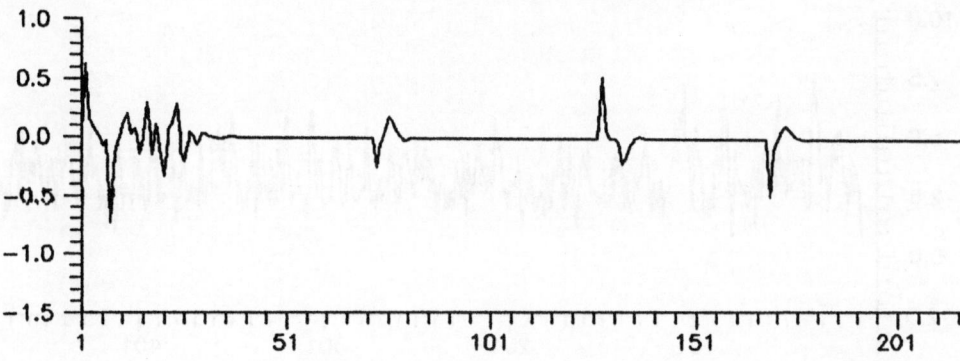

Fig.13. Coefficient "b" of the parabolic
polynomial

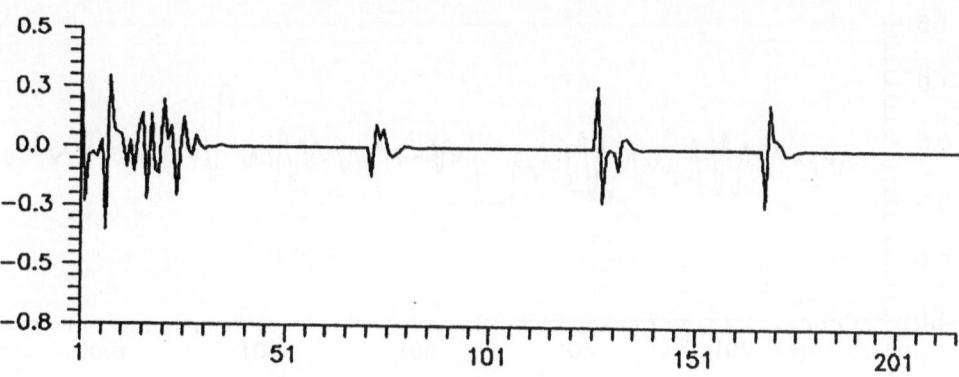

Fig.14. Coefficient "c" of the parabolic
polynomial

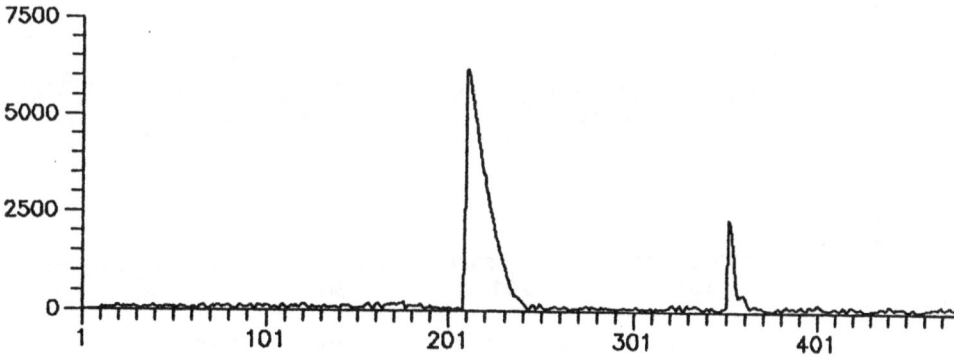

Fig.15. Total deviation

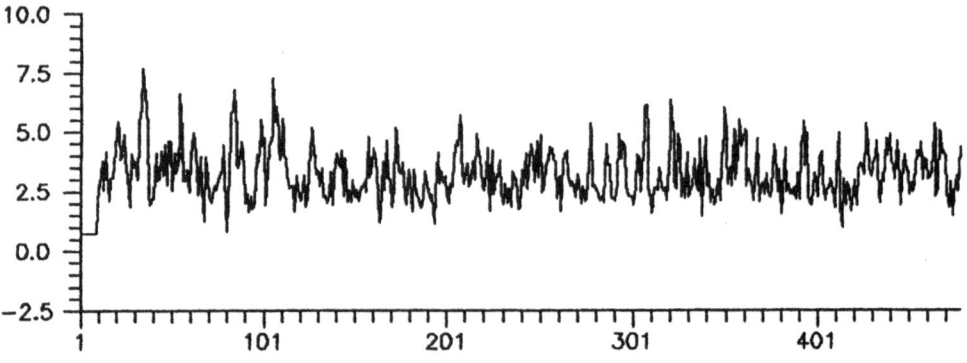

Fig.16. Spline approximation error

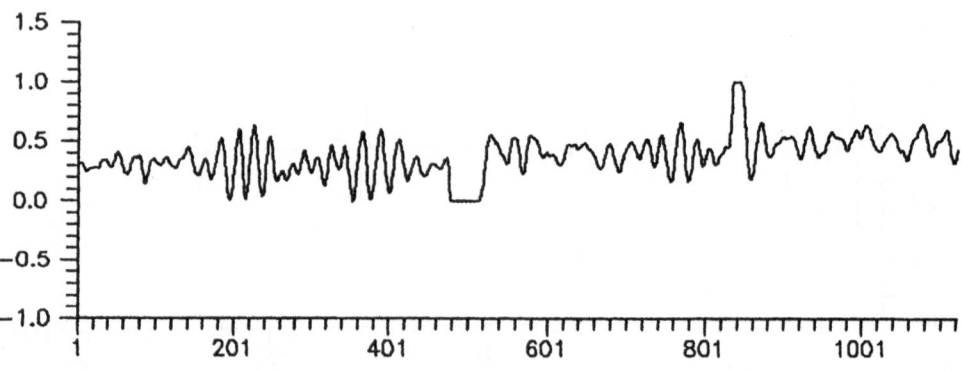

Fig.17. Input signal $u_1(t)$

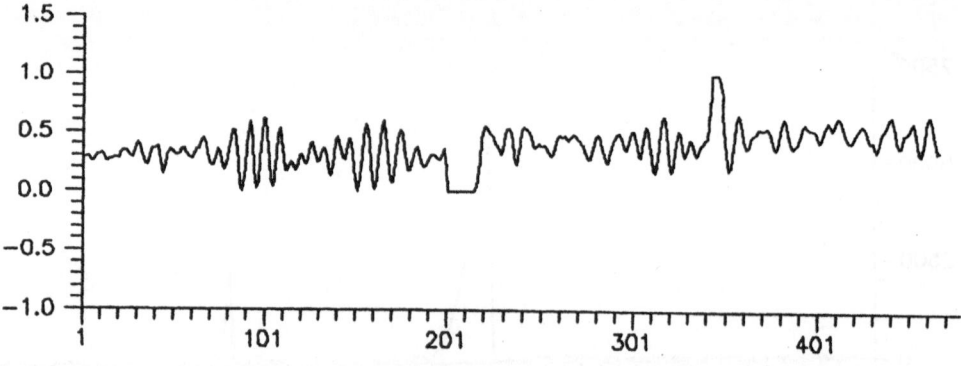

Fig.18. Coefficient "a" of the parabolic
polynomial

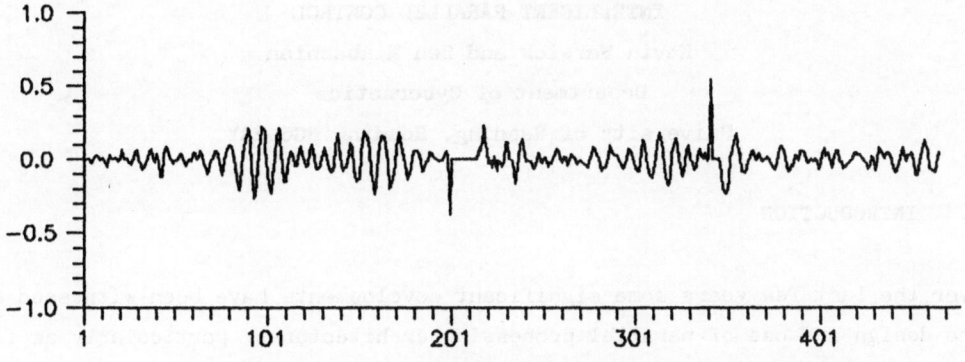

Fig.19. Coefficient "b" of the parabolic
polynomial

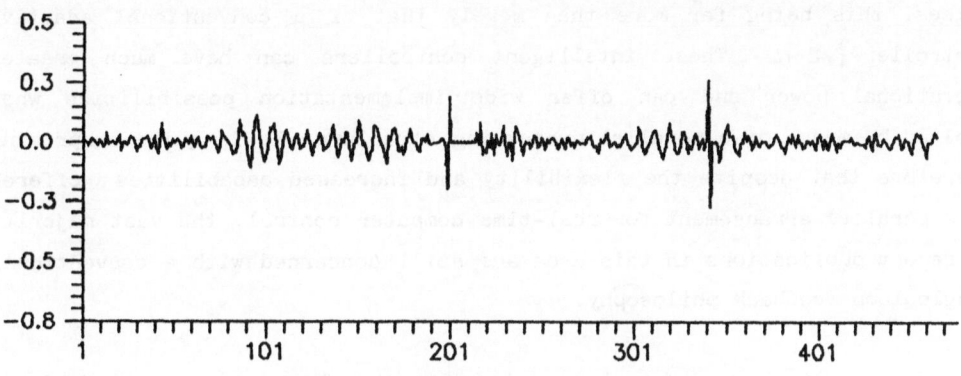

Fig.20. Coefficient "c" of the parabolic
polynomial

INTELLIGENT PARALLEL CONTROL

Kevin Warwick and Ben Minbashian

Department of Cybernetics

University of Reading, Reading RG6 2AY

1.   INTRODUCTION

Over the last few years some significant developments have been witnessed in
the design and use of parallel processing architectures, particularly as far
as control systems are concerned [1].   It is also of importance that
feedback control techniques with an 'intelligence' factor have been
developed, whereby the control action applied is modified as the need
arises, this being far more than merely that of a conventional adaptive
controller [2].   These intelligent controllers can have much greater
operational power and can offer wider implementation possibilities when
applied by means of a parallel processing structure. It is quite surprising
therefore that despite the flexibility and increased capabilities offered
by a parallel arrangement for real-time computer control, the vast majority
of recent publications in this area are still concerned with a conventional
single loop feedback philosophy.

Recently a new approach has been taken for the formulation of a feedback
control structure providing for a novel parallel control scheme [3]. This
approach directly tackles the problem of looking again at the way in which a
basic controller structure is developed, primarily, but not exclusively, for
computer control, such that some of the benefits of present-day technology
can be realised.   The use of a parallel control structure thus brings the
implementation of control systems directly in line with present-day trends.

The implementation of flexible parallel controllers is now only in its
infancy, as essentially much work remains in order to achieve a firm
theoretical basis for study.   Stability, controller convergence, robustness
and performance properties all need to be looked at afresh with regard to
this type of control structure.   It is very straightforward to show that
stability properties of single loop feedback controllers do not map over in
an, as yet, obvious way to parallel controller structures.   Perhaps of more

immediate importance however, is the study of parallel controllers in terms of their implementation in various forms in order to assess where the most positive application areas may lie. At the present time flexible parallel controllers have been introduced and implemented in a limited number of modes, the full extent of possibilities is yet to be found.

The concept of a flexible parallel structure for feedback control is described in this paper and two implementation studies are investigated. The fundamental arrangement for such controllers is, it is felt, a novel way in which feedback control can be arranged, further it is seen to be modern, in that it ties in with present-day technology, and exciting in terms of the potential application areas which can now be investigated.

2.   CONTROLLER STRUCTURE

The introduction of new control techniques directly to industrial processes is always, quite sensibly, much slower than is possible. This is partly due to a natural conservatism and partly due to the need for reliability and efficiency to have been proven for a controller in situ, something that is not ultimately possible unless it is actually employed. A need therefore exists for a control structure which allows for a simple, faster introduction of new control algorithms whilst retaining safety features, efficiency and flexibility.

In many cases, classical control techniques, such as the three term controller, have the distinct advantage of being both familiar and well understood, particularly by those who come into direct contact with the controller, e.g. plant operators. Indeed such a controller may well have been in operation for some time, giving rise to a great deal of resistance when their replacement, with a more modern controller, is considered. Factors causing this resistance are not necessarily based on economical reasons, but are probably very much to do with a lack of confidence in the new controller.

There have been many technological advances over the last decade or so, and going hand-in-hand with a rapid increase in the capability of computers is the development of high performance parallel processing hardware. Some feedback controllers have already been implemented via a parallel processing device [4], however such implementations have been generally obtained simply by chopping up the particular control algorithm into a number of sequential sections, which are then processed in a concurrent fashion. As yet the basic control structure has not been reformulated in a way which makes use of the best parallel features. In this paper a flexible parallel control structure is described, which is intended to be particularly well suited to a parallel implementation structure. As far as advantages of the controller are concerned, it is felt that the scheme goes a long way to solving some of the problems raised in that the technique allows for confidence to be built in a new control algorithm before full implementation occurs, thereby increasing the possible try out of new control methods.

The parallel control structure considered is essentially part of a feedback control path, of the form shown in Figure 1. The plant is shown as a Single Input-Single Output (SISO) system, this is for explanation purposes only, the parallel controller is equally applicable for multi-sensory outputs. The dotted lines in Figure 1 are included merely to show the information gathering capabilities of the upper Control Level/Decision block, all of which happens concurrently with the feedback path operation. Also, all of the control diagram not within the shaded area is assumed to be computational, in that computer control is employed as the parallel controller basis; this is though not a necessity as the same overall structure can be employed by digital or analog means.

Given the computer control arrangement, the implication is that Analog-Digital and Digital-Analog converters are employed at the points at which the signals (plant output and input respectively) cross from the shaded area in Figure 1 to the unshaded area. Also the keyboard input is included in order to allow for information passed from external sources, which can mean interrupts or other devices. The actual feedback control arrangement is described in more detail in the sections which follow.

The main aim behind the control scheme described here is to provide a controller which makes full use of, and is specifically directed towards, the

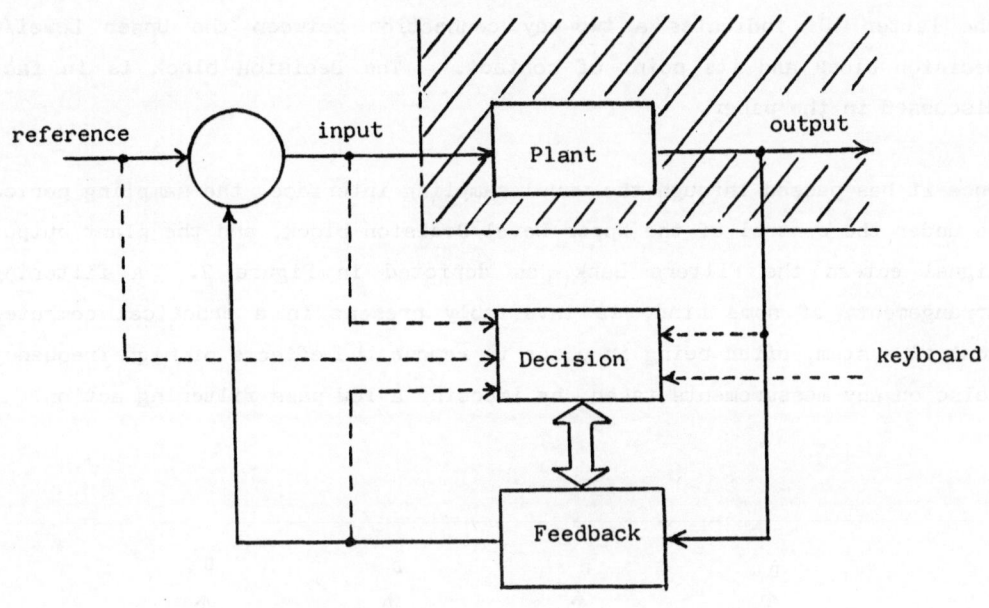

Figure 1   General Controller Scheme

available computer processing facilities, thereby producing a high
performance versatile technique. The controller is a practical, real-world
tool rather than a mathematical exercise and is directly implementable in a
parallel processing hardware framework.   The overall complexity of the
controller depends on the amount of upper level control/decision processes
employed, the complexity of the plant under control and the type of control
employed in the feedback path.   This latter point is considered further in
the following section.

3.   FEEDBACK ARRANGEMENT

It can be seen from Figure 1, that the overall control structure is based on
a low level feedback control structure which is overseen by an upper level
decision block and this directs the operation of the lower level.   The
feedback controller is though more complex than a single loop feedback
approach, and employs a number of subsections, as shown in Figure 2 in which

the letter 'D' indicates a two-way connection between the Upper Level/-Decision block and its point of contact.   The Decision block is in fact discussed in the paper.

Once it has passed through the usual sampling interface, the sampling period is under the control of the Upper Level/Decision block, and the plant output signal enters the Filters Bank, as depicted in Figure 2.   A filtering arrangement, of some kind, is invariably present in a practical computer control system, often being in place to remove the effects of high frequency noise on any measurements taken, by imposing a low pass filtering action.

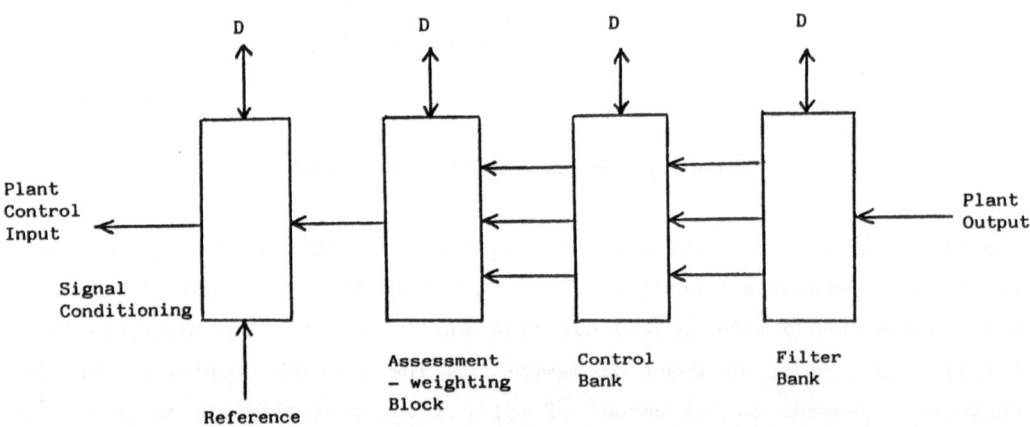

Figure 2   Parallel Feedback Path

In the flexible parallel controller described here, it is at this Filter Bank stage that the measured output signal is split up into a number of parallel feedback paths. It is thus possible for all of the paths to have the same type of filtering action, or more likely each path can have its own unique filtering arrangement. As an example one path may contain a bandpass filter to remove both high frequency noise and any d.c. bias, whilst another path may just use a low pass filter for stochastic noise suppression. The filter bank of Figure 2 is interconnected with the Upper Level/Decision block, which offers the possibility of changing the filtering action on-line in real-time, in one or more of the parallel paths.

On leaving the filter bank the parallel feedback signals, each one being dealt with separately, pass to the Control bank. This Control bank is made up of a number of controllers which operate in a concurrent mode in parallel, the control action in each block having been nominally selected a priori. Each parallel control limb can therefore contain a controller based on a different control objective, i.e. one limb may contain an optimal controller, another a servo controller etc. The selection of individual controllers in a limb is though monitored, and can, if desired, be modified by the upper level of control, due to the decision block interconnection. A special case of the parallel path arrangement is found when all of the limbs contain the same, identical controllers, in which case the method is good for multiple path fault tolerance with majority voting.

The extent of interaction between the upper level/decision block and the feedback controller block depends very much on the type and complexity of the control scheme employed in each limb. For example a fixed gain three term controller would require very little interaction, possibly just a monitoring procedure would be useful, however the employment of an adaptive controller [2] such as a self-tuning algorithm [5], [6] would necessitate a periodic link in which parameter values, directly employed in the controller itself, could be passed between the two levels.

4. CONTROLLER ASSESSMENT

The different control actuation signals, which leave the control block in parallel form, enter the Assessment-Weighting block, remaining in parallel on entry. At this point it may be necessary to synchronize the actuation signals, if required, due to the uneven complexity dependent computation times apparent in each of the limbs. It may well be the case though that such synchronization is not necessary, due to the nature of overall control aimed at. An example is the employment of controllers operating in parallel to deal with different time constants of the plant – one controller operating rapidly to respond to short time constants and one operating slowly to respond to long time contants.

The resultant output from the Assessment-Weighting block is one signal (in the SISO plant case) which consists of a weighted sum of the individual actuation signals found in each of the control bank limbs, a simplified example of which is shown in Figure 3. It should be noted in this figure

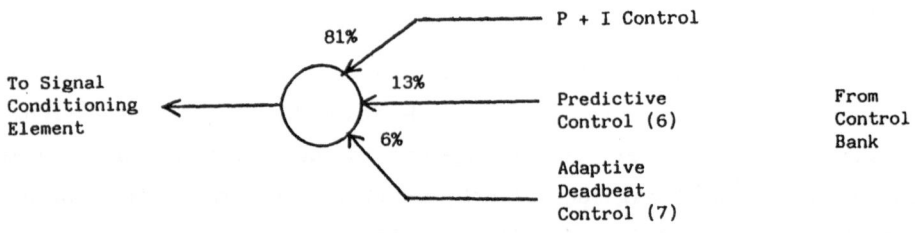

Figure 3 Assessment-Weighting Block (an example)

that no time dependency is indicated, i.e. synchronization of application is assumed.

The weightings on each signal entering the Asssessment-Weighting block can be constant, having been selected a priori, however the true functionality of the parallel control technique is found when the weightings are allowed to be time varying under the control of the upper level. The weightings can therefore be placed under the control of a rule-based hierarchy within the Decision block, in which rules fire in response to such as particular plant condition or keyboard (external) information entry.

A special case of the weighting scheme is encountered if 100% or 0% is applied to a specific control signal, in which case that signal is respectively either the signal actually applied or is discarded completely. This latter event can occur if a controller is found to be malfunctioning whereas if a poor control action appears to be presented it may be sensible for a very low weighting to be awarded to that limb, the weighting selection being made within the upper level. The Assessment-Weighting block can though be used at a point at which error checking is carried out by straightforward signal comparison. This is particularly relevant when a multi-path feedback scheme, employing identical controllers in parallel is used.

The output from the Assessment-Weighting block is passed on for signal conditioning and reference signal inclusion. The signal conditioning may well be such that offset compensation is dealt with.

5. UPPER CONTROL LEVEL

The upper control level is shown in Figure 1 as the Decision block, and this is employed for monitoring the overall performance of the system at many points, processing this information, taking decisions as to courses of action necessary and ensuring that these actions are carried out. It sits on top of the parallel feedback path and is, in that sense, independent of the low level immediate feedback action. Rather the upper level organises the lower level and ensures that the mode of operation is that which is required.

The unit can take on any desired degree of complexity, including various levels of Artificial Intelligence [7], certainly to the extent of a Rule Based Scheduler [8]. The complexity of the plant under control and the type of controllers employed in the feedback path will also directly affect how much information gathering and AI is employed [9]. As an example, a basic need in the upper control level could be for the block to carry out an analysis of signal trends obtained via outcome evaluations based on up-to-date models of the plant and controller being applied, i.e. if a particular course of action is taken, what will be the result? By this means, the validity and performance of a particular control schedule can be tried and tested before actual application. This approach is particularly useful in terms of reducing the effects of faults, or even getting around problem areas altogether.

Another use of the upper level control block is in the operation of adaptive control regimes, particularly self-tuning control, an example of which is given in the following section. If a number of adaptive controllers are employed in parallel, then where an updated model of the plant is used as a basis for any controller calculation [10], it is possible to use the same plant model, which is contained within the upper control level, for all of the adaptive loops, and in this case a number of either separate or usefully related models can be accommodated.

6.   IMPLEMENTATION EXAMPLES

In order to show how the flexible parallel controller can be used in a practical way, a couple of straightforward examples are given in this section.   The first example is based on a feedback loop which consists of a fixed gain three-term controller operating in parallel with a self-tuning (adaptive) pole placement controller.   The second example is based on a feedback loop in which an optimal (minimum variance) controller is operated in parallel with a servo controller.   It is felt that these two examples give an indication of situations in which the use of such a parallel controller has distinct advantages over a conventional single controller arrangement.   The control algorithms have, in each case, been kept in fairly simple terms, the intention being to examine the power of the parallel control method, rather than to claim that any particular control objective, for example optimal control, is either better or worse than any other.

For both examples, the plant under control was considered to be defined by the equation

$$y(t) = \frac{z^{-2}(0.173z^{-1} + 0.104z^{-2})}{1 - 0.992z^{-1} + 0.215z^{-2}} . u(t) \tag{1}$$

where $y(t)$ and $u(t)$ are the plant output and input respectively, at time instant t, also z^{-1} is the unit delay operator, such that $z^{-i}u(t) = u(t - i)$ which corresponds to the control input value at a time i sample periods before t.

The plant was also deemed to be affected by an unmeasurable white noise disturbance directly affecting the plant output signal, this being of zero-mean value with a variance equal to 0.0625. For both the examples which follow, the plant and disturbance were simulated along with the applied control.

Example 1

For this first example, after 20 sample periods had elapsed, it was assumed that the plant characteristics changed from those given in equation (1) to

$$y(t) = \frac{z^{-2}(0.2z^{-1} + 0.1z^{-2})}{1 - 0.6z^{-1} + 0.5^{-2}} \; u(t) \qquad (2)$$

Two controllers were employed in parallel, in the form of a feedback loop, as shown in Figures 1 and 2. The first of these was a three term controller described by equation (3)

$$u_T(t) = [K_p + K_i(1 - z^{-1})^{-1} + K_d(1 - z^{-1})] \, [v(t) - y(t)] \qquad (3)$$

in which K_p, K_i, and K_d are fixed gain values of the three term (Proportional + Integral + Derivative) controller. Also, v(t) signifies the desired reference level for the output signal, referred to as the reference input, and $u_T(t)$ is the control signal, or rather part control signal, which is due to the three term controller action.

From the three term control set up in equation (3), it was found that gain values of $K_p = 1.5$, $K_i = K_d = 1$ produced reasonable results in terms of control performance. These values were therefore retained throughout the whole simulation run.

The three term controller was operated in parallel with a pole-placement self-tuning controller [11], whose control input can be described by equation (4).

$$u_s(t) = -\frac{F}{G} \, y(t) + \frac{C}{G} \, v(t) \qquad (4)$$

in which $u_s(t)$ signifies the control signal, or part control signal, based on the self-tuning element. Also, F, G and C are polynomials which are obtained in the following way.

Firstly, let equation (1) be rewritten as equation (5)

$$y(t) = z^{-2} \frac{B}{A} u(t) \tag{5}$$

Assuming, for a moment, that the applied input $u(t) = u_s(t)$, on substitution of the input defined in equation (4), into equation (5), the closed loop transfer function is given by:

$$y(t) = \frac{z^{-2} BG}{AG + BGz^{-2}} v(t) \tag{6}$$

The denominator of the closed loop system can then be specified to be a pre-selected polynomial T, if the controller polynomials F and G are such that they satisfy

$$AG + BFz^{-2} = T \tag{7}$$

By suitable structural choice for F and G a unique solution to equation (7) is found.

The required steady-state gain of the closed loop system can then be obtained by a suitable selection of C in equation (4). If it is desired, as one might expect, that the requirement is for $y(t)$ to equal $v(t)$ in the steady-state, then from equation (6) this is found if:

$$C = T(1)/B(1) \tag{8}$$

where $T(1)$ signifies the scalar sum of the coefficients of the T polynomial, $B(1)$ being similarly defined such that after 20 sample periods, with reference to equation (2):

$$B(1) = 0.2 + 0.1 = 0.3$$

It was stated earlier that a pole-placement self-tuning controller was employed. Thus far the pole-placement element has been described. The self-tuning scheme means essentially that at every sample period the plant polynomials A and B, defined in equation (5) are recursively estimated such that equation (7) could be evaluated for F and G, at each iteration, by making use of the updated estimated parameter values in the polynomials A and B.

In this example a recursive least squares algorithm [12], was used as part of the self-tuning pole-placement algorithm, with an initial covariance matrix of $10^9 I$, where I is the identity matrix, and a fixed data forgetting factor of 0.99. Also the closed loop poles were selected as z = 0.6 ± j0.24 which is equivalent to a pole polynomial of

$$T = 1 - 1.2z^{-1} + 0.48z^{-2}$$

The control input actually employed though was dependent on how good the self-tuning control part was at any time instant, such a measure being based on the goodness of plant model employed in the self-tuning algorithm. This means that when the plant model, employed within the self-tuning algorithm, is a poor one, the three term controller tends to be used, whereas when the plant model is a good one, the self-tuning controller is used. Such an arrangement is put into practice in the following way:

Let the error between the actual measured output y(t) and the estimated output signal $\hat{y}(t)$, found from the plant model be at each time instant

$$\epsilon(t) = y(t) - \hat{y}(t) \tag{9}$$

In equation (9), $\hat{y}(t)$ is an estimation of the plant output which is based on actual plant data and the most recent estimated plant model parameters.

The control input, u(t), can be written as a combination of the two control signals $u_T(t)$, i.e. that due to the three term controller, and $u_s(t)$, i.e. that due to the self-tuning controller. This can be written as:

$$u(t) = u_T(t) + \alpha[u_s(t) - u_T(t)] \tag{10}$$

in which $0 \leqslant \alpha \leqslant 1$.

The weight α is related to the modelling error ε(t) in a fairly direct way, such that:

when $|ε(t)|$ is high, we want $α → 0$

and when $|ε(t)|$ is low, we want $α → 1$.

Also, γ is specified as the maximum absolute value of $|ε|$ for which confidence is, to any extent, placed in the plant model. Hence the approach was taken

$$ε = 0 \qquad \text{for } |ε(t)| > γ$$

$$α = 1 \quad \frac{|ε(t)|}{γ} \quad \text{for } |ε(t)| \leqslant γ$$

(11)

This is in fact a rather simplistic selection for α and causes α to vary a fair amount with respect to time. One alternative is to consider a weighted window of errors, e.g. 10 ε(t) values, such that some form of $Σ|ε(t)|$ is found and used in equation (11) in place of $|ε(t)|$. An approach of this kind can mean that a number of 'poor quality' control signals may be obtained before actual changes to the applied control input are carried out. So, with a window approach, the self-tuning controller is not switched out as quickly when a poor plant model is encountered. An appropriate trade-off between rapidity of a response and reduction in switching is required for each particular situation. Research is presently being carried out into various selections.

If α is the weight assigned to the input signal $u_s(t)$ attributed to the self-tuning controller, and $β = 1 - α$ is the weight assigned to the input signal $u_T(t)$ attributed to the three term controller, a change from one controller to the other can be found as in Figure 4, and this particular arrangement was employed in Example 1.

Two simulation runs for Example 1 are shown in Figures 5 and 6. Figure 5 shows the operation of a fixed term three term controller, operating on the plant described by equations (1) and (2), such that after 20 samples the

271

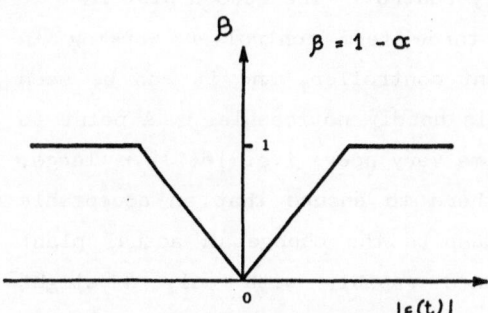
Figure 4 Controller Mixing : Example 1

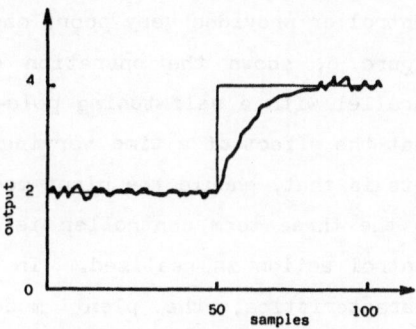
Figure 7 Optimal Controller : Example 2

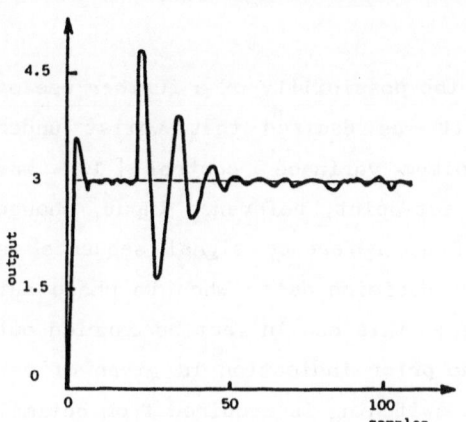
Figure 5 Fixed-Term PID Controller : Example 1

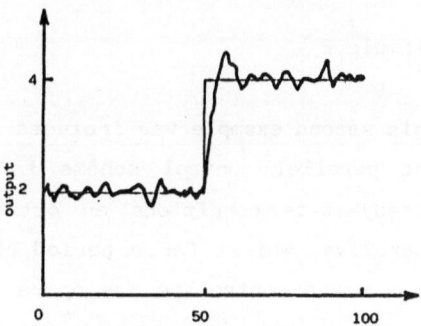
Figure 8 Servo Controller : Example 2

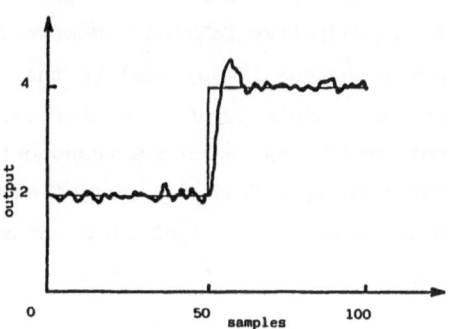
Figure 9 Parallel Controller : Example 2

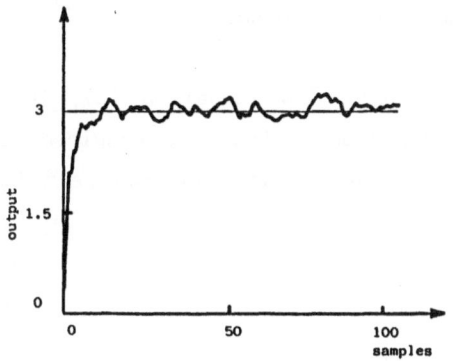
Figure 6 Parallel Controller : Example 1

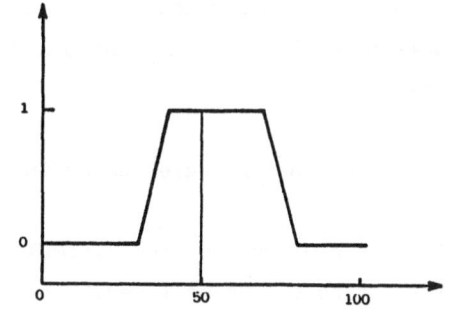
Figure 10 Controller Mixing : Example 2

controller provides very poor, oscillatory control. The second plot in Figure 6, shows the operation of the three term controller working in parallel with a self-tuning pole-placement controller, and it can be seen that the effect of a time varying plant is hardly noticeable. A point to note is that, should the plant model become very poor, i.e. $|\epsilon(t)| \rightarrow$ large, so the three term controller is still there to ensure that an acceptable control action is realised. In fact, despite the change in actual plant characteristics, the plant model retuned rapidly with only a slight temporary increase in $|\epsilon(t)|$ which was well below the specified $\gamma = 0.5$.

Example 2

This second example was included to show the possibility of a further use of the parallel control scheme in which it was desired that whilst under steady-state conditions an optimal, minimum variance, control [13] was operative, whilst for a period around a set-point, reference input, change so a servo controller was operative. If a reference signal sequence is known a priori, it becomes a fairly simple decision as to when to phase out one controller whilst phasing in the other; this can in fact be carried out in a predictive fashion. However, when no prior indication is given of set point changes it may well be that a rapid switching is required from optimal to servo control, as soon as a variation is witnessed. It should be pointed out though that whilst a transport delay between plant input and plant output exists, a finite number of sample periods will occur between the actual occurrence of an input step change, and any effect being witnessed at the plant output. In equation (2), 3 such sample periods occur.

The plant selected for Example 2 was that described by equation (2), this being the case for the entire trial run. The plant output was required to settle at 2 units for the first 50 samples and 4 units for the next 50 samples.

Three controllers were tested in the run:

(i) A minimum variance optimal controller (see Figure 7).
(ii) A pole-placement servo controller (see Figure 8).

(iii) A parallel controller consisting of optimal and pole-placement controllers operating in parallel (see Figure 9).

The minimum variance controller was of the form:

$$u(t) = \frac{-F}{BG} \, y(t) \tag{12}$$

where the controller polynomials F and G are found from the unique solution to the identity

$$1 = AG + z^{-2}F \tag{13}$$

Noting that the system + noise was modelled as

$$Ay(t) = z^{-k}Bu(t) + e(t) \tag{14}$$

where e(t) is a zero-mean white noise disturbance, with a variance of 0.06025.

The pole-placement servo controller was set up as that in Example 1, i.e. as in equation (4) with F and G found from equation (7). Note however that in this example no self-tuning algorithm was introduced, but rather simply a fixed parameter pole-placement controller.

The parallel controller consisted of a mixture of the two controllers described, i.e. the optimal controller of equation (12) and the pole-placement controller of equation (4). Knowledge of the reference input change on 50 sample periods was assumed a priori such that with the applied control input described by

$$u(t) = u_o(t) + \alpha[u_p(t) - u_o(t)] \tag{15}$$

in which $u_o(t)$ is the optimal minimum variance control, $u_p(t)$ is the pole-placement servo control and $0 \leqslant \alpha \leqslant 1$. This means that when $\alpha = 0$ only optimal control is applied, whereas when $\alpha = 1$ only pole-placement control is applied. The switching function for α in Example 2 is shown in Figure 10.

It can be seen from Figure 7 that use of the minimum variance controller produces good regulatory control in the steady-state and very poor servo control during the transient period. The pole placement controller, in Figure 8, meanwhile produces good servo control during the transient period with much worse regulatory control in the steady-state. Results from the parallel controller in Figure 9, show that such control can approach the best of both worlds, i.e. good steady-state and transient properties.

7. CONCLUSIONS

A flexible parallel control scheme has been introduced, which is aimed directly at restructuring feedback control in a way which is immediately applicable within a parallel processing environment, thereby taking advantage of the possibilities offered with such an architecture. In fact the overall computational time for a single time period controller evaluation is, in the worst case, only equivalent to the most time consuming single feedback loop. The parallel control scheme is also useful at making much more of the available computing power whilst adding flexibility to the controller, allowing for a hierarchical control arrangement. Use of a computer, within a computer control scheme, simply for digital control is really felt to be rather like using a sledge hammer to crack a nut.

Within a parallel control implementation, it is of course quite possible to switch to a simple feedback loop if desired, or to achieve robust fault tolerant control. These facilities are off-shoots from the basic scheme and arrive as a natural way of setting up the control. They can be regarded rather as special cases of the general method.

There are many advantages obtained from the implementation of a parallel controller. Firstly, different controllers can be phased in and out, rather than being directly switched. This considerably reduces the possibility of transfer bumps occurring, thereby much reducing transient spikes in the control and output signals. Secondly, the method allows for the cost effective, easy introduction of modern control algorithms, probably including such as self-tuning control, by means of a gradual familiarisation such that confidence can be steadily built up without facing a risky shut down operation. A third point is that the actual cost of implementing a parallel controller scheme is fairly low as control algorithms can be stored in software, providing a portable and readily modifiable controller base.

8. REFERENCES

1. Fleming, P.J. (ed.), 1988, 'Parallel Processing in Control', Peter Peregrinus Ltd.

2. Warwick, K., 1990, Computing and Control Engineering Journal, Vol. 1, 35-39.

3. Warwick, K., 1989, 'Parallelism in Control', Kybernetes, Vol. 18, 20-26.

4. Fleming, P.J. and Jones, D.I., 1988, 'Parallel Processing in Control', Chapter 13 in 'Industrial Digital Control Systems', K. Warwick and D. Rees (eds.), 2nd ed., Peter Peregrinus Ltd.

5. Seborg, D.E., Edgar, T.F. and Shah, S.L., 1986, AIChE Journal, Vol. 32, 881-913.

6. Astrom, K.J., 1983, Automatica, Vol. 19, 471-486.

7. Winston, P.H., 1984, 'Artifical Intelligence', Addison-Wesley.

8. Minbashian, B. and Warwick, K., 1989, 'Decision Rules for Intelligent Parallel controllrs', IEE Colloquium on 'Exploiting the Knowledge Base - Application of Rule Based Control, 1/1-1/14.

9. Peel, D. and Morris, A.J., 1988, 'Artificial Intelligence for Process Regulation and Servo Control', Chapter 15, in 'Industrial Digital Control Systems', K. Warwick and D. Rees (eds.), Peter Peregrinus Ltd.

10. Harris, C.J. and Billings, S.A. (eds.), 1985, 'Self-tuning and Adaptive Control', 2nd ed. revised, Peter Peregrinus Ltd.

11. Wellstead, P.E., Prager, D. and Zanker, P., 1979, Proc. IEE, Part D, Vol. 126, 781-787.

12. Soderstrom, T. and Stoica, P., 1989, 'System Identification', Prentice-Hall International.

13. Astrom, K.J., 1970, 'Introduction to Stochastic Control Theory', Academic Press.

SIGNAL-BASED ALGORITHMS FOR SELF-TUNING REGULATORS EASILY IMPLEMENTABLE ON SINGLE-BOARD MICROCOMPUTERS.

J. Maršík, P. Klán, V.Gorner

Institute of Information Theory and Automation

Czechoslovak Academy of Sciences

May 17, 1990

KEYWORDS

Digital adaptive PID control, sampling period adaptation, single-chip microcomputers, controllers in dual-in-line package

1 ABSTRACT

Some extremely simple identification-free algorithms for self-tuning PID controllers realizable on single-board or single-chip microcomputers have been developed. The algorithms are based on the fact that single PI or PID control loops behave like a second or third order system, regardless of the controlled plant order. This enables us to characterize any control process by its oscillation or damping index and a certain global time lag only. These two parameters, which are evaluated form the control error, are sufficient for continual adaptation, the former adjusting the common gain, the latter the time lags and leads inclusively the sampling period. The adaptation loops work like a usual control loop, no additional test signals are used.

2   THE PRINCIPLE OF SELF-TUNING

As it follows from the Abstract, the control loop is capable of "lowering" the system order to the values three or two. This also implies the similarity of optimum adjusted control processes which practically differ from each other in time and amplitude scale factors only.

Nevertheless this welcome property can hardly be proved exactly. Thus our approach could not been built upon a pure mathematical basis - except the rule of three which has proved to be very useful, as it will be shown at once: Suppose that we have simulated an optimum control loop and we wish to change the time scale factor of the process k-times. It means we must change k-times all time lags and leads of the plant as well as those of the controller not to affect the optimum adjustment. Consequently the above mentioned global time lag will change k-times, too. Therefore, making all controller lags and leads proportional to the global lag, the condition of adaptivity to time scale changes is fulfilled. However, this property alone may not ensure the optimum performance unless suitable coefficients of that proportionality are chosen. Fortunately, due to the resemblance of optimum control processes, we succeeded in experimental finding uniform "optimum" coefficients for a wide class of controlled systems.

2.1   DETERMINATION OF THE GLOBAL LAG

Using the frequency response approach, the global time lag can be defined very simply as

$$\tau = \frac{1}{\omega} = \frac{1}{2\pi f} \tag{1}$$

where f is the frequency and ω is the corresponding angular frequency Eq. (1) implies

$$\omega\tau = \omega^2\tau^2 = \omega^3\tau^3 \tag{2}$$

If the relations (1) or (2) are met for any frequency, then the desired adaptivity to time scale changes is provided.

Applying this to the Fourier transform of the PID control law

$$U(j\omega) = E(j\omega)\left(\frac{C_I}{j\omega\tau} + C_P + C_{\dot{D}}j\omega\tau\right), \tag{3}$$

we obtain for any single frequency ω

$$U(j\omega) = E(j\omega)\left(\frac{C_I}{j} + C_P + jC_{\dot{D}}\right), \tag{4}$$

Here C_P, C_I and C_D are the coefficients of the P, I and D components, E is the control error, U is the controller output.

(4) tells that the adaptation of the global process lag according to (1) makes the amplitude of each component equal to its respective coefficient, regardless of the frequency (i.e. regardless of the time scale factor, too), C_I, C_P and C_D remain unchanged. Naturally, these considerations are valid for single frequencies only, serving as a simple illustration of the adaptation principle which must be extended to more general processes than a sine wave. The generalization can be carried out in the time domain where the following relations between mean absolute values of the control error derivatives (or differences) correspond to (2):

$$\tau \overline{\left| \frac{de}{dt} \right|} = \tau^2 \overline{\left| \frac{d^2 e}{dt^2} \right|} \tag{5}$$

or in the realizable discrete form

$$\tau \overline{\left| \frac{\Delta e}{T_s} \right|} = \tau^2 \overline{\left| \frac{\Delta^2 e}{T_s^2} \right|} \tag{6}$$

(T_s denotes the sampling period). From (6) we get the global time lag

$$\tau = T_s \frac{\overline{|\Delta e|}}{\overline{|\Delta^2 e|}} \tag{7}$$

The global lag is then used in the PID control law in the following manner:

$$u = \frac{c_I T_s}{\tau} \Sigma g e + g c_P e + g c_D \tau \frac{\Delta e}{T_s}, \tag{8}$$

or in case of the more advantageous incremental form

$$\Delta u = g \frac{T_s}{\tau} \left(e + c_P^* \tau \frac{\Delta e}{T_s} + c_D^* \tau^2 \frac{\Delta^2 e}{T_s^2} \right) \tag{9}$$

where g is the common gain which will be discussed later, as well as suitable values of the other constants).

2.2 ADAPTATION OF THE SAMPLING PERIOD

As was said before, the sampling period T_s is made proportional to the global lag τ , too. This seems to be reasonable not only for the sake of adaptivity to time scale factor changes. In addition, there exists an optimum sampling period which is comparatively long if the plant order is high. In that case the PID controller has an insufficient structure to perform e.g. the time optimum control so that the control speed cannot be higher than a certain limit corresponding to a critical frequency. The useful frequency range of any control loop is limited just by the critical frequency. Within this range the feedback is negative and able to compensate disturbances of lower frequencies than the critical one. Outside this range the control is quite ineffective because the feedback is positive. Since the critical frequency f_{crit} is the highest one which can be generated by the feedback loop, then, according to the sampling theorem (Shannon, Kotelnikov) the necessary minimum sampling frequency is

$$f_s = 2f_{crit} \tag{10}$$

Simultaneously with the sampling theorem, the relation (10) defines the maximum critical frequency which can be generated in the control loop sampled with frequency f_s (it is the case when the control error changes its sign at every sampling instant). The real relation is always

$$f_{crit} < \frac{f_s}{2} \tag{11}$$

or expressed in terms of periods

$$T_{crit} > 2T_s \tag{12}$$

because even if $T_s \rightarrow$, then always $T_{crit} > 0$ Experiments have shown that there exists a minimum of T_{crit} at $T_s \gg 0$ (not at $T_s = 0$, as it might seem). The question of the best sampling period under normal work conditions had to be solved experimentally, too. It has been found to be convenient if

$$T_s = \tau \tag{13}$$

Substituting (13) into (7) gives

$$\frac{\overline{|\Delta e|}}{\overline{|\Delta^2 e|}} = 1 \tag{14}$$

This implicit relation with the unknown variable T_s prescribes that.

$$\overline{|\Delta e|} - \overline{|\Delta^2 e|} = 0 \tag{15}$$

and our task is to find a suitable value of T_s which fulfils this condition. (15) can be solved iteratively in the following way

$$\Delta T = cT \text{sign}\left(|\Delta e| - |\Delta^2 e|\right) \tag{16}$$

where $0.01 < c < 0.05$ This approach is analogous to an integral "on-off" control maintaining the right-hand side of (16) (=control error) zero. Since available microprocessors are not equipped with programmable, continuously variable sampling, the control error itself must be sampled with a constant, sufficiently short basic period T_0 . Then T_s will be an integer multiple of T_0. The main advantage of the sampling period adaptation is saving of computing and simplification. For example, owing to the equality $T_s = \tau$, the control lag (9) becomes much simpler. The operation of division could be left out entirely. In addition, the said similarity of control processes with different plants became even better because the sampling period T_s is long enough to wipe off some details in plant dynamics. The performance optimum with respect to the sampling period is flat so that the adjustment is not crucial. Simultaneously a good noise filtering is provided for control as well as for adaptation.

2.3   GAIN ADAPTATION

It has been ascertained that if the plant gain equals to one, then the suitable values of the common gain g (see (8) or (9)) fluctuate within a rather narrow interval $0.5 < g < 1$. In that case g need not be adapted and can serve as a supplementary parameter for correction of the control process shape according to the user's demands. Otherwise the plant gain can be identified and its reciprocal used for compensation of the total loop gain. This brings the same effect as if the virtual plant gain were equal to one. The plant gain can be estimated as a ratio of plant output y and input u

$$g_p = \frac{y}{u} \tag{17}$$

provided that the plant is linear and the mean values of y and u are non-zero and predominating over their variances. If the static plant characteristic has the form (which is supposed to be roughly known)

$$y = a + ku \tag{18}$$

then

$$g_p = \frac{y - a}{u} = k \tag{19}$$

The control law then takes the final form (compare with (9))

$$\Delta u = g\frac{u}{y-a}\left(e + c_P\Delta e + c_D\Delta^2 e\right) \tag{20}$$

where $c_P \approx c_D \approx 0.5$, $g \approx 0.5$ for plants with a transport lag, or $c_P \approx c_D \approx g \approx 1$ for plants without transport lag. These values have been found by numerous experiments. This approach, though very primitive, proved to be superior to some theoretically perfect but much more complicated ones. Thanks to its rapidity it readily compensates even great and quick plant gain changes so that instability may not arise. The control law (20) comprises the operation of division. It can be avoided, howerer, at the cost of loss of speed. The gain compensating term from (20)

$$Z = g\frac{u}{y-a} \tag{21}$$

can be comnputed by solving the corresponding implicit equation

$$gu - Z(y-a) = 0 \tag{22}$$

using similar method as in (15) and (16):

$$\Delta Z = cZ\mathrm{sign}(gu - Z(y-a)) \tag{23}$$

where $0.1 < c < 0.3$ The main advantage of this method is its resistance to situations when the values of u or $y - a$ might be zero. (Note that in (20) the computation sticks when $u = 0$ or breaks down as soon as $y - a = 0$. Therefore it must be protected against such events). Besides the above mentioned methods of gain adaptation some different ones have been developed. They are based on new criteria of damping or oscillatory behaviour of the control process. Since they have been published already, merely two representatives of them will be briefly treated in this paper, the former using the "oscillation index", the latter the "damping index", both having a heuristic character. It must be premised that any attempt to make out those criteria mathematically must fail. They have been made "to measure" utilizing the said similarity of control processes. They comprise inherently the necessary properties for this kind of processes only. Each of these criteria can easily be calibrated by known processes in order to find the "optimum" reference value which is practically uniform-applicable in any case.

2.3.1 GAIN ADAPTATION BASED ON THE OSCILLATION INDEX

Roughly speaking, the oscillation index Q indicates the relative level of a sine wave in the control process. One variant is defined as follows:

$$Q = \overline{\text{signe}(k-1)\text{sign}\Delta^2 e(k)} \tag{24}$$

For a pure sine wave it gives $Q = -1$, for an exponential function $Q = +1$. Optimum processes give $Q \approx -0.3$ The algorithm for adaptation of the common gain g has the form:

$$\Delta g = cg(\text{sign } e(k-1)\text{sign } (e(k) - 2k + (k-1) + e(k-2)) + 0.3) \tag{25}$$

where the reasonable extent for c is $0.03 < c < 0.1$

2.3.2 GAIN ADAPTATION BASED ON THE DAMPING INDEX

The damping index is the inverse of the oscillation index. For its formulation the same components can be used as for the oscillation index, only its structure is different:

$$D = \overline{|\text{signe}(k-1) + \text{sign}\Delta^2 e(k)|} \tag{26}$$

For a sine wave $D = 0$, for an exponential $D = 2$, for an optimum process $D \approx 0.7$ The corresponding adaptation algorithm is:

$$\Delta g = cg(|\text{signe}(k-1) + \text{sign}(e(k) = 2e(k-1) + e(k-2))| - 0.7) \quad \text{with} 0.03 < c < 0.1 \tag{27}$$

Remark

It can be proved that both these algorithms are mutually "complementary" because for any process

$$D - K = 1 \tag{28}$$

Thus, irrespective of which algorithm has been used, the adaptation has equivalent properties. The algorithms can be used either independently or in combination with the plant-gain compensation (23).

3 THE COMPLETE ALGORITHM

The algorithm of the adaptive PID controller consists of the following few formulae:

(I.) $\quad \Delta e(k) \quad = \quad e(k) - e(k-1)$

(II.) $\quad \Delta^2 e(k) \quad = \quad e(k) - 2e(k-1) + e(k-2)$

(III.) $\quad u(k+1) \quad = \quad u(k)g\frac{u(k)}{y(k)-a}(e(k) + 0.5\Delta e(k) + 0.5\Delta^2 e(k))$
$\quad\quad\quad\quad\quad\quad\quad$ see (20) up to (23)

(IV.) $\quad T_s(k+1) \quad = \quad T_s(k)\mathrm{sign}(|\Delta(k)| - |\Delta^2 e(k)|)$ see (16)

(V.) $\quad g(k+1) \quad = \quad g(k) + 0.03g(k)(\mathrm{sign}e(k-1)\mathrm{sign}(e(k)-$
$\quad\quad\quad\quad\quad\quad\quad -2e(k-1) + e(k-2)) + 0.3)$ see (25) or (27)

Formula (V) need not be used in case of linear plants. Formulae of adaptation (IV) and (V) must be skipped in the program if $|e| \to 0$ or if u exceeds its preseribed limits ("Do not adapt if you need not or cannot control!"). According to the saturation of the actuator u must be limited from above and from below; u and $y - a$ must not be zero. These arrangements are necessary to avoid the wind-up effects in control as well as in adaptation.

4 THE CONTROLLERS HARDWARE

The construction of the controllers is made in the form of a "black box" integrated in a dual-in-line package. It is based on eight-bit microcomputers (INTEL 8748, 8751 or perspectively 8798). Each controller consists of the following parts:

The process interface working as the link between the central processing unit and process. It performs signal conversion and matching, receiving process data from sensors (through the voltage range 0-10V) and returning a feedback signal to the process (in the same voltage range).

The central processing unit controls the processing sequence and performs the desired processing functions.

The communication components enable the process station or human operator to observe the process and intervene as necessary (by means of the RS 232 serial port).

5   THE SOFTWARE

The real-time operating system performs administrative functions for the organization of program runs, interactive communication as well as checking function within the controller.

The modular function programs (firmware) consist of function-oriented modules. They contain the control algorithm, logical operations, arithmetic units (fast fixed point arithmetic) etc. The controllers have been designed to have the following properties:

The controller is a basic unit of a process station which performs the functions of data input as well as control command output and implements automation functions such as closed-loop control (automatic mode), open-loop control (manual mode) and monitoring (display mode).

The controller can serve as a real-time measuring and communication unit for connection to a personal computer (e.g. IBM PC XT/AT) in order to utilize CAD program systems. The serial communication (at the rate of 2400 Bd) with the PC serves for initial adjustment of controller parameters (sampling period, setpoint, gain) and for graphic monitoring of controller operation to check its work. The serial communication ability of these controllers can be employed in real-time experiments in CAD for control systems package such as SIC ("Simulation, Identification and Control" developed in our Institute), MATLAB, etc.

6   SIMULATION TESTS AND APPLICATIONS

A series of tests was conducted simulating a wide range of processes and comparing the performance of the above described self-tuning PID controllers with optimum statically tuned ones. The results were very good. Good results have been obtained in varions industrial applications, such as temperature control of greenhouses, glass and quartz melting furnaces, wood-drying kilns, a sodium iodide crystallizer etc. Successful experiments have been made with power control of a technological laser.

7 CONCLUSION

Extremely simple algorithms for self-tuning PID controllers were developed and implemented on readily available single-chip microcomputers. Our aim was to solve the problem as simply as possible, at minimum cost and maximum user-friendly. The self-tuning controllers of this type might substitute the statically tuned ones used in the industry.

References

[1] Maršík, J.: A New Conception of Digital Adaptive PSD Control. Problems of Control and Information Theory 12(4), 1983, pp. 267-279.

[2] Maršík, J., Strejc V.: Heuristic Adaptive Process Computer Control. Proc. of the 7th IFAC/IFIP/IMACS Conference on Digital Computer Applications to Process Control, Vienna 17-20 Sept. 1985, pp. 387-392.

[3] Klán P., Maršík, J., Drozen J.: Adaptive PID Controllers on a Single-chip Microcomputer. IFAC Symposium LCA 86 on Components, Instruments and Techniques for Low Cost Automation and Applications, Valencia, Spain, 27-29 Nov. 1986, pp. 349-352.

[4] Maršík, J., Strejc V.: Application of Identification-free Algorithms for Adaptive Control. Proc. of the 10th World Congress IFAC on Automatic Control. Preprints Vol. 3, pp.15-21, 27-31 July 1987, Munich.

[5] Maršík, J., Strejc V.: Application of Identification-free Algorithms for Adaptive Control. Automatica Vol. 25, No.2, pp.273-277, March 1989.

[6] Klán P., Maršík, J.: Implementation and Application of Single-Board Microcomputers-Based Auto-Tuners. IFAC Workshop on Ecaluation of Adaptive Control in Industrial Applications, 16-20 Oct. 1989 Tbilisi.

[7] Maršík, J.: Robust Adaptive PID Controller with Automatic Adjustment of Sampling Period. Problems of Control and Information Theory, Vol. 19(3) pp. 243-252, 1990.

Self-Tuning Control of Robot Manipulators

Karam Z. Karam

Department of Cybernetics
University of Reading, UK.

Introduction

The increasing use of robotic manipulators in various areas such as industrial automation and remote handling is being accompanied by a greater need to improve the performance of the underlying controller elements, this is particularly the case for more modern lightweight and high speed manipulator specifications.

Manipulators generally consist of a number of linked mechanical components each driven by some form of actuator. Individual actuator loads are therefore highly variable in character depending on the overall manipulator configuration, movement of other mechanical components in the system and the nature of the manipulated load. Simple fixed control structures as are used in most current commercial manipulators, cannot provide consistent performance over a wide range of robot motions and loads. Consequently speed of operation, load bearing capacity, positioning accuracy and stability are usually compromised.

Recently the use of adaptive or self-tuning controllers to circumvent such problems has been increasingly studied [1-5]. Such control philosophies provide the possibility of adapting controller gains continuously in order to accommodate the varying robot dynamics and thus increase the performance consistency. However there remain various problems associated with adaptive or self-tuning schemes in the context of robot control, for example due to the high speed dynamics and their rapid variation and the difficulty in providing rigorous theoretical analysis of stability and overall performance.

In order to highlight the problems involved, this paper will first briefly review the general nature of robot dynamics and will then discuss typical fixed control topology. A description of some more recent adaptive ideas will then be presented and this is followed by a more detailed look at a particular self-tuning implementation with an illustration of some results using computer simulations.

Robot Dynamics

A mathematical description of the dynamic characteristics for a particular robot can be derived in a variety of ways eg. using Lagrangian mechanics or Newton-Euler formulations [6] and although the precise relationships will vary for different robots, the general form relating joint actuation forces and link motions is given below :

$$\tau = D(q)\ddot{q} + C(q,\dot{q}) + h(q) + b(\dot{q}) \tag{1}$$

where the terms have the following physical interpretations : τ represents joint actuation forces, $D(q)\ddot{q}$ represents inertia associated with the distribution of mass, $c(q,\dot{q})$ represents interaxis velocity coupling due to centrifugal and coriolis forces, $h(q)$ represents gravity loading forces and $b(\dot{q})$ represents the effects of friction.

For any but the most simple of robots the dynamic model is highly complex due to the numerous terms existing in the above matrices and vectors. For a typical n-link manipulator the model will consist of a set of n highly coupled non-linear, time varying second order differential equations. This combination of factors makes the high performance control of a typical robot a very challenging task indeed.

Robot Control

Most current commercial robot controllers take advantage of a number of simplifying conditions in order to achieve reasonable performance. One of the more complex terms in the dynamic model is that of the velocity coupling terms $(c(q,\dot{q}))$, however since most commercial robots are operated at modest speeds the contribution of this term is substantially reduced relative to other dynamic effects. Furthermore other aspects inherent to existing robot designs such as substantial gear ratios and gravity compensating devices, reduce the effects of variable components of inertia, gravity loading and velocity coupling as seen at the link actuator mechanisms. Finally, a majority of industrial applications involve repetitive manipulation tasks with infrequent and usually well defined changes in the task requirements.

Under these circumstances it becomes sufficient to design relatively simple controllers whose parameters can be adjusted manually to suit a particular job. However it must be pointed out that the advent of new designs using more efficient and accurate direct drive actuators and the use of robots in less structured environments for truly flexible applications will increasingly highlight the limitations of such controllers.

Single-axis fixed PID control

A common approach to the control problem and one which is used in most current commercial robots, is the use of single-axis PID control. Here each robot axis is individually controlled disregarding the effects of other robot axes and using a simple servoing strategy based on well tested designs adopted from other control applications. Many such controllers rely on traditional analogue electronic implementations but more recent designs are increasingly employing a digital (sampled signal) approach.

A typical continuous time model for the single-axis PID controller is given by the following :

$$u(t) = K_P e(t) + K_I \int_0^t e(t)dt + K_D e(t) \qquad (2)$$

where $u(t)$ is the signal applied to the actuator, K_P, K_I and K_D are the Proportional, Integral and Derivative gains respectively and $e(t)$ is the reference tracking error which is defined by :

$$e(t) = r(t) - q(t) \tag{3}$$

where $r(t)$ is the reference or required link position and $q(t)$ is the measured position.

From (2) it is easy to see that the function of the control structure is to apply control action proportional to the error (in order to reduce it), proportional to the time integral of the error (to remove offsets) and proportional to the rate of change of the error (to dampen oscillatory responses). Since the remainder of the discussion will only be concerned with discrete time mathematical representations suitable for implementation on digital computers, a particular discrete time (z-domain) form of the PID structure is given below :

$$Iu(k) = Se(k) \tag{4}$$

here, $u(k)$ and $e(k)$ are the sampled signals at sample no. k,
$S = s_0 + s_1 z^{-1} + s_2 z^{-2}$,
$I = 1 - z^{-1}$
and z^{-1} is the backward shift operator ie. $z^{-1}e(k) = e(k-1)$.

The coefficients in S have the following relationship with the more familiar PID gains :

$$s_0 = -K_p + K_I + K_D, \quad s_1 = K_p + 2K_D, \quad s_2 = -K_D.$$

Note that the above represents only one form of the discrete PID controller. In practice slight variations are used to overcome undesirable effects such as 'set point kick'. Fig. 1 below depicts a single axis servoing scheme incorporating a general fixed parameter digital PID controller.

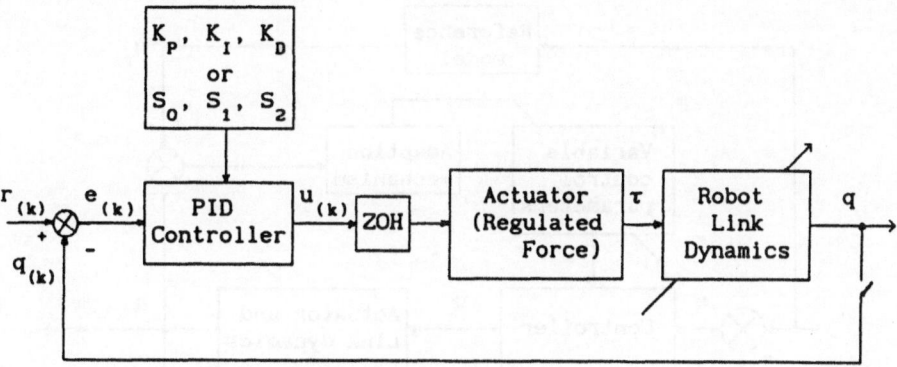

Figure 1. Digital single axis PID control scheme

Typically, rough values of K_P, K_I and K_D (or S_0, S_1, S_2) are selected by the robot manufacturer based on a knowledge of the robot mechanics. In some cases these parameters may be further adjusted on site to provide better characteristics for a particular application, however this retuning procedure usually relies on operator experience or on a trial and error basis, and more often than not is a task not entirely suitable for the faint of heart!

Adaptive control schemes

Adaptive control techniques range from simple gain scheduling to full blown multivariable optimising algorithms, but whilst many of these have been put to good use in various controlled mechanisms, their incorporation into the area of robot control has been somewhat hindered by the complexity of the dynamics and by the high sample rates required. Nonetheless research in this area continues with the aim of overcoming these problems and indeed this paper presents one possible solution.

In the context of robot control, three popularly cited adaptive schemes are depicted in figures 2,3 and 4. The model reference controller shown in figure 2 attempts to obtain the same servo response from the real robot as that specified by a user defined reference model. Any difference or error between the two responses is used to adjust the controller parameters in order to eliminate that error.

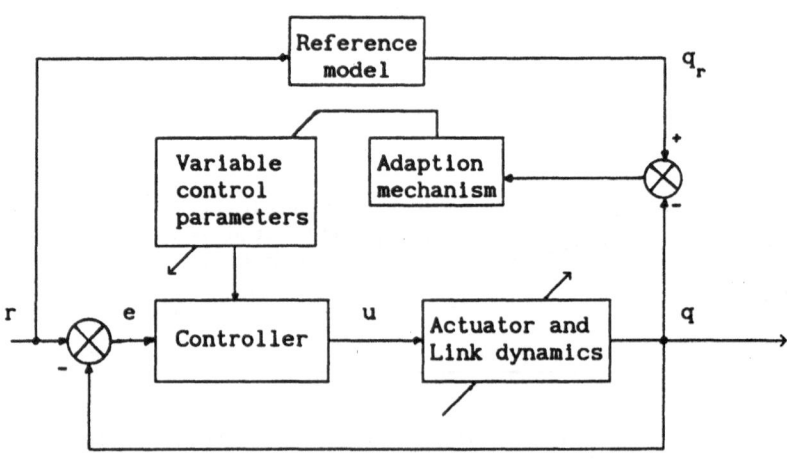

Figure 2. Model reference adaptive control

Figure 3 depicts a somewhat different approach; in this case a dynamic
model of the robot mechanism is used to predict the forces required for a
particular movement and these are fed forward directly to the actuators.
Any remaining error between the reference signal and the actual robot
response (due to model mismatch or unknown disturbances) is then regulated
by a secondary optimal controller. Non-linearities or time variations in
the nature of the error signal are further tracked via a parametric
estimator and the optimal controller parameters are adjusted accordingly in
order to maintain overall control consistency.

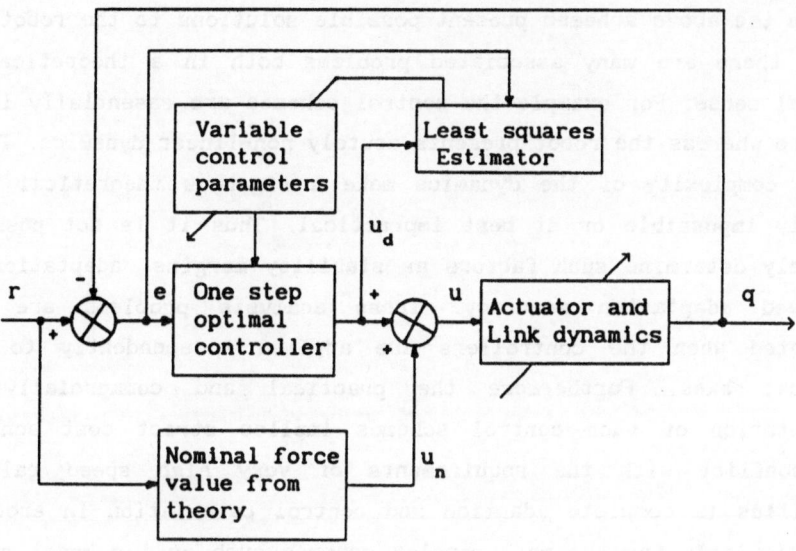

Figure 3. Differential adaptive controller

In figure 4, the adaption mechanism relies on estimating a model of the robot dynamics which is then used to calculate controller parameters in order to effect a particular control behaviour. If the estimation procedure is only invoked upon commissioning the robot then this simply results in a self-tuning controller, however if the estimation procedure is carried out continuously then adaptive behaviour is achieved.

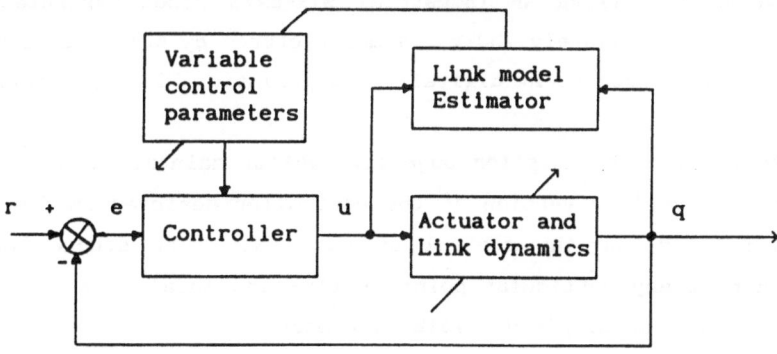

Figure 4. Explicit self-tuning control

Although the above schemes present possible solutions to the robot control problem there are many associated problems both in a theoretical and a practical sense. For example the control schemes are essentially linear in structure whereas the robot presents acutely non-linear dynamics. This fact and the complexity of the dynamics make a rigorous theoretical analysis virtually impossible or at best impractical. Thus it is not possible to accurately determine such factors as stability margins, adaptation speeds or indeed adaptation accuracy. These analysis problems are further complicated when the controllers are applied independently to control individual axes. Furthermore the practical and commercially viable implementation of such control schemes implies strict cost constraints which conflict with the requirements of very high speed calculating capabilities (a complete adaption and control calculation in around 5-10 ms) particularly for the more complex schemes such as the model reference or differential controllers.

A Fast Single-Axis Adaptive Controller

Notwithstanding the theoretical analysis problems mentioned above a simple structure single axis self-tuning/adaptive controller is now presented which is practically implemented on a low cost microprocessor based system and yet provides the required control rates. The impetus behind the controller development arose from a requirement to assess the possibility of adaptively controlling an industrial six-axis robot manipulator. The robot weighing approximately 750kg, is manufactured by KUKA and is normally controlled by six independent digital fixed parameter PID type controllers.

In order to achieve the adaption objective whilst maintaining a reasonable cost a simple structure version of the controller depicted in figure 4 is used independently for the control of each axis. The scheme begins by assuming that at any particular point of time the local axis dynamics can be represented by the simple discrete time model:

$$Aq(k) = bu(k-1) + d(k) \tag{5}$$

where $A = 1 + a_1 z^{-1} + a_2 z^{-2}$, b is a scalar and d(k) is an unknown disturbance.

The structure of the controller block is simply that of a particular form of the PID algorithm, given by :

$$u(k) = u(k-1) + S^1 r(k) - Sq(k) \tag{6}$$

where $S = S_0 + S_1 z^{-1} + S_2 z^{-2}$, and $S^1 = S_0 + S_1 + S_2$ ie. $S|_{z=1}$.

The equation for the closed characteristics can be derived from (5) and (6) and is given below:

$$q(k) = \frac{z^{-1} b S^1 r(k) + (1-z^{-1}) d(k)}{(1-z^{-1})A + z^{-1} bS} \tag{7}$$

A pole assignment strategy [8] is employed as the control objective, thus controller parameters (coefficients in S) are computed in such a way as to ensure that the denominator expression in equation (7) becomes equal to a user defined polynomial T, whose roots are chosen to give the desired response characteristics ie. :

$$(1-z^{-1})A + z^{-1} bS = T, \qquad (T = 1 + t_1 z^{-1} + .. + t_3 z^{-3}) \tag{8}$$

Solving (8) gives the following relationships :

$$\begin{aligned} S_0 &= (t_1 - a_1 + 1)/b \\ S_1 &= (t_2 - a_2 + a_1)/b \\ S_2 &= (t_3 + a_2)/b \end{aligned} \tag{9}$$

Rewriting the closed loop expression using (7) and (8) gives :

$$q(k) = \frac{z^{-1} T^1}{T} r(k) + \frac{(1 - z^{-1})}{T} d(k) \tag{10}$$

where $T^1 = 1 + t_1 + t_2 + t_3$.

It is clear from the above that the desired control objective is achieved assuming a correct dynamic model. However for most robot axes the simple linear model is not sufficiently correct over a wide range of robot movement and in any case must be evaluated at least upon commissioning, it is therefore necessary to use a parameter estimation algorithm in order to achieve model parameterisation and subsequent adaption. Bearing in mind the requirements for speed the following computationally efficient approximate form of the least squares algorithm is used :

Defining the following parameter estimate ($\hat{\theta}$) and data (X) vectors:

$$\hat{\theta}(k) = (\hat{a}_1, \hat{a}_2, \hat{b})^T$$
$$X(k) = (-q(k-1), -q(k-2), u(k-1))^T$$

Then, a prediction q_p, of the axis position is given by:

$$q_p(k) = X^T(k)\hat{\theta}(k-1) \tag{11}$$

A recursive form of the approximate least squares algorithm [7] is now used to continuously update parameter estimates in $\hat{\theta}$ so as to reduce the prediction error ε, as follows :

$$\varepsilon(k) = q(k)_p - q(k),$$

$$\hat{\theta}(k) = \hat{\theta}(k-1) + \gamma(k-1)\varepsilon(k),$$

$$\gamma(k) = \frac{P(k-1)X(k)}{\beta + X^T(k)P(k-1)X(k)}, \tag{12}$$

$$P(k) = DIAG\{(I - \gamma(k)X^T(k))P(k-1)/\beta\}.$$

where P is a 3x3 diagonal matrix, I is the unit matrix and β is a 'forgetting factor' which ensures that old data is gradually phased out. Typically P is initialised as $100I$ and β is given a value of 0.98.

Microprocessor implementation

In order to achieve the necessary computation speeds at commercially economical costs the algorithm was implemented using the 16 bit M68000 microprocessor from Motorola. Cost constraints also precluded the use of a floating point co-processor so a set of high speed semi-precision floating point routines were developed in assembly language. The structure of the numerical representation (which is chosen to fit into one processor long word of 32 bits) is shown below :

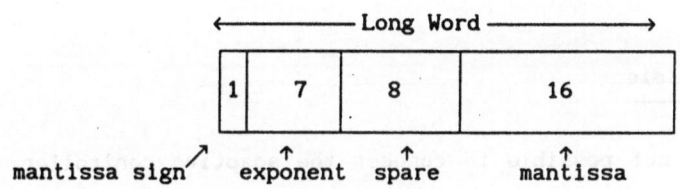

Figure 5. Floating point representation

The precision offered by this representation is approximately 5 decimal places and the absolute range is from 10^{-39} to 10^{38}. This accuracy is more than sufficient for the application [4].

In order to further increase computation speed the main core of the control algorithm was written in assembly language. This infact was quite an easy task due to the simplicity of the scheme and the powerful processor instruction set. The remainder of the program which included initialisation blocks, safety jacketing, various diagnostic routines and the user interface was implemented in a high level language.

Summary of adaption/control sequence

1 - Collect/update signal values q, u and r.
2 - Generate control signal u.
3 - Update parameter estimates in the linear axis dynamic model, using
    the recursive approximate least squares algorithm.

4 - Calculate new set of PID parameters S_0, S_1, S_2, using results of step 3.

5 - Wait for end of sample period.

6 - Goto step 1.

For the implementation described above, the complete computation sequence (steps 1-4) could be achieved in a time span of around 2ms using an 8MHz clock. This was actually faster than the control update rate of the robot's own internal fixed parameter digital PID controllers, which took around 15ms.

Simulation Trials

Since it was not possible to connect the adaptive controller to the real robot at the time, testing was achieved by application to a comprehensive non-linear time domain simulation of the six-axis robot dynamics running on a separate mini-computer. The performance was assessed in comparison with that of the fixed parameter controllers used on the real robot.

Figures 6, 7 and 8 summarise some typical results obtained via the simulation trials. In this particular simulation the manipulator is asked to move from its zero position to a position defined by the following axis angles (joint co-ordinates) : A1 = -110°, A2 = 30°, A3 = -20°, A4 = 0°, A5 = 160°, A6 = 0°, in a period of 2 seconds.

Figure 6 depicts the case for the internal fixed PID controllers, and it is clear that significant inaccuracies exist both in end-effector path following as well as in the final rest position. The adaptive controller on the other hand shows much improved performance as can be seen in figure 7. Furthermore, examination of the progression of parameter estimates in the linear axis model for one of the six adaptive controllers used on the robot shows that adaption is taking place during the movement, indicating that the controllers are rapidly tracking the changing dynamics. This behavior can be seen in figures 8(a) & (b), which show the progression of parameters estimates in the $\hat{A}$ polynomial for the controller attached to axis 3 (which is responsible for moving link 3 on the robot).

Note that in this particular case the adaptive controllers were initialised with no information about the axis dynamics ie. the initial axis model parameters were set to zero, resulting in the initial erratic estimates which can be seen in figure 8. In practice this situation would not occur since the controllers would always be initialised with values from a previous run, and even upon commissioning it is possible to run the estimators alone for a while during a run using the internal fixed parameter PID controllers, and thus obtain model information prior to switching on the adaptive controllers.

Conclusions

Adaptive control techniques are becoming increasingly important in the field of robot control in order to maintain consistency and accuracy in the face of changing robot dynamics particularly in applications involving unstructured environments, high speeds or lightweight manipulators. Several adaptive schemes are available but it is very difficult to provide theoretical analysis for their performance within a robot environment. Furthermore many adaptive schemes involve large amounts of computation which makes them difficult to implement economically using current computer technology. However this paper presents a scheme which satisfies cost constraints whilst providing high control rates. This is achieved using a combination of simple control structure, a high efficiency adaption algorithm, a low cost microprocessor together with assembly level coding and limited precision floating point arithmetic. The resultant controller provides computation times down to 2ms per adaption/control cycle, and has shown significant improvements over the fixed parameter equivalent, when compared on a simulation of an industrial robot.

References

1 - Koivo A.J. and Guo T.H., 'Adaptive linear controller for robotic manipulators', IEEE Trans. AC, AC-28, 2, pp162-170, Feb. 1983.

2 - Lee C.S.G. and Chung M.J., 'An adaptive control strategy for mechanical manipulators', IEE Trans. AC, AC-29, pp873-890, 1984.

3 - Warwick K. and Pugh A. (eds.), 'Robot Control: theory and applications', Peter Peregrinus Ltd., 1988.

4 - Karam K.Z., 'Fast adaptive control algorithms and their application to industrial robots', Ph.D. Thesis, University of Newcastle Upon Tyne, UK, 1988.

5 - Karam K.Z. and Warwick K. 'An adaptive control simulator for industrial robot applications', Proc. IMA conf. on Robotics, LUT, UK, July 1989.

6 - Fu K.S., Gonzalez R.C. and Lee C.S.G., 'ROBOTICS Control, Sensing, Vision, and Intelligence', McGraw-Hill, 1987.

7 - Farsi M., Karam K.Z., Warwick K., 'Simplified recursive identifier for ARMA processes', Elect. letters, 20, p913-915, 1984.

8 - Wellstead P.E., Prager D. and Zanker P., 'Pole assignment self-tuning regulator', Proc. IEE, Vol. 126, p781-787, 1979.

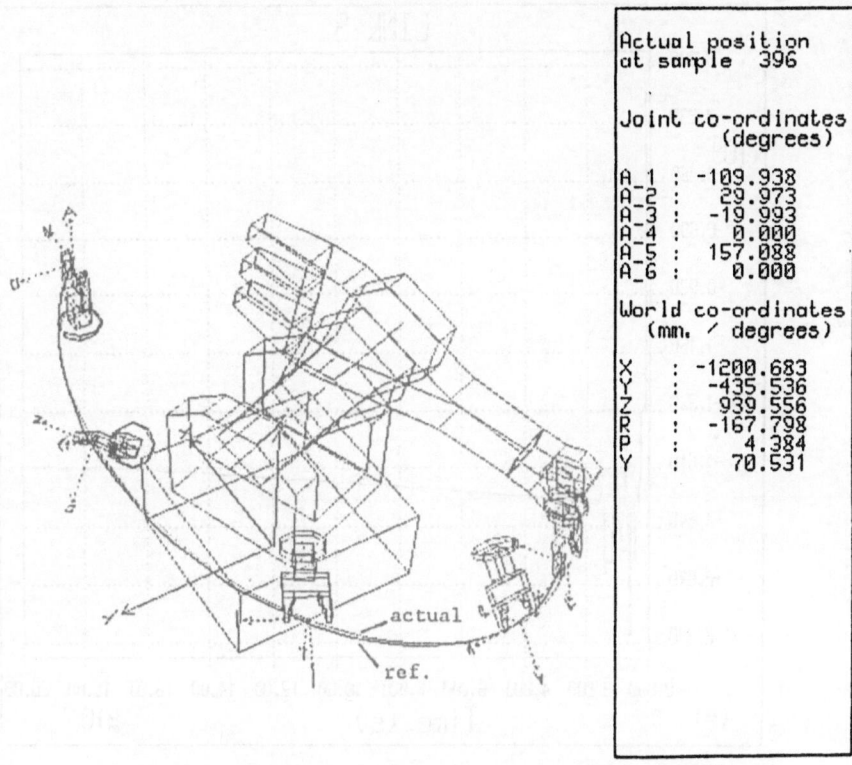

Figure 6. Simulated response using fixed parameter controllers

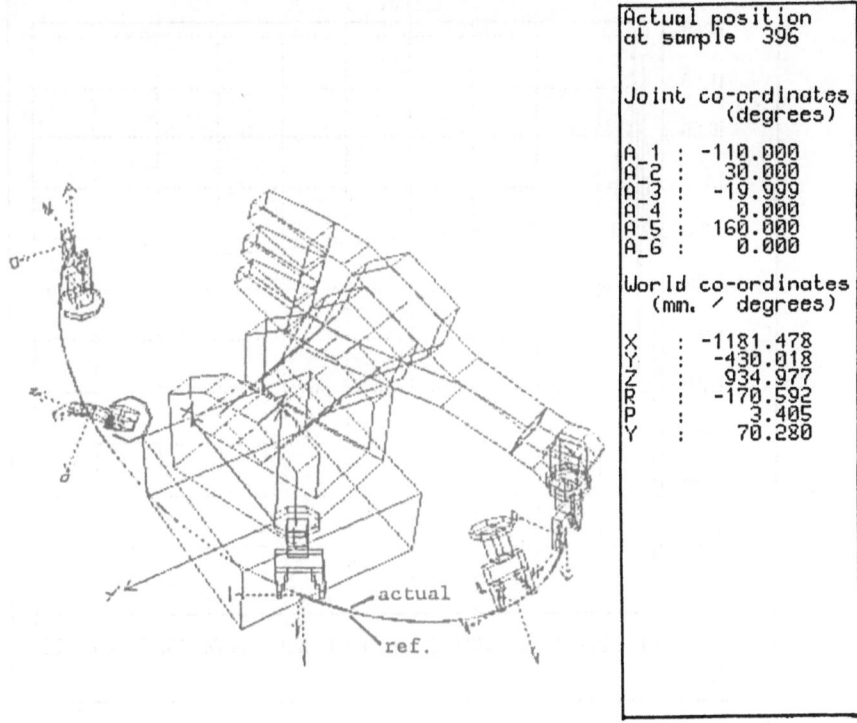

Figure 7. Simulated response using adaptive controllers

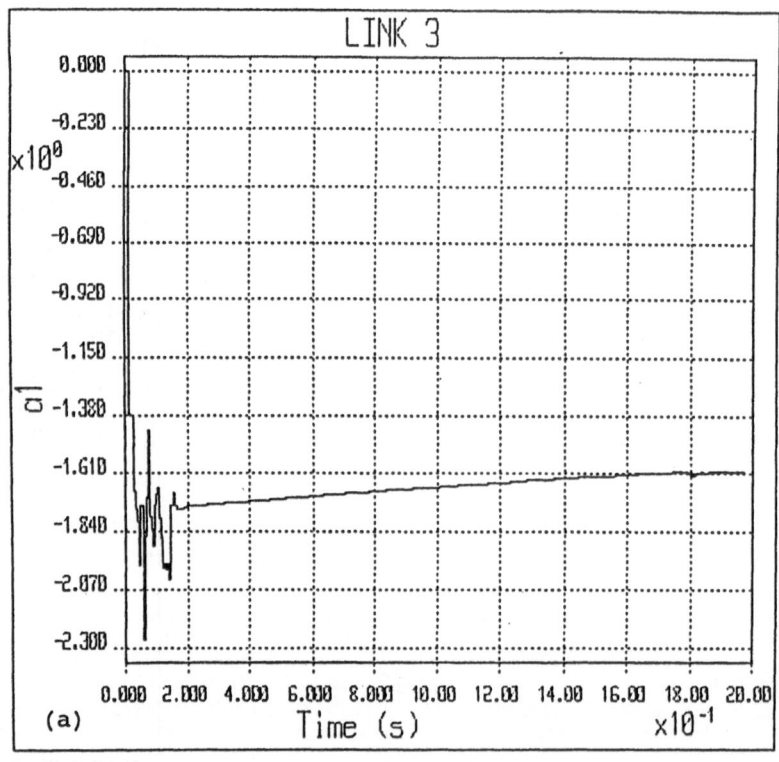

(a)

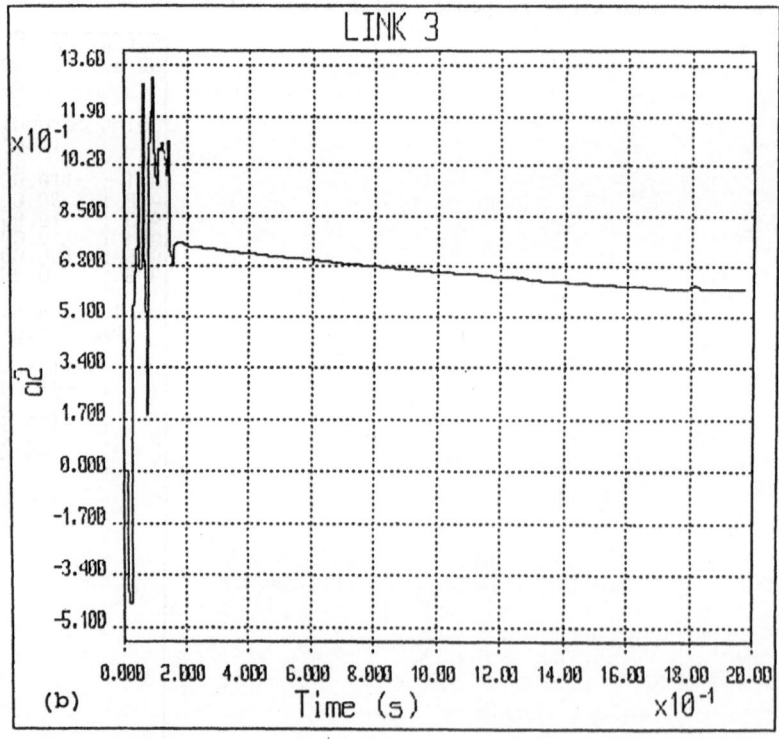

(b)

Figure 8.   Parameter estimates in model of axis 3

PRELIMINARY TUNING OF SELFTUNERS

M. Kárný, A. Halousková

Institute of Information Theory and Automation,
Czechoslovak Academy of Sciences
182 08 Prague 8, Czechoslovakia

1 Introduction

Computers are natural tools for tuning of computer controllers. Paradoxically, preliminary tuning of selftuners is not supported enough: neither theoretically nor by a suitable software.

The paper describes strivings to overcome this situation which contributes to unsatisfactory extent of industrial applications of selftuners. The activities should result into the *complete* set of tools which support all steps of the selftuner design. It means that available solutions of particular steps have to be joined and existing holes filled up. The dreamed up result is expected to bring new quality in selftuning area, namely,

– substantial decrease of demands on the user who is not forced to understand deeply the theory behind the selftuner in detail;

– efficient exploitation of the available prior information which increases a chance for high performance of the selftuner (including its quiet start up).

The algorithms and tools described in the paper are tailored to the multivariate selftuner which minimizes *multistep quadratic* criterion (Q) assuming the controlled system to be describable by *linear* (L) *gaussian* (G) *regression* model (the selftuners are referred to as LQG type for simplicity). The same approach can be (and should be) elaborated for other types of selftuners.

The results presented in the paper have roots in the long lasting research of V. Peterka and his co-workers. The thesis [2] can be taken as starting point of the research branch described here. The other relevant references are [3,4,8,13].

The explanations are organized as follows:

– background of the tuned selftuner is recalled;

– user's preliminary options are listed;

– the tools supporting particular choices are sketched;

– full preliminary design is illustrated on a simulated example;

– open problems and intended ways of the solution conclude the paper.

The paper itself has two objectives: to inform potential users about the design tools available and to stimulate discussion whether the right direction of the research has been chosen.

2   The tuned selftuner

The controller uses m_y-dimensional output of the controlled system $y(t)$, m_u-dimensional input $u(t)$ and m_v-dimensional measured disturbance $v(t)$ (of course, $v(t)$ need not be present). Their values are labelled by the discrete time $t = 1, 2, ..$ which counts control periods, i.e. time istants at which the inputs can be modified by the assumed feedback.

Control quality   optimized by the controller is quantified by the expected value of the *quadratic loss*

$$K[W] = \frac{1}{T}\{(z(T) - z_0(T-1))'S_z(z(T) - z_0(T-1))+$$
$$+ \sum_{t=1}^{T}[(y(t) - y_0(t-1))'Q_y(y(t) - y_0(t-1)) +$$
$$+ (u(t) - u_0(t-1))'Q_u(u(t) - u_0(t-1))]\}, \tag{1}$$

with a long control horizon T. In the loss function (1), W is common name for the penalizations $W = \{Q_y, Q_u, S_z\}$ which weight particular deviations from reference values (marked by the subscript $_0$).

Model of the controlled system   used for predicting behaviour of the controlled system is multivariate *linear* regression (*gaussian*) model

$$y(t) = P'z(t) + e(t) = \tag{2}$$
$$= \sum_{i=1}^{l_y} P_{yi}y(t-i) + \sum_{i=0}^{l_u} P_{ui}u(t-i-n_u) + \sum_{i=1}^{l_v} P_{vi}v(t-i-n_v) + P_1 + e(t).$$

For $x = y, u, v$, P_{xi} denotes matrix coefficient at the x value delayed by i steps. The number of coefficients is determined by "orders" l_x as well as by transport delays n_u, n_v. P_1 is a (vector) absolute term. $\{e(t)\}_{t\geq 1}$ is sequence of mutually independent m_y-dimensional random variables having common gaussian distribution with zero mean and a constant covariance matrix R (so called gaussian white noise).

The comparison of the "sum" form and of the compact form (all coefficients are compressed into a single matrix P, ' denotes transposition) explains the definition of the regression vector z used also in the loss function (1).

Model of the measured disturbance v adopted is the autoregression

$$v(t) = \sum_{i=1}^{l_{av}} P_{avi}v(t-i) + P_{a1} + e_a(t), \tag{3}$$

in which all symbols have direct counterpart in the definition of the regression model.

Admissible control strategies of the controller are determined by informational and generated-input restrictions. Selftuning LQG controller (approximately) minimizes expectation of the quadratic loss (1) under the causality condition: the input $u(t)$ is selected using data $d(1), d(2), .., d(t-1)$ where data item $d(t)$ means

$$d(t) = (y(t), u(t), v(t)). \tag{4}$$

Values of $u(t)$ have to be in the interval

$$u(t) \in [ul(t), uu(t)] \tag{5}$$

and their increments $\Delta u(t) = u(t) - u(t-1)$ in

$$\Delta u(t) \in [\Delta ul(t), \Delta uu(t)]. \tag{6}$$

Transformation of measured data for controlling is needed as a rule. The signals used by the selftuner are often sampled more frequently than the inputs are changed by the feedback. The signals are usually selected from a wider set of variables measured on the controlled process. Thus, the measured signals have to be adapted to the chosen control period and closed-loop signals selected from the measured set.

Let (for simplicity) the *control period* T_f be integer multiple of a fixed sampling period and $\bar{D}(\tau)$ be full data record sampled at sampling moments labelled by $\tau = 1, 2, .., L$. By our version of the selftuner, the sampled data

$$\bar{D}(\tau), \quad \tau = (t-1)T_f + 1, .., tT_f$$

are compressed into a single representant $D(t)$ (filtered value)

$$D(t) = \alpha[\bar{D}((t-1)T_f + 1), .., \bar{D}(tT_f)]T_f + \beta[\bar{D}((t-1)T_f + 1), .., \bar{D}(tT_f)]$$
$$t = 1, 2, ... \tag{7}$$

whose subset is used by the feedback. The vector coefficients $\alpha[.]$, $\beta[.]$ determine regression straight line fitted to the data measured within the control period [1,9].

The selection of the data items $d(t)$ (4) from $D(t)$ (7) which are used in the specific closed loop (they enter into the loss function (1) and into the models (2), (3)) is formalized by the relations

$$x_i(t) = D_{ip_x(i)}(t), \quad i = 1, 2, .., m_x, \quad x = y, u, v, \quad t = 1, 2, ... \tag{8}$$

where the (selection) vectors ip_x ($x = y, u, v$) have appropriate integer entries.

Principle of the assumed selftuner is quite standard. The selftuner estimates recursively the unknown coefficients of both models (2), (3) and (approximately) minimizes the expectation of the quadratic loss (1) restricting the inputs $u(t)$ or their changes to the pre-selected intervals (5), (6) using data gained according to the formulae (7), (8).

3 User's preliminary options

The description of the tuned controller implies characteristics to be chosen before controller start up. Three groups of problems have to be solved:

Structure estimation concerns integer quantities determining signals and array dimensions used by the controller. The choice of the controlled outputs (described by the selection m_y-vector ip_y) is usually determined by technical requirements. On the contrary, there is often freedom in selecting of inputs and measured disturbances. The appropriate selection of signals as well as of all remaining integer parameters of both models (2), (3) is called *structure estimation*.

For given m_y, ip_y, the structure is described by the collection of integer variables

$$s = \{ip_u, m_u, ip_v, m_v, n_u, n_v, l_y, l_u, l_v, l_{av}\} \tag{9}$$

which are known to be in pre-specified ranges.

Parameter estimation relates to estimating unknown real quantities occuring in the assumed models. The controller need not to know parameters of both models (2), (3), i. e. coefficients and noise covariances, as it estimates them when controlling. However, its bumpless start-up requires *their initial estimates and their uncertainties* to be adapted to the specific controlled system.

Let us re-write both models in a more compact form in order to specify more precisely the required initial conditions:

$$y(t) = P'z(t) + e(t), \tag{10}$$

$$v(t) = P'_a z_a(t) + e_a(t). \tag{11}$$

These versions of models are related to the original ones as follows

$$
\begin{aligned}
P' &= [P_{u0}, P_{y1}, P_{u1}, .., P_{yl_y}, .., P_{ul_u}, P_{v1}, .., P_{vl_v}, P_1], \tag{12}\\
z(t) &= [u'(t - n_u), y'(t - 1), u'(t - 1 - n_u), .., y'(t - l_y), .., u'(t - l_u - n_u),\\
&\qquad v'(t - 1 - n_v), .., v'(t - l_v - n_v), 1]', \qquad\qquad dim(z) = i_z,\\
P'_a &= [P_{av1}, .., P_{avl_{av}}, P_{a1}],\\
z_a(t) &= [v'(t - 1), .., v'(t - l_{av}), 1]', \qquad\qquad\qquad\qquad dim(z_a) = i_a.
\end{aligned}
$$

The selftuner estimates the *unknown parameters* $\Theta = \{P, P_a, R, R_a\}$ by the algorithm which coincides formally with recursive least squares. It starts from *point estimates* $\hat{\Theta} = \{\hat{P}, \hat{P}_a, \hat{R}, \hat{R}_a\}$. It needs *uncertainties* of the coefficients described by symmetric positive definite matrices $C > 0$ type (i_z, i_z), $C_a > 0$ type (i_a, i_a). C determines together with $\hat{R}$ mutual correlation of P entries.

Uncertainties of noise covariances R, R_a depend on $\hat{R}, \hat{R}_a$ and they are inversely proportional to positive scalars ν, ν_a, called *degrees of freedom*.

Prediction of control quality is the key novel tool in the described tripple. The penalization matrices (W) specifying the loss function (1) are to be chosen beforehand, too. These penalizations influence ranges of particular signals incorporated into the loss. It is desirable to *transform the required ranges* into proper *values of penalties* in order to meet technical requirements on the closed loop behaviour.

The penalties can be re-tuned in the control course. However, manual tuning can be overdemanding for slow systems and/or unexperienced users.

Moreover, the penalty S_z on the terminal regressor $z(T)$ in (1) influences strongly rate of convergence of the control design performed in real time. Thus, it is desirable to find an estimate $\hat{S}_z$ of the value which maximizes this rate.

4 Informational sources for preliminary tuning of the selftuner

Let us discuss pieces of information about the control problem which can be used for preliminary tuning of the controller.

User's objective and knowledge form decisive external inputs to the design. The user has to supply control objective (e. g. some quantity should be kept at zero level), restrictions on the controller complexity (e. g. computer memory available) and input actions (e. g. admissible ranges) as well as prior knowledge about the controlled system (e. g. potential inputs, upper bound on model complexity etc.). The current solution is oriented to the most common knowledge in order to decrease burden on the user. Extensions which exploit more detailed description of prior knowledge are planned.

Measured data are the key information source for the preliminary design. It relies mostly on the data gained in preliminary identification experiment. Their quality is decisive for its success. Up to now, there is no fully satisfactory experiment design for our problem. Consequently, the quality of used data is the weakest point of the approach. Common rules of thumb have to be used under these circumstances.

The evaluations described below use L records of rough data $\{\bar{D}(\tau)\}_{\tau=1}^{L}$.

5 Outline of the theoretical backgrounds for solving key steps of the preliminary design

The complete preliminary design consists of structure estimation, parameter estimation and tuning of penalties in such a way that the predicted quality of the closed loop is close to the user's requirements. The tasks are referred to as STR, RLS and EVA, respectively. They can be used independently and independently will be described. The more common is the algorithm the shorter its description is.

5.1 Structure estimation, STR

For the selected control period T_f, the task STR exploits $L_f = L/T_f$ filtered data $\{D(t)\}_{i=1}^{L_f}$ gained according to the formula (7). The user has to select the controlled outputs $y(t)$, *potential inputs* $U(t)$ and *potential measured disturbances* $V(t)$ from the full data record $D(t)$:

$$D(t) = (y(t), U(t), V(t)). \tag{13}$$

The task STR estimates the structure s (9) of the regression model (2) and of the autoregression model (3). These models describe (for chosen inputs) dependence of a data item $d(t)$ — formed by a fixed selection (8) from the representants $D(t)$ (13) — on delayed values $d(t - i)$, $i \geq 0$.

The user has to choose an integer ORDE which refers to the "oldest" data item $D(t-\text{ORDE})$ an entry of which could be incorporated into regressors of the final models (2), (3). Let us introduce *maximal data vectors*

$$
\begin{aligned}
DY(t) &= [D'(t); D'(t-1); ..; D'(t-\text{ORDE}); 1]' = \tag{14}\\
&= [y'(t), U'(t), V'(t); y'(t-1), U'(t-1), V'(t-1); ..;\\
&\qquad y'(t-ORDE), U'(t-ORDE), V'(t-ORDE); 1]'\\
DV(t) &= [V'(t), V'(t-1), .., V'(t-\text{ORDE}), 1]' \quad t = 1, 2, ...
\end{aligned}
$$

Structure estimation is formulated as testing of the finite (but huge) amount of the following hypotheses, indexed by possible structures s (9):

Appropriate regressors are formed by entries selected from the maximal data vectors (14) in such a way that the models (2), (3) have the structure described by the collection of integer variables $s = \{ip_u, m_u, ip_v, m_v, n_u, n_v, l_y, l_u, l_v, l_{av}\}$.

The following hypotheses illustrate the concept:

empty regressor $\leftrightarrow$ no entry in $DY(t)$ contributes to predicting the controlled output $y(t)$;

full regressor $\leftrightarrow$ all entries in $DY(t)$ should be used for predicting the controlled output $y(t)$.

A lot of methods have been created for testing of finite amount of the hypotheses of the above type. Intuitively acceptable procedure should weight:

predictive abilities of the models of the inspected structure;

precision of the parameter estimates of the models having the particular structure;

number of estimated *parameters* and *amount* of available *data*.

All these features are well balanced in posterior probability of hypothesis gained by the straightforward application of the bayesian methodology [13].

Overwhelming amount of possible hyphotheses $2^{dim(DY)} + 2^{dim(DV)}$ (14) prevents us to evaluate fully the posterior distribution on them. For this reason, an efficient search for its maximizer has been proposed in [4]. The task STR is based on this repeatedly verified successful algorithm. In order to find the global maximizer, it starts automatically the search from the empty and the full regressor. If the results coincide they are taken as the solution otherwise the user is asked for his guess of the maximizing structure.

5.2 Parameter estimation, RLS

The task RLS exploits $L_f = L/T_f$ data items $\{d(t)\}_{i=1}^{L_f}$ selected from the representants $D(t)$ adapted to the chosen sampling period. The selection of the controlled output $y(t)$, the system input $u(t)$ and the measured disturbance $v(t)$ forming $d(t)$ is performed either by the task STR or by the user.

Recursive least squares used for estimation of the unknown parameters Θ of the models (2) and (3) with a fixed structure are also interpreted within bayesian framework [13]. In this way, appealing interpretation of the point estimates $\hat{\Theta} = \{\hat{P}, \hat{P}_a, \hat{R}, \hat{R}_a\}$ and their uncertainties — given by the positive definite matrices C, C_a, by the point estimates $\hat{R}$, $\hat{R}_a$ and by the positive scalars ν, ν_a — is gained.

Specifically, let $\mathcal{E}[.|t]$ denote expectation conditioned on the observed data $d(1), .., d(t)$. Then (cf. (10)–(12))

$$\hat{P}(t) = \mathcal{E}[P|t], \qquad \hat{P}_a(t) = \mathcal{E}[P_a|t],$$
$$\hat{R}(t) = \mathcal{E}[R|t], \qquad \hat{R}_a(t) = \mathcal{E}[R_a|t], \tag{15}$$

$$C_{ik}(t)\hat{R}_{jl}(t) = \mathcal{E}[(P_{ij} - \hat{P}_{ij}(t))(P_{kl} - \hat{P}_{kl}(t))|t], \tag{16}$$
$$j, l = 1, .., m_y, \qquad i, k = 1, .., i_z,$$
$$C_{a,ik}(t)\hat{R}_{a,jl}(t) = \mathcal{E}[(P_{a,ij} - \hat{P}_{a,ij}(t))(P_{a,kl} - \hat{P}_{a,kl}(t))|t],$$
$$j, l = 1, .., m_v, \qquad i, k = 1, .., i_a.$$

For simplicity, the covariances of the noise covariances are not listed. Qualitatively, they decrease with increasing degrees of freedom $\nu(t)$, $\nu_a(t)$.

Important consequence of the above interpretation is the conclusion that the results of the estimation in the preliminary stage are suitable initial conditions for identification performed by the selftuner.

This conclusion — similarly as whole background of the preliminary design — is valid when parameters of the assumed models are time invariant. The standard extension to slowly time-varying parameters through some sort of forgetting is, of course, possible.

5.3 Preliminary evaluation of loss function, EVA

The task EVA uses point parameter estimates $\hat{\Theta} = \{\hat{P}, \hat{P}_a, \hat{R}, \hat{R}_a\}$ (15) and their uncertainties C, C_a, ν, ν_a (16) (the structure s (9) of both models (2), (3) is fixed).

These data are gained as the outputs of the tasks STR and RLS. In principle, they can be gained from other — measured data independent — sources and used, for instance, for analysis of projected systems!

The task EVA predicts ideal closed loop behaviour using uncertain models of the controlled system (2), (3). It admits to modify systematically penalties in the loss (1) in such a way that the controlled system inputs and their changes remain with high probability within prespecified ranges.

This part of the preliminary design extends results given in [2,3]. Its description is the most thorough as it is not a standard part of the current state of the art within the specialists' community.

5.3.1 Prediction of the expected loss

Let Θ be fixed parameters of the models (10), (11)

$$\Theta = \{P, P_a, R, R_a\} \tag{17}$$

and W be fixed penalization matrices in the loss (1)

$$W = \{Q_y, Q_u, S_z\}. \tag{18}$$

The selftuner design has to specify control law L, i. e. mapping which assigns the system input to the observed data

$$L : d(1), d(2), .., d(t-1) \rightarrow u(t), \quad t = 1, 2, ... \tag{19}$$

Let us assume temporarily that the expectation of the loss is minimized within the class of control laws which can exploit value of the parameters Θ. Thus, such control laws are given by mappings

$$L(\Theta) : \Theta, d(1), d(2), .., d(t-1) \rightarrow u(t). \tag{20}$$

It can be shown simply that it holds

$$\min_{L(\Theta)} \mathcal{E}\{K[W]\} = \mathcal{E}\{\min_{L(\Theta)} \mathcal{E}[K[W]|\Theta, L(\Theta)]\}. \tag{21}$$

The minimum of the expected loss attained by any selftuner is greater than the value (21) as the parameters Θ are unknown for it. A successful selftuner should converge to the controller which knows the parameters Θ. Thus, the value $\min_{L(\Theta)} \mathcal{E}[K[W]|\Theta, L(\Theta)]$ — from which the outer expectation is taken in (21) — serves as useful lower bound on control quality with the well tuned selftuner.

The task EVA analyses $\min_{L(\Theta)} \mathcal{E}[K[W]|\Theta, L(\Theta)]$ for long control horizon T.

5.3.2 Prediction of the controller performance

The scalar quadratic loss (1) — used for the controller design — expresses the summarized behaviour of all closed-loop signals. User's view point is, as a rule, multicriterial: he evaluates behaviour of particular signals and judges their mutual relations. The task EVA helps even in this respect.

For a fixed weights W, let $L(W, \Theta)$ be control law minimizing the expected loss in the class (20) and $q(W, \Theta)$ corresponding value of the inner expectation (21)

$$
\begin{aligned}
q(W, \Theta) &= \mathcal{E}[K[W]|\Theta, L(W, \Theta)] = \\
&= \frac{1}{T} \mathcal{E}\{(z(T) - z_0(T-1))'S_z(z(T) - z_0(T-1)) + \\
&\quad + \sum_{t=1}^{T}[(y(t) - y_0(t-1))'Q_v(y(t) - y_0(t-1)) + \\
&\quad + (u(t) - u_0(t-1))'Q_u(u(t) - u_0(t-1))]|\Theta, L(W, \Theta)\} = \\
&= q_z(W, \Theta) + tr[Q_v q_v(W, \Theta)] + tr[Q_u q_u(W, \Theta)].
\end{aligned}
\tag{22}
$$

The third equality in (22) defines some quantities needed in the sequel. The scalar $q_z(W, \Theta)$ denotes

$$
q_z(W, \Theta) = \frac{1}{T} \mathcal{E}\{(z(T) - z_0(T-1))'S_z(z(T) - z_0(T-1))|\Theta, L(W, \Theta)\}.
\tag{23}
$$

For $x = y, u$, the symmetric positive semidefinite matrices $q_x(W, \Theta)$, of the type (m_x, m_x), are used in the trace symbol $tr[.]$. Their entries are defined for $i, j = 1, .., m_x$ as follows

$$
q_{x,ij}(W, \Theta) = \frac{1}{T} \sum_{t=1}^{T} \mathcal{E}[(x_i - x_{0,i})(x_j - x_{0,j})|\Theta, L(W, \Theta)].
\tag{24}
$$

The term $q_z(W, \Theta)$ converges to zero for the assumed long horizon and stabilizing control law $L(W, \Theta)$ (because of the factor $\frac{1}{T}$).

The entries $q_{x,ii}(W, \Theta)$ are expectations of the expected squared deviations — related to a single control period — of the i-th x-entry from the corresponding reference value $x_{0,i}$. The off-diagonal entries $q_{x,ij}(W, \Theta)$ describe mutual correlations of the i-th and j-th deviations.

The task EVA analyses and changes mutual relations of these contributions — called *partial criteria* — and consequently the overall expected loss (criterion). It projects the uncertainty of the parameters Θ into the uncertanties of the partial criteria and it changes penalties W in order to reach desirable compromise among entries $q_{x,ij}$.

5.3.3 Conceptual flow chart of the task EVA

1. An initial guess of a suitable penalty W (18) is chosen.

2. Independent samples of the parameters $\{\Theta(\tau)\}_{\tau=1}^{N}$ ($N >> 1$) are generated with the common distribution determined by the point estimates $\hat{\Theta} = \{\hat{P}, \hat{P}_a, \hat{R}, \hat{R}_a\}$ and the uncertainties C, C_a, ν, ν_a of the unknown parameters Θ gained by the task RLS or specified by the user.

3. The optimal control law $L(W, \Theta(\tau))$ and partial criteria $q_{x,ij}(W, \Theta(\tau))$ for $i, j = 1, .., m_x$ and $x = y, u$ are determined for each sample $\Theta(\tau)$, $\tau = 1, .., N$. The (overall) criterion $q(W, \Theta(\tau))$ is determined as the properly weighted sum of the partial criteria.

4. The gained independent samples of the criteria are used for constructing the distribution of the partial criteria $q_{z,ij}(W, \Theta)$ and of the criterion $q(W, \Theta)$.

5. The right-hand-side quantiles of the partial criteria

$$q^{\gamma}_{z,ii}(W), \quad i = 1, .., m_z, \quad x = y, u \tag{25}$$

and of the overall criterion are determined at a pre-selected level $\gamma < 1$. Recall that the quantil of a random variable is defined as the number under which the variable lies with the chosen probability γ. In the case assumed, it means

$$\text{Probability}\{q_{z,ii}(W, \Theta) \leq q^{\gamma}_{z,ii}(W)\} = \gamma$$

6. The least favourable confidence intervals of the inspected deviations are determined

$$\left[-k\sqrt{q^{\gamma}_{z,ii}(W)} + x_{0,i}, k\sqrt{q^{\gamma}_{z,ii}(W)} + x_{0,i}\right],$$
$$k \in [1, 2], \quad i = 1, .., m_z, \quad x = y, u. \tag{26}$$

The confidence intervals (26) are based on the normality of the closed loop signals in steady state. Consequently, the coefficient k determines well defined quantil. For $k = 1$, 67% of the signal values is expected to be in the interval, for $k = 2$ approximately 97%.

7. Automatic tuning of penalization matrices (18) in the loss (1) is based on fitting of the above confidence interval (26) to the restrictions (5). The fitting is performed by a version of the golden cross method which proposes new penalties W and procedure is repeated from the step 2.

The algorithm stops if the confidence intervals (26) are satisfactory from the user's view point.

The algorithm is able to tune penalization matrices Q_y, Q_u. The penalization of the terminal regressor guarantees closed loop stability and its value is irrelevant for long horizon. For short horizons, which are used in real time because of time shortage, the penalization matrix S_z should extend continuously the infinite horizon optimization. Such matrices (so called Riccati) $S_z(W, \Theta(\tau))$ are gained for particular parameter samples $\Theta(\tau)$ when evaluating the corresponding optimal control law. The sample mean $\hat{S}_z$ is proposed as an estimate of the suitable value of S_z

$$\hat{S}_z = \frac{1}{N} \sum_{\tau=1}^{N} S_z(W, \Theta(\tau)) \tag{27}$$

which is determined for the final penalizations Q_y, Q_u.

5.3.4 Algorithmization of particular steps

Random samples $\Theta(\tau)$ are generated by a suitable transformation of independent normal variables. Details can be found in [2,3]. The cited results are enriched by generating the parameters of the measured disturbance model.

The **optimal control law** $L(W, \Theta(\tau))$ is found by using factorized version of dynamic programming which is numerically stable version of solving Riccati equation.

The **overall criterion** $q(.)$ is taken as simple sum of partial criteria and values of the criterion for a fixed (typically zero) control law are generated for comparison.

The **distribution of criteria** is nonparametrically estimated in [2]. Essentially, histogram is evaluated. The evaluations are substantially speeded up by assuming log-normal distribution of the "diagonal" (with indeces $_{ii}$) partial criteria. This distribution has been found by inspecting results of the nonparametric estimation. The logarithms of the independent samples are described by the simple models

$$\ln(q_{z,ii}(W, \Theta(\tau))) = \mu_{z,ii} + e_{z,i}(\tau), \quad i = 1, .., m_z, \quad x = y, u, q \tag{28}$$

where $\mu_{z,ii}$ is an unknown constant and $\{e_{z,i}(\tau)\}_{\tau \geq 1}$ is white gaussian noise with an unknown dispersion $\omega_{z,i}^{-1}$.

The i-th model, which well describes the observed samples, is a simple special case of the regression model. Its identification reduces to an elementary version of least squares. The probability density function of the searched for distribution has the form (with simplified notation $q = q_{z,ii}(W, \Theta)$, $q(\tau) = q_{z,ii}(W, \Theta(\tau))$)

$$p(q|q(1), .., q(N)) = \frac{c}{\lambda^{\frac{1}{2}} q} \left[1 + \frac{(ln(q) - \mu)^2}{(N+1)\lambda} \right]^{-\frac{N+1}{2}} \tag{29}$$

with

$$
\begin{aligned}
c &= \frac{\Gamma(\frac{N+2}{2})}{\sqrt{N\pi} \Gamma(\frac{N+1}{2})} \approx \sqrt{\frac{N+1}{2\pi N}}, \\
\mu &= \frac{1}{N} \sum_{\tau=1}^{N} ln(q(\tau)), \\
\lambda &= \frac{1}{N} \sum_{\tau=1}^{N} ln^2(q(\tau)) - \mu^2.
\end{aligned}
\tag{30}
$$

$\Gamma(.)$ denotes Euler function which is well approximable by Stirling formula [10].

Evaluation of the quantil of the probability density function (29) is another important step which has not been published yet. It motivates the more detailed description.

Let $p(q)$ be probability density function (abbr. p.d.f.) (29) of a quantity $q \geq 0$ (a diagonal partial criterion (24): indices, arguments and condition are suppressed for clarity).

For $\gamma \in (0, 1)$ ($\gamma \approx 1$), the right-hand side quantil $q^\gamma \geq 0$ is defined by the equation

$$\gamma = \int_0^{q^\gamma} p(q) \, dq, \tag{31}$$

i. e. probability assigned by the p.d.f. $p(q)$ to the interval $[0, q^\gamma]$ is γ.

Let $q^{j,\gamma}$ be the j-th approximation of the q^γ. Then the new approximation is searched for by Newton method, which can be derived using first order Taylor expansion of the right-hand side in (31) at the point $q^{j,\gamma}$:

$$q^{j+1,\gamma} = q^{j,\gamma} + \frac{\gamma - F(q^{j,\gamma})}{p(q^{j,\gamma})} \tag{32}$$

where the distribution function is used

$$F(q^{j,\gamma}) = \int_0^{q^{j,\gamma}} p(q)\, dq. \tag{33}$$

The iterations (32) — which always converge with high rate due to monotonicity of $F(.)$ — take place until the equality (31) is fulfilled with sufficient precision.

Values of the distribution function (33) are needed for these evaluations. For log-normal model (28), bayesian identification supplies the needed p.d.f. $p(q)$ (29). The related distribution function $F(.)$ (33) can be written in the form:

$$F(q^{j,\gamma}) = \mathcal{T}_N \left[\frac{\sqrt{N}(ln(q^{j,\gamma}) - \mu)}{\sqrt{(N+1)\lambda}} \right] \tag{34}$$

where $\mathcal{T}_N[.]$ denotes Student distribution function with N degrees of freedom. This distribution has the p.d.f.

$$p_T(q) = \frac{\Gamma[\frac{N+1}{2}]}{\Gamma[\frac{N}{2}]\sqrt{N\pi}} \left(1 + \frac{q^2}{N} \right)^{-\frac{N+1}{2}}. \tag{35}$$

By the relation (34), the problem reduces to the determination of the probability assigned by the Student distribution to the given intervals. For a real ξ, the following approximation [14] is chosen

$$\mathcal{T}_N(\xi) \approx 1 - \frac{p_T(\xi)(N+1)}{H(\xi)N} \left[1 - \frac{1}{2} \left(\frac{\bar{H}(\xi)}{H^2(\xi)} + \frac{1}{N+1} \right) \right]. \tag{36}$$

The auxiliary quantities $\bar{H}(\xi)$, $H(\xi)$ are defined as follows

$$H(\xi) = \frac{\xi(N+1)}{N+\xi^2}, \quad \bar{H}(\xi) = \frac{(N+1)(N-\xi^2)}{(N+\xi^2)^2}. \tag{37}$$

The approximation (36), (37) is precise enough if $\mathcal{T}_N \in (0.5, 1)$. In order to stay within this range even when iterating we
- choose the highest sample value of q as the first approximation of the quantil $q^{0,\gamma}$;
- find the point $\bar{q}$ maximizing the inspected p.d.f. $p(q)$;
- deviate from the Newton method (32) and choose

$$q^{j+1,\gamma} = \bar{q} + \frac{q^{j,\gamma} - \bar{q}}{2} \tag{38}$$

if $q^{j+1,\gamma}$ determined by (32) is smaller than $\bar{q}$.

The algorithm converges safely and with high rate.

Automatic choice of penalties is restricted at the moment to diagonal matrix Q_u, fixed Q_y and symmetric restrictions of inputs or their increments $ul = -uu$. The algorithm rests on conjecture that the width of the influenced interval of expected deviations decreases monotonically with increasing value of the corresponding penalty. Under this assumption, the algorithm is a simple modification of the golden cross method.

6 Illustrative example

This section should introduce the implemented version — called DESIGNER — of the tuning system described above, and illustrate a typical result of its use.

6.1 The program system DESIGNER

DESIGNER has been designed to provide the adaptive controller with appropriate parameters and suitable initial conditions for bumpless start and reliable function.

The DESIGNER is an extensive conversational system, which consists of programs solving the three main tasks: structure determination (STR), parameter estimation (using recursive least squares – RLS), control evaluation (EVA), and supplies several auxiliary programs for user's convenience. The tasks can be solved separately or sequentially (results of the previous task are used automatically as initial data for the following one). Only brief summary of necessary inputs and main outputs can be given here, a broader description has been published in [2,3].

Input information needed for the DESIGNER:

STR – measured data file (sufficient length and excitation)
– maximal order of the regression models assumed
– rough estimate of lower bound on transport delays in the models
RLS – results of solving the STR task or an equivalent structure specification
– specification of forgetting the obsolete information (fixed or variable)
EVA – results of solving the RLS task or an equivalent structure and parameter specification
– desired limits of the input signal or its increments

Most important outputs of the DESIGNER

STR – specification of the chosen regression model for both the output and the measured disturbance
– posterior probability of the structure (a measure of certainty of the final choice)
– ordering of the particular regression elements according to their relative importance
– probabilities of the substructures nested in the best one
– mean values, variances of signals
RLS – point parameter estimates and their uncertainties for both models
EVA – appropriate penalization for meeting input limits
– result files with all data needed for bumpless start of the adaptive control
– graphs of estimated probability densities for the squared outputs, inputs and expected loss

6.2 Example of DESIGNER use

To show typical results of DESIGNER use, a simple simulated example is given in the following. The analyzed data are generated by regression models (10) and (11) using the KOS-SIC conversational system for simulation, identification and control, developed in the Institute of Information Theory and Automation of the Czechoslovak Academy of Sciences [12].

The simulated system

$$y(t) = P'z(t) + e(t)$$
$$P' = [.1, 1.8, .7, -.81, .2, .3, 0.], \quad R = .01$$
$$z'(t) = [u(t), y(t-1), u(t-1), y(t-2), u(t-2), v(t-1), 1]$$
$$v(t) = P'_a z_a(t) + e_a(t)$$
$$P'_a = [.8, 0.], \quad z'_a(t) = [v(t-1), 1], \quad R_a = .1$$

Data supplied to the DESIGNER

— Data file – 400 records of input, output, disturbance

— Maximal order allowed – $ORDE = 3$

— Lower bound on delay estimate – $n_u = 0, n_v = 1$

— Type of forgetting – no forgetting

— Input limits – $u(t)\epsilon[-1., 1.]$

Selected DESIGNER results

— Output model structure specification – given as a vector I_d of binary significance indeces for the elements of the maximal data vector $DY(t)$ ($ORDE = 3$ assumed in advance):
$$DY'(t) = [y(t), u(t - n_u), v(t - n_v), y(t - 1), u(t - n_u - 1), v(t - n_v - 1), ...,$$
$$....., y(t - 3), u(t - n_u - 3), v(t - n_v - 3), 1]$$
$$I'_d = [1, 1, 1, 1, 1, 4, 1, 1, 4, 4, 4, 4, 4]$$
$I_d(i) = 1$ means accepting the element in the regression model
$I_d(i) = 4$ means rejecting the element in the regression model

— Ordering of the regression elements according to the importance (numbers correspond to the sequence in $DY(t)$ vector – 4 7 5 8 3 2

— Disturbance model structure specification (meaning similar to that for the output model)
$$DV'(t) = [v(t), v(t - 1), v(t - 2), v(t - 3), 1]$$
$$I'_{ad} = [1, 1, 4, 4, 4]$$

— Mean values and variances for all variables measured

— Point parameter estimates
$$\hat{P}' = [.0963, 1.79, .705, -.799, .210, .249, .0]$$
$$\hat{P}'_a = [.789, .0]$$
$$\hat{R} = .0249, \ \hat{R}_a = .107$$

— Input penalization recommended – $Q_u = .184$

— File of relevant data stored for the controller initialization

6.3 Cross – validation

The influence of DESIGNER use on the selftuner performance is illustrated in the following figures (using the example described above). The standard initialization (controller default) we have used for comparison is often quite satisfactory; nevertheless, in multivariate cases and in case of input limitations (as in the illustration) the start of control can be substantially improved. Moreover, the danger of substantial mismodelling error with its well-known consequences is decreased.

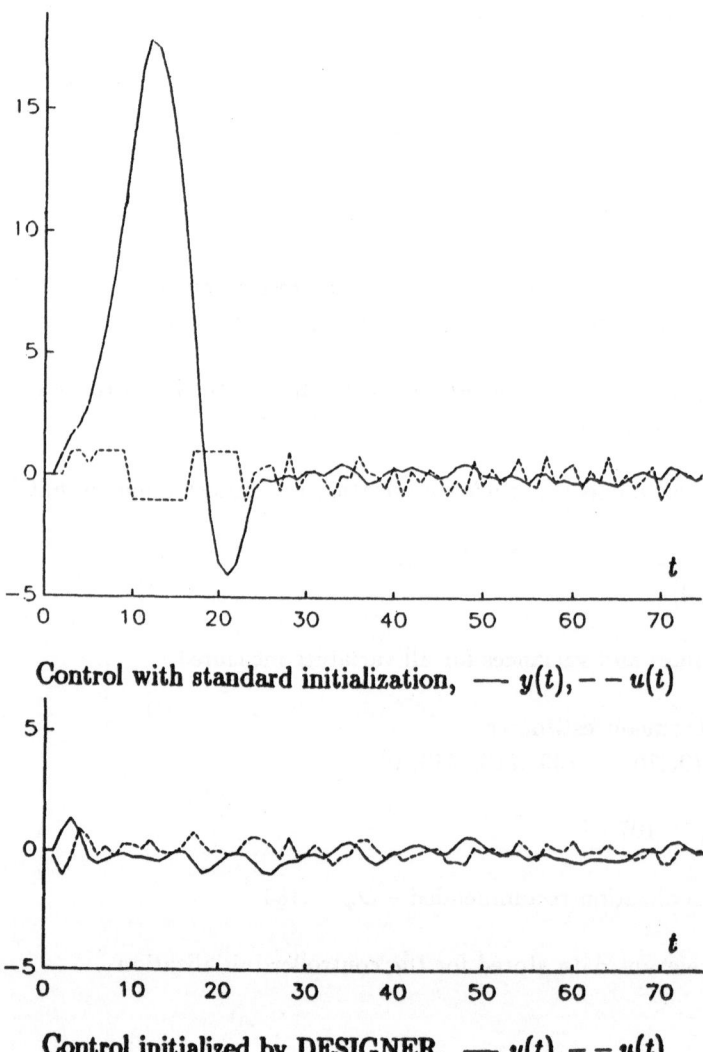

Control with standard initialization, $\quad$ — $y(t)$, $--$ $u(t)$

Control initialized by DESIGNER, $\quad$ — $y(t)$, $--$ $u(t)$

7 Conclusions

The problem of the computerized preliminary tuning of selftuners is formulated in the paper. The computerized tuning decreases user's effort when implementing this powerful type of controller. Moreover, it increases chance for the most satisfactory tuning and bumpless start up of the selftuner. Key steps of the problem, i. e.

- determining the structure of the selftuner
- generating the sensible initial conditions for parameter estimation
- shaping of loss function in order to reach desirable closed loop performance

are solved in the special case of LQG selftuners.

Limited experience with the resulting software is encouraging but the solution is far to be complete. The following improvements are prepared and/or foreseen:

- automatic tuning of control period (theoretical solution is available but it is not incorporated into overall design philosophy);
- automatic choice of non-diagonal entries of the penalization matrices Q_y, Q_u (1) (the solution is straightforward with some programming effort);
- decrease of the time requirements when optimizing the weights W (a conceptual solution based on appropriate sequential stopping rules when generating random samples is proposed)
- experiment design (remains open).

The above list supports our feeling that the preliminary tuning of selftuners is a new and promising research area. We believe that it is even more that it can be quite useful when implementing selftuners. The feedback of practical users will be, however, decisive. The paper tries to stimulate it.

References

[1] Hebký Z. (1984). Multiple sampling in computer control (In Czech). Automatizace, **27**, 6, 142-145.

[2] Jeníček T. (1988). Prior analysis of the expected control quality for selftuning controllers (In Czech). Thesis, ÚTIA ČSAV, Praha.

[3] Kárný M., T. Jeníček, W. Ottenheimer (1990). Contribution to prior tuning of LQG selftuners. Kybernetika, **26**, 2.

[4] Kárný M. (1983). Algorithms for determining the model structure of a controlled system. Kybernetika, **19**, 2, 164-178.

[5] Kárný M. (1984). Quantification of prior knowledge about global characteristics of linear normal model. Kybernetika, **20**, 5, 376-385.

[6] Kárný M., A. Halousková, J. Böhm, R. Kulhavý, P. Nedoma (1985). Design of linear quadratic adaptive control: theory and algorithms for practice. Kybernetika, **21**, Supplement to numbers 3, 4, 5, 6.

[7] Kárný M., R. Kulhavý (1988). Structure determination of regression-type models for adaptive prediction and control. In: Spall J. C. (ed.): Bayesian Analysis of Time Series and Dynamic Models. Marcel Dekker, New York.

[8] Kárný M. (1990). Estimation of sampling period for selftuners. Accepted for 11th IFAC World Congress, Tallinn.

[9] Kárný M. (1990). Local filter robust with respect to outlying data. Research report No. 1644, ÚTIA ČSAV, Praha.

[10] Korn G. A., T. M. Korn (1968). Mathematical Handbook for Scientist and Engineers. McGraw-Hill Book Company, New York, San Francisco, Toronto, London, Sydney.

[11] Kulhavý, R. (1987). Restricted exponential forgetting in real-time identification. Automatica 23,5,598-600.

[12] Kulhavý R. (1989). SIC - Package for Simulation Identification and Adaptive Control of Stochastic Systems: User's Guide. Research report No. 1592, ÚTIA ČSAV, Praha.

[13] Peterka V. (1981). Bayesian system identification. In: Eykhoff P. (ed.): Trends and Progress in System Identification. Pergamon Press Oxford, 239-304.

[14] Vajda I. (1982). Theory of information and statistical decision (In Slovakian). Alfa, Bratislava.

ADAPTIVE CONTROL ALGORITHM FOR TIME-VARIANT PROCESSES

Vladimír Velička

Research Institute of Chemical Equipment

Křižíkova 70, 602 00 Brno, ČSFR

1 Introduction

In industrial applications of adaptive control algorithms following restrictions often arise:

- on-line identification of controlled system model has to be performed in closed loop

- parameters of controlled system model are time-varying

- there are periods when control system is little excited and controller adaptation conditions became worse

- some disturbances are directly or indirectly measurable but with different accuracies

- variations in closed loop dynamics should be minimized

- system to be controlled includes a time delay

- there are engineering and technology constraints in position of control valve and in its maximum difference between two subsequent control actions.

Multiple-effect falling-film evaporator of sugar solution is the apparatus having above described properties. The evolution of algorithm for output fluid density control through input fluid flow manipulating is described in following sections. The aim of adaptive control algorithm synthesis is to cope with foregoing problems that are not specific only for heat-exchanging processes.

2 System model and closed loop identification

A linear single-input single-output discrete second-order model of controlled system with time delay is used:

$$y(k) = B1\, u(k-d-1) + B2\, u(k-d-2) - A1\, y(k-1) - A2\, y(k-2) \qquad (1)$$

where $u(k), y(k)$ are input and output signals respectively and k is an integer $k = 0, 1, 2, \ldots . B1, B2, A1, A2$ are (a priori) unknown coefficients the values of which have to be estimated. The time delay d is an integer number of sample intervals. The output signal $y(k)$ is reconstructed using the difference between measured process output $ym(k)$ and estimated output $n(k)$ of noise filter:

$$y(k) = ym(k) - n(k) \tag{2}$$
$$n(k) = D1 \, v(k-h-1) + D2 \, v(k-h-2) - C1 \, n(k-1) - C2 \, n(k-2) \tag{3}$$

The measured noise $v(k)$ is used in order to lower input-output cross-correlation and to improve the closed loop on-line identification. The values of coefficients $D1, D2, C1, C2$ and of the time delay h are considered to be a priori known.

A recursive least-squares performance criterion is used in on-line identification of coefficients $B1, B2, A1, A2$ in 1. As these coefficients are time-dependent a variable forgetting factor F is used the value of which is obtained by stabilisation of the Euclidean norm $SM(k)$ of the covariance matrix diagonal vector:

$$SM(k) = \sqrt{MC_{1,1}(k)^2 + MC_{2,2}(k)^2 + MC_{3,3}(k)^2 + MC_{4,4}(k)^2} \tag{4}$$

where $MC_{i,i}(k)$, $i = 1, 2, 3, 4$ are elements of covariance matrix diagonal vector. The equation used for determining of F has a simple form:

$$F(k) = \sqrt[H]{EW/SM(k)} \tag{5}$$

where EW is a desired value of SM and H is an integer number $3 < H < 10$. The higher is the predetermined value of EW the smaller is the influence of $u(k), y(k)$ on variations of $B1, B2, A1, A2$. When a closed loop is persistently excited the values of $SM(k)$ are rather higher than EW and $F(k) < 1$. After excitement disappearing $SM(k)$ goes down to EW and F converges to unity. Speed of this convergence is dependent on the value of H. In practice $H = 5$ was used in general.

3 Digital adaptive controller

Basic second-order digital controller takes the form:

$$u(k) = S0 \, e(k) + S1 \, e(k-1) + S2 \, e(k-2) - R1 \, u(k-1) - R2 \, u(k-2) \tag{6}$$

where $e(k)$ is control error. The values of $S0, S1, S2, R1, R2$ are determined under several conditions:

- controller incorporates an integrator to ensure zero steady-state tracking error:

$$1 + R1 + R2 = 0 \tag{7}$$

- poles of controlled system discrete transfer function are cancelled by zeros of controller transfer function:

$$S1 = S0 \, A1 \tag{8}$$
$$S2 = S0 \, A2 \tag{9}$$

- poles $T1, T2$ of closed loop discrete transfer function are assigned in accordance with process technology requirements concerning closed loop behaviour:

$$T1 = S0\,B1 + R1 \tag{10}$$
$$T2 = S0\,B2 + R2 \tag{11}$$

The equations 7 to 11 consequently enable the adaptation of coefficients in control law 6. A similar set of equations can be easily derived for the third-order controlled system model and related adaptive controller.

When a time delay $d > 0$ is present in model 1 a simple reconfiguration of the equation 6 is done in order to keep the dynamic properties of closed loop. Following modified equation is in fact a digital form of Smith predictor:

$$u(k) = S0\,e(k) + S1\,e(k-1) + S2\,e(k-2) - T1\,u(k-1) - T2\,u(k-2) \tag{12}$$
$$+ S0\,B1\,u(k-d-1) + S0\,B2\,u(k-d-2) \tag{13}$$

The equations 8, 9 remain valid and $S0$ (as well as for 6) is determined from :

$$S0 = (1 + T1 + T2)/(B1 + B2) \tag{14}$$

4 Control law at valve saturation

Using control law 6 when control valve is in marginal position in step $k-1$ generally results in an undesired behaviour of both manipulated and controlled variable. One way often used to solve this problem is to incorporate real (limited) value of previous control action $u(k-1) = U_{lim}$ into control law 6 instead of the calculated value $u(k-1)$ exceeding the manipulated variable limit. But in this case larger difference $u(k) - U_{lim}$ can appear which can be in many processes inadmissible. When we are able to determine maximum usable difference $UD = u(k) - U_{lim}$ which is possible to apply within closed loop we make use of equation 6 and modify it into form:

$$u(k) = S0\,e(k) + S1\,e(k-1) + S2\,e(k-2) - R1\,u_r(k-1) - R2\,u_r(k-2) \tag{15}$$

When the control action in step $k-1$ was limited then we use following value of $u(k-1)$ reduced for the purpose of determining $u(k)$:

$$u_r(k-1) = (UD - U_{lim} + (S0 + S1)\,e(k-1) + S2\,e(k-2) - R2\,u(k-2))/R1 \tag{16}$$

Equations 15, 16 are used for each k whenever control valve is saturated in step $k-1$. Relatively small changes in controlled variable within control valve saturation interval are presumed.

When in 15 $u(k-1)$ instead of $u_r(k-1)$ is used then a larger controller reset windup occurs. That is why equations 15, 16 can be looked at as a compromise between separate solving either problem of controller reset windup or problem of excessive control actions after control valve saturation. A similar modification can be made for Smith predictor 13 too.

5 Sugar solution density control

Sugar solution density control used in a multiple-effect evaporator is important for subsequent crystallization stage efficiency. As stated above, the inflow of sugar solution represents the manipulated variable u. The measured error v is the steam pressure in the second separator of a quadruple- effect falling film evaporator. The approximations 2 and 3 are used because external consumption of part of second-stage steam significantly affects remaining two stages function.

The incentive for adaptive control usage is the time variability of evaporator heat-transfer dynamics. This is caused mostly by changing of heat-transferring surface quality and by varying of steam condensation conditions on heating tubes. In nominal state of operation preliminary values of model 1 constants ($B1 = -.0522, B2 = -.0432, A1 = -1.51, A2 = .568, d = 4$) were estimated and used for start-up of on-line identification with $EW = 40$. The input density of sugar solution was from 12 to 13 mass percent of sugar. The desired values of the output solution concentration were set between 55 and 60 mass percent of sugar. The inflow of sugar solution was manipulated within the range from 8 to 15 kg/s .The constants of error filter 3 were estimated as follows: $D1 = -.4, D2 = .395, C1 = -.9, C2 = 0, h = 1$. The second separator steam pressure nominal value was 50 kPa absolute. The sampling interval was 1 minute for all control runs.

6 Conclusions

One class of digital adaptive control algorithm convenient to control of heat exchanging processes has been introduced. In particular, it has several specific features: variable forgetting factor is used in discrete model on-line identification; controller has pole assignement tuning and integral action; controller extended form is suitable for time-delayed systems control and a control law modification proposed to use after control valve saturation makes a good compromise between reset windup effect and excessive control action immediately after control valve saturation. It is important to use better identification of disturbance filter parameters because their values are also dependent on the external consumption of the second stage steam.

Numerous experiments using laboratory simulator and microcomputer-controller have indicated the robustness of both on-line identification and adaptive control. A series of output fluid density adaptive control runs was made on a quadruple-effect falling-film evaporator in a sugar works. Comparing of control runs was influenced by input steam pressure variations which affected the function of the whole system. After experiments evaluation further experiments will be performed in an industrial-scale equipment.

ALGORITHMS FOR CONVERSION BETWEEN THE CONTINUOUS-TIME MODELS OF LINEAR SYSTEMS AND THE DISCRETE-TIME ONES

Jan Ježek

Institute of Information Theory and Automation,
Czechoslovak Academy of Sciences
182 08 Prague 8, Czechoslovakia

Algorithms for conversion of the continuous-time polynomial model $B(s)/A(s)$ of a linear system to the discrete-time model $b(d)/a(d)$ and vice versa are described. Here s is the derivative operator and d is the delay one. The algorithms are based on the computation of exponential and logarithm function in the algebra of polynomials.

1 Introduction

There are two approaches to system control: the continuous-time one and the discrete-time one. Each of them has a different basic principle of work: the continuous-time control works unintermittently and is implemented by analogue elements, the discrete-time one works in periodic instants of time and is implemented by digital elements.

Each of these two approaches builds its own mathematical models. This paper is devoted to polynomial models: $B(s)/A(s)$ for continuous-time, s being the derivative operator, $b(d)/a(d)$ for discrete-time, d being the delay operator.

Often it is necessary to perform conversion from B/A to b/a or vice versa. For example, the continuous-time system is to be controlled discretely, or the result of discrete-time identification is to be controlled continuously. The numerical algorithms for conversion are subject of the paper.

In section 2, the continuous-time and the discrete-time signals are described and properties of the conversion between them are investigated. After a brief survey of existing numerical conversion methods in section 3, a method using congruences of polynomials is described in section 4. The conversion of signals being only a preparatory matter, the main theme is the conversion of transfer functions, section 5,6. Finally, some modifications of the algorithms are mentioned in section 7 and the implementation is descibed in section 8.

2   Discretization of signals

The continuous-time signals $x(\tau)$ are functions of $\tau \geq 0$. Their Laplace transform is

$$\bar{X}(s) = \int_0^\infty X(\tau)\mathrm{e}^{-s\tau}\mathrm{d}\tau.$$

For rational signals of order n, it is

$$\bar{x}(s) = \frac{B(s)}{A(s)} = \frac{B_0 + \cdots + B_{n-1}s^{n-1}}{A_0 + \cdots + A_n s^n}.$$

We suppose $\deg B < \deg A$ not to allow Dirac impulses.

Given a sampling period $T > 0$ and the relative shift ε, $0 \leq \varepsilon < 1$, the sampling instants $t = 0, 1, 2, \ldots$ are defined by $\tau = (t + \varepsilon)T$. The sequence

$$x_t = D_{T,\varepsilon}\{X(\tau)\},$$

the discretization of $X(\tau)$, is defined by sampling the function $x(\tau)$. Its d-transform is

$$\bar{x}(d) = \sum_{t=0}^\infty x_t d^t.$$

The symbol D is used also for transforms:

$$\bar{x}(d) = D_{T,\varepsilon}\{\bar{X}(s)\}.$$

Discretization of rational $\bar{X}(s)$ is rational in d:

$$\bar{x}(d) = \frac{b(d)}{a(d)} = \frac{b_0 + \cdots + b_{n-1}d^{n-1}}{a_0 + \cdots + a_n d^n}$$

It is always $a_0 \neq 0$, the point $d = 0$ is never a pole (causality). It is also $\deg b < \deg a$, the function $\bar{x}(d)$ is strictly proper, its polynomial part is zero.

There exist some signals with $\deg b \geq \deg a$ but they do not correspond to any rational continuous-time signals. They may correspond to transcendental signals, e.g. the transport delay $\bar{X}(s) = \mathrm{e}^{-T_d s}$.

An illustration of rational signal is

$$X(\tau) = \mathrm{e}^{-\alpha\tau}, \quad \bar{X}(s) = \frac{1}{s + \alpha}. \tag{1}$$

Its discretization is

$$x_t = \mathrm{e}^{-\alpha T(t+\varepsilon)}, \quad \bar{x}(d) = \frac{\mathrm{e}^{-\alpha T\varepsilon}}{1 - \mathrm{e}^{-\alpha T}d}. \tag{2}$$

A multiple root case:

$$x(\tau) = \frac{\tau^k}{k!}\mathrm{e}^{-\alpha\tau}, \quad \bar{X}(s) = \frac{1}{(s + \alpha)^{k+1}}. \tag{3}$$

Its discrete-time analogy (for $\varepsilon = 0$) is

$$x_t = \frac{t^{(k)}}{k!}e^{-\alpha T t}, \quad \bar{x}(d) = \frac{e^{-\alpha T k}d^k}{(1 - e^{-\alpha T}d)^{k+1}} \tag{4}$$

where $t^{(k)} = t(t-1)\ldots(t-k+1)$ is the "factorial power", a function more suitable than the power t^k. The multiple-root case is somewhat more complicated than the simple root case: the discretization of (3) is not (4) but it can be expressed as a sum of such terms.

We can see some properties of rational $\bar{x}(d)$ from the fractions (1)-(4). The pole $s_0 = -\alpha$ is mapped to $d_0 = e^{-\alpha T}$. The strip

$$-2\pi/T < \text{Im } s_0 < 2\pi/T \tag{5}$$

in the s-plane is mapped to the whole d-plane with the exception of d real, $d \leq 0$. With condition (5) for all poles, the conversion D is bijective and we can define

$$\bar{X}(s) = D_{T,\varepsilon}^{-1}\{\bar{x}(d)\}.$$

The real poles $d < 0$ are excluded; they are possible in the discrete-time signals but they have no continuous-time counterpart of the same order.

3 Numerical methods of discretization

The numerical problem of discretization consists in computing b_i, a_i, given $n, B_i, A_i, T, \varepsilon$. Several algorithms were designed for it. The standard procedure uses the partial fraction expansion: we compute the roots of $A(s)$, convert every fraction according to (1)-(4) and finally we sum the converted fractions. A bit of complication occurs for multiple roots or complex roots. Nevertheless, it is possible to construct a general algorithm. But the numerical experience says that the precision reached is not high due to computation of roots, fraction composition and decomposition.

It is natural to look for an algorithm where the root computation is not required. In [1], the discretization is performed by numerical computation of the exponential function of a matrix:

$$D_{T,\varepsilon}\{Ce^{A\tau}B\} = C\Phi^t e^{AT\varepsilon}B$$

where $\Phi = e^{AT}$ with square matrices A, Φ, a column vector B and a row vector C. The conversion from polynomial models to matrix ones and vice versa is well known. The numerical experience is good.

4 Congruences of polynomials

The algorithm which uses the exponential function was further modified by replacing the matrix algebra by the algebra of congruences of polynomials. In this way, the number of computational operations needed was reduced. Moreover, the solution is formulated in polynomial terms and fits to the polynomial theory of linear systems. Let us mention only the basic ideas, more details can be found in [5], [6].

A separate part is conversion $A(s)$ to $a(d)$. Let us consider a linear space $\mathcal{L}$ of functions $x(\tau)$ satisfying the differential equation

$$A(s)x = (A_0 + A_1s + \cdots + A_{n-1}s^{n-1} + s^n)x = 0.$$

Its dimension is n. On $\mathcal{L}$, only the operators $1, s, \ldots, s^{n-1}$ are linearly independent, the operator s^n is a linear combination of them:

$$s^n = -A_0 - A_1s - \cdots - A_{n-1}s^{n-1}.$$

The same holds for any polynomial operator $F(s)$: it can be replaced on $\mathcal{L}$ by a polynomial of degree at most $n - 1$, namely by its remainder from division by $A(s)$. The algebra $\mathcal{A}$ (the ring) of polynomial operators on $\mathcal{L}$ is the algebra of congruence classes modulo $A(s)$.

The algebra $\mathcal{A}$ can be also equipped with a norm and topology, we can use power series, iterative processes etc. We can construct functions similarly to function of matrices. The delay operator d is defined by the exponential function $d = e^{-sT}$ and can be computed in $\mathcal{A}$ numerically by Taylor series and similar tricks. As the dimension of $\mathcal{A}$ is n, the powers of d are linearly dependent:

$$a_0 + a_1d + \cdots + a_nd^n = 0.$$

The coefficients a_i can be found numerically; they yield the discrete-time denominator $a(d)$.

Having computed $a(d)$, the computation of $b(d)$ can be performed via discretization of the impulse response $h(\tau)$:

$$h_t = \frac{1}{2\pi i} \int \frac{B(s)}{A(s)} e^{sT(t+\epsilon)} ds.$$

Such expressions of the form

$$I = \frac{1}{2\pi i} \int \frac{\Phi(s)}{A(s)} ds$$

can be computed by congruences of polynomials: the function $\Phi(s)$, even the transcendental one, can be replaced by its remainder $\Phi_R(s)$ modulo $A(s)$. Now, having computed $h(d)$, we can easily compute $b(d)$ from $h(d) = b(d)/a(d)$.

5 Conversion of the plant transfer function

Up to now, we have dealt with conversion of signals or of their transforms. The conversion of transfer functions needs something more. A linear system is supposed in the form

$$Y = \frac{B}{A}U + \frac{C}{A}E$$

with output Y, input U and white noise E. This scheme may be continuous-time or discrete-time, polynomials A, B, C in s or in d. Each of transfer functions B/A and C/A is converted in a different way.

The continuous-time transfer function $B(s)/A(s)$ is the Laplace transform of the response to Dirac impulse $\delta(\tau)$. In digital control, the input is not $\delta(\tau)$ but is held constant for $0 \le \tau < T$ and then reset. The discrete-time transfer function is

$$H(d, \varepsilon) = (1 - d)D_{T,\varepsilon}\left\{\frac{H(s)}{s}\right\},$$

ε being the relative shift between the samples of the input signal and those of the output one.

6 Conversion of the noise filter

When looking for a discrete-time equivalent of the noise filter $C(s)/A(s)$ we want the discretely generated noise to have the same statistical properties as the continuously generated one. We require that their autocorrelation functions coincide for $\tau = Tt$. In transforms, the spectral densities are converted

$$\frac{c(d)c(d^{-1})}{a(d)a(d^{-1})} = D_{T,\varepsilon}\left\{\frac{C(s)C(-s)}{A(s)A(-s)}\right\}.$$

The autocorrelation function is two-sided symmetric: $R(\tau) = R(-\tau)$. Our discretization D is defined for one-sided functions ($\tau \ge 0$) only, so we discretize one side of it only. The algorithm is as follows:

1) Given A, C, the continuous-time spectral density is decomposed into G/A:

$$\frac{C(s)C(-s)}{A(s)A(-s)} = \frac{G(s)}{A(s)} + \frac{G(-s)}{A(-s)}.$$

It is performed by solving a symmetric polynomial equation.

2) $G(s)/A(s)$ is converted to discrete-time $g(d)/a(d)$. A special care must be taken for $\tau = 0$; this point is common to $G(s)/A(s)$ and $G(-s)/A(-s)$. We must take only half of it:

$$\frac{g(d)}{a(d)} = D_{T,0}\left\{\frac{G(s)}{A(s)}\right\} - \frac{1}{2}\left[\frac{G}{A}\right]_{\tau=0}.$$

3) The resulting discrete-time spectral density is factorized to obtain $c(d)/a(d)$:

$$\frac{g(d)}{a(d)} + \frac{g(d^{-1})}{a(d^{-1})} = \frac{c(d)c(d^{-1})}{a(d)a(d^{-1})}.$$

It is performed by solving the polynomial equation for spectral factorization.

7 Modifications

Besides of the discretization, the backward conversion $b(d)/a(d)$ to $B(s)/A(s)$ was also solved. It uses the logarithm function $s = -(\ln d)/T$. The computation of it in the algebra $\mathcal{A}$ uses Taylor series, iterative processes and similar tricks known from the real number analysis.

Conversion of the discrete model $b(d)/a(d)$ from one sampling period to another can be performed by the power function

$$d_1 = e^{-sT_1}, \quad d_2 = e^{-sT_2}, \quad d_2 = d_1^{\frac{T_2}{T_1}}.$$

An efficient algorithm for computing this function is described in [2]; it can be used in $\mathcal{A}$, too.

Another idea is to replace the delay operator d by the finite difference operator Δ:

$$\Delta x_t = \frac{x_{t+1} - x_t}{T}.$$

The conversion algorithm can be modified for this case, the exponential and the logarithm being replaced by a slightly more complicated functions. The reason for using Δ is the unification of the continuous-time theory with the discrete-time one: the derivative is a special case of the difference for T approaching 0. For more details, see [3].

8 Implementation and experiences

The algorithms have been programmed in Fortran and implemented in PC computers. They are included in POLPACK, the package of mathematical routines for polynomial operations. The routines run under a conversational monitor. No other software package is known which could offer such a capability for linear system conversion.

The routines have been tested by many examples. The various conversion routines can be cross-checked mutually. For reasonable data, the routines work reliably. Troubles can occur when T is too small or too big. For small T, d-fractions break down numerically and the information is lost but Δ-fractions behave well and converge to continuous-time ones. For big T, troubles may occur when an oscillatory mode is present. Approaching the border of the region (5), the numerical behaviour gets ill and the backward conversion breaks down. Another source of difficulties for the backward conversion is approaching the pole $d = \infty$. Such a case occurs for T big compared with time constants of the plant.

9 References

1. Ježek,J.,1968: Algorithm for computing the discrete time transfer function of a linear dynamic system (in Czech). Kybernetika 4,3,246-259.

2. Ježek,J.,1988: An efficient algorithm for computing real power of a matrix and a related matrix function. Aplikace matematiky 33,1,22-32.

3. Ježek,J.,1990: Conversion of the polynomial continuous-time model to the delta discrete-time and vice versa. Accepted to 11th IFAC World Congress, Tallinn, U.S.S.R.

4. Nagy,I. and Ježek,J.,1986: Polynomial LQ control synthesis for delta operator models. 2nd IFAC Workshop on adaptive systems, Lund,Sweden, pp.323-327.

5. Vostrý,Z.,1977: Congruence of analytical functions modulo a polynomial. Kybernetika 13,12,116-137.

6. Vostrý,Z.,1978: Polynomial approach to conversion between Laplace and Z transforms. Kybernetika 14,4,292-306.

Lecture Notes in Control and Information Sciences

Edited by M. Thoma and A. Wyner

Lecture Notes in Control and Information Sciences

Edited by M. Thoma and A. Wyner

Lecture Notes in Control and Information Sciences

Edited by M. Thoma and A. Wyner

For information about Vols. 1-96 please contact your bookseller or Springer-Verlag